Lecture Notes in Control and Information Sciences

Edited by M. Thoma and A. Wyner

Vol. 62: Analysis and Optimization
of Systems
Proceedings of the Sixth International
Conference on Analysis and Optimization
of Systems
Nice, June 19–22, 1984
Edited by A. Bensoussan, J. L. Lions
XIX, 591 pages. 1984.

Vol. 63: Analysis and Optimization
of Systems
Proceedings of the Sixth International
Conference on Analysis and Optimization
of Systems
Nice, June 19–22, 1984
Edited by A. Bensoussan, J. L. Lions
XIX, 700 pages. 1984.

Vol. 64: Arunabha Bagchi
Stackelberg Differential Games
in Economic Models
VIII, 203 pages, 1984

Vol. 65: Yaakov Yavin
Numerical Studies
in Nonlinear Filtering
VIII, 273 pages, 1985.

Vol. 66: Systems and Optimization
Proceedings of the Twente Workshop
Enschede, The Netherlands, April 16–18, 1984
Edited by A. Bagchi, H. Th. Jongen
X, 206 pages, 1985.

Vol. 67: Real Time Control of Large Scale Systems
Proceedings of the First European Workshop
University of Patras, Greece, Juli 9–12, 1984
Edited by G. Schmidt, M. Singh, A. Titli,
S. Tzafestas
XI, 650 pages, 1985.

Vol. 68: T. Kaczorek
Two-Dimensional Linear Systems
IX, 397 pages, 1985.

Vol. 69: Stochastic Differential Systems –
Filtering and Control
Proceedings of the IFIP-WG 7/1 Working Conference
Marseille-Luminy, France, March 12-17, 1984
Edited by M. Metivier, E. Pardoux
X, 310 pages, 1985.

Vol. 70: Uncertainty and Control
Proceedings of a DFVLR International Colloquium
Bonn, Germany, March, 1985
Edited by J. Ackermann
IV, 236 pages, 1985.

Vol. 71: N. Baba
New Topics in Learning Automata
Theory and Applications
VII, 231 pages, 1985.

Vol. 72: A. Isidori
Nonlinear Control Systems:
An Introduction
VI, 297 pages, 1985.

Vol. 73: J. Zarzycki
Nonlinear Prediction
Ladder-Filters for Higher-Order
Stochastic Sequences
V, 132 pages, 1985.

Vol. 74: K. Ichikawa
Control System Design based on
Exact Model Matching Techniques
VII, 129 pages, 1985.

Vol. 75: Distributed Parameter
Systems
Proceedings of the 2nd International
Conference, Vorau, Austria 1984
Edited by F. Kappel, K. Kunisch,
W. Schappacher
VIII, 460 pages, 1985.

Vol. 76: Stochastic Programming
Edited by F. Archetti, G. Di Pillo,
M. Lucertini
V, 285 pages, 1986.

Vol. 77: Detection of
Abrupt Changes in Signals
and Dynamical Systems
Edited by M. Basseville,
A. Benveniste
X, 373 pages, 1986.

Vol. 78: Stochastic
Differential Systems
Proceedings of the 3rd Bad Honnef
Conference, June 3–7, 1985
Edited by N. Christopeit, K. Helmes,
M. Kohlmann
V, 372 pages, 1986.

Vol. 79: Signal
Processing for Control
Edited by K. Godfrey, P. Jones
XVIII, 413 pages, 1986.

Vol. 80: Artificial Intelligence
and Man-Machine Systems
Edited by H. Winter
IV, 211 pages, 1986.

For information about Vols. 1–61 please contact your bookseller or Springer-Verlag.

Lecture Notes in Control and Information Sciences

Edited by M. Thoma and A. Wyner

130

W. A. Porter, S. C. Kak
J. L. Aravena (Editors)

Advances in Computing and Control

Springer-Verlag Berlin Heidelberg GmbH

Editors

William A. Porter
Subhash C. Kak
Jorge L. Aravena

Department of Electrical
and Computer Engineering
Louisiana State University
Baton Rouge, LA 70803
USA

ISBN 978-3-540-51425-1   ISBN 978-3-540-46260-6 (eBook)
DOI 10.1007/978-3-540-46260-6

2161/3020-543210   Printed on acid-free paper.

PREFACE

This volume is a collection of papers selected out of those presented at the 1988 International Conference on Advances in Communication and Control Systems. This conference was held in Baton Rouge, Louisiana during October 19-21, 1988 and it was sponsored by National Science Foundation, Louisiana State University, Southern University, and International Federation of Information Processing.

The aim of this conference was to bring together researchers in the fields of communications, control, signal processing, and computing to explore emerging common themes and report advances in research. A total of 131 papers were presented at the conference. Sixty-eight of these papers were selected for inclusion in two Springer-Verlag LNCIS volumes. These volumes are titled <u>Advances in Communications and Signal Processing</u> and <u>Advances in Computing and Control</u>.

We would like to thank A.V. Balakrishnan, Nirmal Bose, Zdzislaw Bubrucki, Romano DeSantis, Thomas Dwyer, Erol Gelenbe, Don Halverson, Manju Hegde, Edward Kamen, Hong Lee, Ruey-wen Liu, Mario Lucertini, Steve Marcus, Sanjit Mitra, Mort Naraghi-Pour, Dave Neuhoff, Roberto Triggiani and Nick Tzannes for helping with the organization of the sessions and selection of contributors.

We also thank Margaret Brewer and Rachel Bryant for their enthusiastic help with the preparation of the volumes.

William A. Porter
Subhash C. Kak
Jorge L. Aravena
Editors

TABLE OF CONTENTS

1. A. El-Amawy and K.R. Dharmarajan
   A VLSI Array for Stage Matrix Inversion Using Gauss-
   Jordan Diagonalization   1

2. R.A. Lincoln and K. Yao
   Kalman Filtering Systolic Array Design Using Dependence
   Graph Mapping   11

3. William A. Porter and Jorge L. Aravena
   Generalized Distributive Memory Arrays   23

4. D.L. Gray, A.N. Michel and W. Porod
   Design and Implementation of a Class of Neural Networks   38

5. F. Pourboghrat and M.R. Sayeh
   Neural Network Models for the Learning Control of
   Dynamical Systems with Application to Robotics   50

6. Alexander Skavantzos and Thanos Stouraitis
   Comples Multiplicaiton Using the Polynomial
   Residue Number System   61

7. Tein Y. Chung, Suresh Rai and Dharma P. Agrawal
   On Routing and Performance Analysis of the Doubly
   Connected Networks for MANs &n LANs   71

8. M.C. Stinson and S.C. Kak
   Bicameral Neural Computing   85

9. D. Prados and S.C. Kak
   Non-Binary Neural Networks   97

10. C.H. Youn and S.C. Kak
    New Learning and Control Algorithms for Neural
    Networks   105

11. H. Barada and A. El-Amawy
    A Fixed-Size Systolic System for the Transitive
    Closure and Shortest Path Problems   117

12. Jorge L. Aravena
    Systematic Design of Fast Discrete Controllers   128

13. Eyad H. Abed
    Nonlinear Stabilizing Control of High Angle of
    Attack Flight Dynamics   140

14. A.S.C. Sinha
    Controllability of Large-Scale Nonlinear Systems
    with Distributed Delays in Control and States   150

15. P.S. Min
    Robust Application of Beard-Jones Detection Filter   162

16. D.L. Lukes
    Lie Groups Underlying Fault Avoidance in Dynamical
    Control Systems   174

17. W.H. Bennett
    Computation and Implementation of Precision Control
    for Flexible Space Structures   182

18. R.M. DeSantis
    On PI and PID/Sliding Mode Controllers   195

19. M. Bodson
    Tuned Values in Adaptive Control   206

20. E. Fernandez-Gaucherand, A. Arapostathis and S.I. Marcus
    On the Adaptive Control of a Partially Observable
    Binary Markov Decision Process   217

21. E.W. Kamen, T.E. Bullock, and C.-H. Song
    Adaptive Control of Linear Systems with Rapidly-
    Varying Parameters   229

22. K.L. Moore, M. Dahleh and S.P. Bhattacharyya
    Learning Control for Robotics   240

23. C-L. Lee and M.W. Walker
    Instantaneous Trajectory Planning For Redundant
    Manipulator in the Presence of Obstacles   252

24. P-Y. Woo and C.N. Dorny
    Coordinated Control in a Multi-Manipulator Workcell   264

25. H.G. Lee, A. Arapostathis and S.I. Marcus
    Digital Control of Linearizable Systems   276

26. C.F. Martin
    Representations of Systems with Finite Observations   281

27. R. Malhame, S. Kamoun and D. Dochain
    On-Line Identification of Electrical Load Models
    for Load Management   290

28. J.E. Lagnese
    Uniform Stabilization of a Thin Elastic Plate
    By Nonlinear Boundary Feedback   305

29. W. Littman
    A Generalization of a Theorem of Datko and Pazy   318

30. R. Datko
    The Destablizing Effect of Delays on Certain
    Vibrating Systems   324

31. W. Desch, K.B. Hannsgen and R. L. Wheeler
    Feedback Stabilization of a Viscoelastic Rod   331

32. J.A. Goldstein and G.R. Rieder
    Continuous Dependence of the Asymptotics
    in Discrete and Continuous Dynamical Systems   338

33. W.A. Sethares and C.R. Johnson, Jr.
    Persistency of Excitation and (lack of) Robustness
    in Adaptive Systems   340

34. M.K. Sain, J.L. Peczkowski and B.F. Wyman
    On Complexity in Synthesis of Feedback Systems   352

35. Michael K.H. Fan, John C. Doyle and Andre L. Tits
    Robustness in the Presence of Parametric
    Uncertainty and Unmodeled Dynamics   363

A VLSI ARRAY FOR STABLE MATRIX INVERSION
USING GAUSS-JORDAN DIAGONALIZATION

A. El-Amawy and K.R. Dharmarajan
Louisiana State University

1.   Introduction: The high throughput requirements of computationally intensive operations in fields like signal processing and image processing have far exceeded the capabilities of general purpose multiprocessor machines.   Moreover, the real time nature of these operations imposes a severe constraint on speed.   In recent years, a number of special purpose architectures have been developed to perform such computationally demanding operations.   The need for such special purpose architectures has been complemented by rapid developments in VLSI technology.   Recent advances in VLSI technology make such architectures an attractive proposition.

Many of the computationally demanding contemporary applications involve operations on matrices. Matrix operations like triangularization and inversion are vital to the simplification and solution of many problems.   A number of VLSI architectures have been proposed for performing matrix computations [1]-[4]. These architectures exploit the parallelism that is inherent in most matrix operations. This study deals entirely with the problem of matrix inversion on systolic or VLSI arrays.

Several VLSI architectures have been proposed earlier for performing matrix inversion [1]-[3].   These architectures are based on the Gaussian elimination or LU-factorization techniques (or a variant).   Most of these architectures do not implement pivoting of any sort. To minimize the effects of rounding error, it has been proved [5]-[6] that it is essential to implement pivoting of some sort. Without implementing pivoting, these inversion algorithms ensure numerical stability only for diagonally dominant or positive definite matrices.

The paper presents a fast, efficient parallel implementation of matrix inversion on a VLSI array, based on the Gauss-Jordan diagonalization method.   The parallel algorithm is implemented on a bilinear systolic array, that employes only $4n+1$ PEs.   The array has a completion time of $n^2 + 3n-1$.   The bilinear aray has been supplemented with a buffer array, to eliminate the need for inter-iteration I/O.   The architecture implements the Gauss-Jordan diagonalization method with partial pivoting.   The architecture utilizes the pivoting strategy impelmented in [4].   This pivoting technique, apart from enhancing the numerical quality of the diagonalization process, also eliminates the need for any global communication.

The Gauss-Jordan diagonalization method is similar to the Gaussian elimination process, but elimination is performed above and below the diagonal in each iteration. The algorithm starts with the system (A I), where A is the input matrix, and through a series of operations, transforms it into the system (I A^{-1}). All operations in the algorithm are performed on the augmented n x 2n matrix (A,I).

Section 2 reviews the Gauss-Jordan diagonalization technique and other salient features of the algorithm. In Section 3, the systolic architecture, its control structure and the PE functions have been described in detail. Section 4 summarizes the conclusions drawn from this study.

2. The Algorithm: As stated earlier, to ensure the numerical stability of the Gaussian elimination process or a variant, it is desirable to implement pivoting of some sort [5]-[6]. There are different types of pivoting strategies namely complete pivoting, partial pivoting, neighbor pivoting, etc. Partial or maximal column pivoting, one of the more popular pivoting strategies, involves selecting the element, in the same column, on or below the diagonal that has the largest absolute value. For performing stable matrix inversion, the Gauss-Jordan diagonalization method, which is a variant of the Gaussian elimination process, with partial pivoting has been chosen.

Let us consider the following linear system of equations:

$$A\bar{x} + \bar{y} = 0$$

$$\text{i.e. } A\bar{x} + I\bar{y} = 0 \tag{1}$$

$$\text{Hence }, \quad \bar{x} + A^{-1}\bar{y} = 0 \tag{2}$$

So, we have to start with (1), do a sequence of operations and end up with the equivalent system (2). In the process, we would have computed A^{-1}. If we start with the n x 2n matrix (A I) and perform Gaussian elimination on it, we would get the system

$$U\bar{x} + G\bar{y} = 0 \tag{3}$$

where U is upper triangular. But since U is not the identity matrix or a diagonal matrix, (3) is not equivalent to (2). So we would have to convert U to a diagonal matrix, to get a system of equations equivalent to (2). This suggests the following variation of the Gaussian elimination process: eliminate above the diagonal as well as below in each step of Gaussian elimination.

The k-th step in Gaussian elimination is

$$a'_{ij} = a_{ij} - (a_{ik}/a_{kk})\, a_{kj}, \quad \text{for } i > k \text{ and all } j,$$

$$a'_{ij} = a_{ij}, \quad \quad \quad \text{for } i \leq k \text{ and all } j.$$

But for the Gauss-Jordan process, the k-th step would be:

$$a'_{ij} = a_{ij} - (a_{ik}/a_{kk})\, a_{kj}. \quad \text{for } i \neq k \text{ and all } j,$$

$$a'_{kj} = a_{kj}, \qquad\qquad \text{for all } j.$$

Thus the Gaussian elimination process is performed for all rows except the pivot row. The pivot element a_{kk} is selected as in Gaussian elimination with partial pivoting, that is, the element in the pivot column on or below the diagonal, with the largest absolute value, is chosen. It should be noted that all operations are performed on the augmented n x 2n matrix (A,I).

After n pivoting steps, we would get the following system:

$$D\bar{x} + H\bar{y} = 0 \qquad\qquad (4)$$

where D is a non-singular diagonal matrix and H is non-singular. Comparison of (4) with (2) gives $A^{-1} = D^{-1}H$. Since D is diagonal, A^{-1} can be obtained by just dividing each row of H by the corresponding diagonal entry of D.

In the above algorithm, since the last iteration alone involves a division of each row of H by the corresponding diagonal entry of D, the last step is different from the earlier steps. For implementation on a systolic array, it is imperative that the operations be made regular and identical in all the iterations. This objective can be achieved by using a variant of the Gauss-Jordan pivoting algorithm [7]. In this variant, the elements of the pivot row are divided by the pivot element in each iteration. The k-th step in this variant would be:

$$a'_{ij} = a_{ij} - (a_{ik}/a_{kk}) \cdot a_{kj}, \qquad \text{for } i \neq k \text{ and all } j,$$

$$a'_{kj} = a_{kj}/a_{kk}, \qquad\qquad \text{for all } j.$$

All iterations of the algorithm are identical in this variant and this makes it feasible to implement the diagonalization algorithm on systolic arrays. This modification does not affect the stability of the algorithm, since the partial pivoting strategy is implemented as in the original version of the algorithm. After n steps, the system (1) would be directly transformed into the following system of equations:

$$I\bar{x} + A^{-1}\bar{y} = 0$$

i.e. $$\bar{x} + A^{-1}\bar{y} = 0$$

It should be noted that at the k-th iteration step, the operations performed on the elements of the k-th row are different from the operations performed on other rows. So the control logic of the parallel algorithm would have to detect the k-th row in the k-th iteration and modify the operation performed appropriately.

2.1   Stability of the Algorithm: As stated in [8], the Gauss-Jordan diagonalization process with partial pivoting can be split into two distinct stages:

(i) Reduction to upper triangular form by the standard Gaussian elimination algorithm with partial pivoting and

(ii) Further reduction of the triangular system to a diagonal system by an elimination process without pivoting.

In the Gauss-Jordon process, the search for the pivot is done only on or below the diagonal.   Hence, the magnitude of the multipliers $m_{ik} = a_{ik}/a_{kk}$ is always less than unity below the diagonal ($i > k$).   Thus, the growth of errors below the diagonal is bounded.   If the Gaussian elimination technique is used, after step (i) the resulting triangular system would be solved by back-substitution.   But in the Gauss-Jordan process, step (i) is followed by a further reduction of the system to diagonal form by elimination.   It has been shown by Peters and Wilkinson [8] that the absolute error obtained in step (ii) of the diagonalization process is strictly comparable to that obtained in back-substitution.   Hence, the numerical stability of the Gauss-Jordan diagonalization method with partial pivoting is comparable to that of Gaussian elimination with partial pivoting plus back-substitution.

3. The Architecture: The systolic array structure and the input sequence for a 3 x 3 input matrix are shown in Figure 1. A bilinear array with (4n + 1) PEs is used to implement the Gauss-Jordan diagonalization algorithm with partial pivoting. The structure of the bilinear array is similar to the one proposed in [4]. The bilinear array is supplemented with an array of buffers to eliminate the need for costly inter-interation I/O. Each column of PEs (excluding the first column) receives a column of the augmented n x 2n matrix (A I), where A is the input matrix and I is the identity matrix.

As in [4], the function performed by each row of PEs can be clearly demarcated as follows: the PEs in the second row search for the pivot in each iteration and after the selection of the pivot, transmit the elements of the pivot's row to the first row of PEs. The pivot is selected using the partial pivoting strategy. The PEs in the first row update the elements of the matrix in each iteration. The operations performed by both the rows are maximally overlapped to optimize the computation time.

As can be observed from Figure 1, the bilinear array has 2n columns of PEs (excluding P_{11}). The first n columns each receive a column of the matrix to be inverted as input and the second n columns each receive a column of the identity matrix (of order n) as input, in a skewed fashion. The input elements are fed

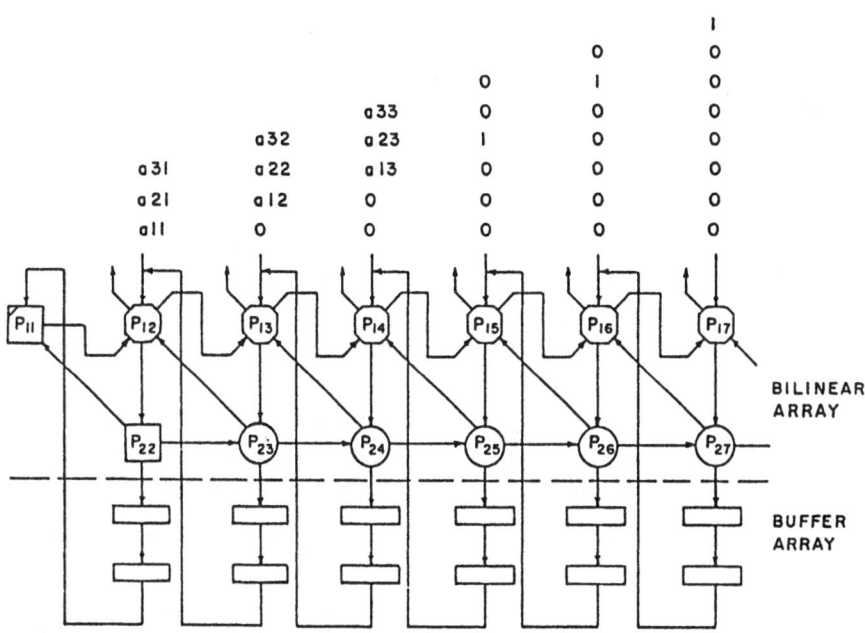

Figure 1. The bilinear inversion array (n = 3).

directly into the second row of PEs. By appending an extra bit, the first row of
PEs can be made to act as simple delay buffers for the south-bound input elements,
during the first iteration alone. After n iterations, the outputs (the elements of
A^{-1}) are retrieved from the first row of PEs.

In the Gauss-Jordan process, as stated in Section 2, operations performed on
the k-th row (pivot row) in the k-th iteration are different from the operations
performed on the other rows. Moreover, elimination has to be performed above and
below the diagonal. To accommodate these two distinguishing features of the
algorithm mentioned above, the control logic of the architecture has to make two
things possible:

(i) The PEs in the first row (P_{1j}, j > 1) should have the ability to detect when
     the modified updating operation for the pivot row is to be performed.
(ii) It is essential that the elements flow out of the first row of PEs in perfect
     sequential order, i.e., if P_{1j} receives the i-th column of the matrix, then
     the elements have to come out of P_{1j} in the following order: a_{11},
     a_{21},......., a_{n1}.

In the proposed array, the two objectives outlined above are accomplished by
appending a tag bit 't' with certain south-bound outputs from P_{1j} (these would be
inputs for the next iteration). A tag 't' associated with an input element would

indicate that the element is above the pivot row and hence has to be sent downwards (after being updated by P_{1j}) before the pivot row. Thus, when any P_{1j} (j > 1) receives the first south-bound input without the tag 't' attached to it, it knows that the modified "update" operation for the pivot row has to be performed. Thus the tag 't' serves the purpose of identifying the pivot row and also ensures that the elements flow out of the first row of PEs in sequential order. A second type of control bit (a token) '*' is also used by the architecture. Presence of the token '*' identifies the last element for that particular iteration.

The PE functions are described in Figure 2. As stated earlier, P_{22} searches for the pivot in each iteration, i.e., it compares the south-bound input element with the element stored in it and performs an interchange if necessary. P_{22} systolically transmits the result of the search to the other PEs in the second row. When a south-bound input with token '*' is received, the search for the pivot is stopped. The PEs in the second row transmit the elements of the pivot's row diagonally to the first row. The PEs in the first row store the pivot row elements, with P_{11} storing the pivot element (a_{kk}). Using the stored pivot element, P_{11} computes and transmits eastward the Gaussian elimination multipliers (a_{ik}/a_{kk}), corresponding to each row. These multipliers are used by each P_{1j}(j > 1) to modify its south-bound inputs.

P_{1j} (j > 1) performs the normal updating process (the Gaussian elimination process) for south-bound input elements with a tag 't'. When it receives the first input element without a 't', it performs the modified "update" operation for the pivot row. It modifies the stored pivot row element (i.e. computes a_{kj}/a_{kk}) and transmits it on its south-bound output. The tag 't' is also attached to this output since this element would be above the pivot row for the next iteration. For succeeding elements without the tag 't' (i.e. for elements below the pivot row), the normal updating process is performed.

It should be noted that, in each iteration, a 'dummy' row of 0's is transmitted on the south-bound outputs of P_{2j}, instead of the elements of the pivot row. These 'dummy' row elements become the south-bound inputs to P_{1j} in the next iteration. In every iteration, the first south-bound input element without a tag 't' received by P_{1j}(j>1), is an element belonging to the 'dummy' row. P_{1j}(j>1) ignores this 'dummy' row input and performs the modified 'update' operation using the stored pivot row element, as stated earlier. It is also noted that an output element is transmitted on each of the north-west bound outputs of P_{1j} in every iteration. But these outputs represent valid output elements (i.e. the elements of A^{-1}) only after the first n iterations.

Since the search for the pivot is performed only on or below the diagonal, P_{22} transmits elements with tag 't' unchanged, i.e. without any comparison or interchange. For such inputs, it also makes $g_{out} = 0$, so that all P_{2j} (j > 2)

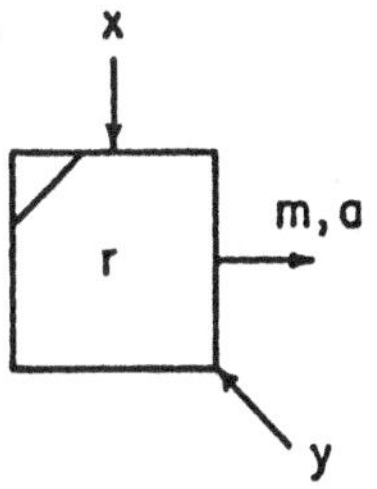

$$x$$

m, a

r

$$y$$

P_{11}

if 'x' on SE-input (with y) then r ← y ;

m ← x / r ;   a ← r ;

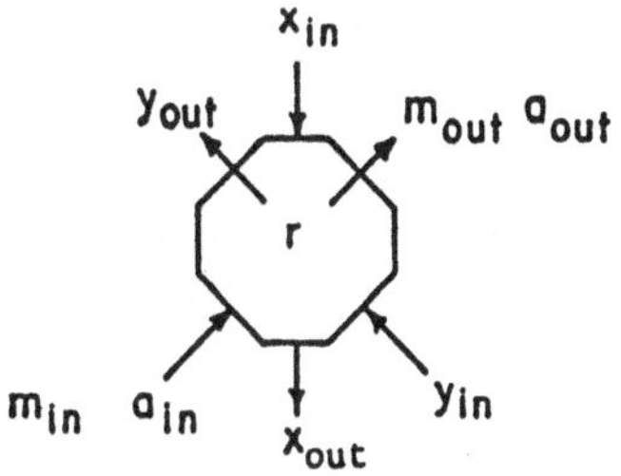

x_{in}

y_{out} m_{out} a_{out}

r

m_{in} a_{in} y_{in}

x_{out}

P_{1j} (j > 1)

if 'x' on SE-input then r ← y_{in} ;

if ('t' on N-input) or (flag set) then

$$\left[x_{out} \leftarrow (x_{in} - m_{in} \cdot r) ; y_{out} \leftarrow (x_{in} - m_{in} \cdot r) ; \right]$$

else

begin

flag ← 1;

x_{out} ← r / a_{in} , t;

y_{out} ← r / a_{in} ;

end

a_{out} ← a_{in} ;

m_{out} ← m_{in} ;

if 'x' on N-input then flag ← 0 ;

x_{out} carries 'x' and 't' if x_{in} carries them.

Figure 2. PE functions. Continued.

also transmit these elements without an interchange. Elements without a 't' are on or below the diagonal and hence are searched as usual for the pivot.

It is noted that all operations performed on A are also performed on the identity matrix I. Thus, at the end of n iterations, A would be transformed to I and I would be transformed to A^{-1}.

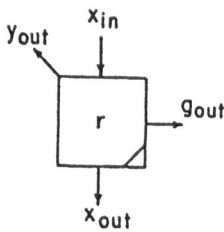

P_{22}

Mode I ('*' on N-input).

if $|x_{in}| \geqslant |r|$ then
 begin
  $y_{out} \longleftarrow x_{in}$, * ;
  $x_{out} \longleftarrow r$, * ;
  $g_{out} \longleftarrow I$;
 end
else
 begin
  $y_{out} \longleftarrow r$, * ;
  $x_{out} \longleftarrow x_{in}$, * ;
  $g_{out} \longleftarrow 0$;
 end
$r \longleftarrow 0$;

Mode 2 (No '*' on N-input)

if 't' on N-input then
 $[x_{out} \longleftarrow x_{in}$, t ;
  $g_{out} \longleftarrow 0$;   ]
else
 if $|x_{in}| \geqslant |r|$ then
  begin
   $x_{out} \longleftarrow r$;
   $r \longleftarrow x_{in}$;
   $g_{out} \longleftarrow I$;
  end
 else
  begin
   $x_{out} \longleftarrow x_{in}$;
   $g_{out} \longleftarrow 0$;
  end ;

Figured 2. Continued

3.1 Completion Time

As shown in Figure 1, the proposed array is supplemented with an array of buffers, with (n-1) buffers in each column. Hence, in each iteration, an input element would have to pass through the 2 PEs in that particular column and the (n-1) buffers below that column of PEs.

Thus, time for each iteration = 2 + (n-1) = n + 1.

Total number of iterations = n.

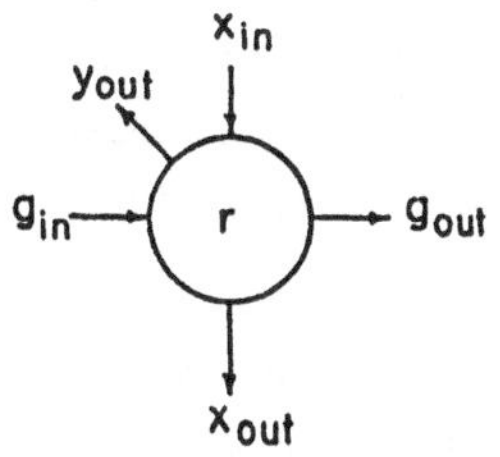

P_{zj} (j > 2)

Mode I ('*' on N- input)

if (g_{in} = I) then

  begin

    $y_{out} \leftarrow x_{in}$, * ;

    $x_{out} \leftarrow r$, * ;

  end

else

  begin

    $y_{out} \leftarrow r$, * ;

    $x_{out} \leftarrow x_{in}$, * ;

  end

$g_{out} \leftarrow g_{in}$;

r $\leftarrow$ 0 ;

Mode 2 (No '*' on N- input)

if (g_{in} = I) then

  begin

    $x_{out} \leftarrow r$;

    r $\leftarrow x_{in}$;

  end

else

    $x_{out} \leftarrow x_{in}$;

    x_{out} carries 't'

    if x_{in} carries 't'.

Figure 2. Continued

Hence, total computation time = n.(n + 1) + (2n - 1)

The (2n-1) time units are added to account for the skewing of the output. Thus, the completion time for the array = $n^2 + 3n - 1$.

4. <u>Conclusions</u>: A purely systolic architecture for computing the inverse of a matrix has been presented. The architecture implements an inversion algorithm based on the Gauss-Jordan diagonalization method with partial pivoting. The architecture employs 4n + 1 PEs and has a time complexity of $O(n^2)$. Thus the area-time complexity is $O(n^3)$, which matches the performance of the fastest systolic implementation of matrix inversion (numerically unstable in most cases), reported to date.

A regular and continuous data flow is maintained within the array. The bi-linear array has been supplemented with a buffer array to eliminate the need for costly inter-iteration I/O. Thus, the total number of I/O operations has been minimized and I/O operations are needed only to input the matrix to be inverted and to retrieve its inverse.

<u>Acknowledgement</u>: This work was supported in part by the State of Louisiana under Grant LEQSF-RD-A-17. The authors are with the Department of Electrical and Computer Engineering, Louisiana State University, Baton Rouge, LA 70803.

References

1. El-Amawy A., "High speed inversion of dense matrices on an optimal systolic array", CompEuro Conf. Proc., Hamburg, W. Germany, pp. 707-711, May 1987.
2. Chern M. and Murata, T., "A fast algorithm for LU decomposition and matrix inversion", Proc. Int'l. Conf. on Parallel Processing, Columbus, OH, pp. 79-86, Aug. 1983.
3. Comon P. and Robert Y., "A Systolic array for computing BA^{-1}", IEEE Trans. on Accoustics, Speech and Signal Processing, Vol. ASSP-35, No. 6, June 1987.
4. El-Amawy, A. and Barada, H., "Efficient Linear and Bilinear Arays for Matrix Triangularization with Partial Pivoting", 21st Asilomar Conf. Signals, Systems, and Computers, Pacific Grove, CA, pp. 757-761.
5. Wilkinson, J.H., "The Algebraic Eigenvalue Problem', Oxford University Press, 1965.
6. Stewart, G.W., "Introduction to Matrix Computations", Academic Press, 1973.
7. Rau, N., "Matrices and Mathematical Programming", St. Martin's Press, New York, 1981.
8. Peters, G. and Wilkinson, J.H., "On the stability of Gauss-Jordan elimination with pivoting", Communications of the ACM, Vol. 18, pp. 20-24, Jan. 1975.

KALMAN FILTERING SYSTOLIC ARRAY DESIGN USING DEPENDENCE GRAPH MAPPING

Robert A. Lincoln
University of California, Los Angeles

Kung Yao
University of California, Los Angeles

Introduction. In many control and communication applications, a signal is to be extracted from noisy measurements. Such systems can frequently be described as a discrete time-varying first order vector-valued dynamic system with additive colored system and observation noise given by

$$x(K+1) = F(K)x(K) + w(K+1),$$ (1)

$$y(K) = C(K)x(K) + v(K).$$

An estimate of the N-dimensional system vector, $x(K)$, is the signal of interest, but only the M-dimensional measurement vector, $y(K)$, is observed. The noise vectors, $w(K)$ and $v(K)$, are zero mean uncorrelated vectors with known covariance matrices, $R_W(K)$ and $R_V(K)$ respectively. The Kalman filter (KF) is a recursive approach for the solution of the minimum variance estimator or predictor for $x(K)$ [1]. The KF method was not only an important contribution to statistical estimation theory but has widespread use in modern aerospace and avionic guidance and control applications. The original KF recursively updates the error covariance matrix. Various alternatives such as updating the square-root covariance matrix, information matrix, and square-root information matrix have since been formulated [2]-[3]. In particular, Paige and Saunders [4] proposed a modified square-root information filter (SRIF) for Kalman filtering which transforms the problem into a least squares computation at each filter iteration or recursion.

Triangular systolic array implementation of the measurement updating equations based on the square-root covariance approach was considered by Jover and Kailath [5]. Using the SRIF approach, Chen and Yao [6]-[9] presented a Kalman filter systolic array which performed both the time and measurement updating equations by solving the least squares computation by QR decomposition method. Heuristic manipulation of a known QR decomposition systolic array led to a systolic array for the in-place updating of the time and measurement equations. Kung and Hwang [10]-[11] observed that by taking advantage of the special form of a submatrix in the QR decomposition, the number of computations necessary for the decomposition is reduced. Dependence graph mapping techniques were then used to develop an array for the first part of the QR decomposition. Based on heuristic duality observations, this array was then adapted for the rest of the algorithm. Gaston and Irwin [13] proposed a systolic array formulated by making similar use of the submatrix form and heuristically restructuring the QR decomposition systolic array. In this paper, using the dependence graph mapping methodology of Rao [12], Kalman filter systolic arrays are derived in a completely systematic manner. In addition to providing an alternative approach in deriving the Kung-Hwang array, a new array is derived which has a greater speed-up factor and processor utilization. The Gaston and Irwin array is also developed using dependence graph mapping techniques. Comparisions of the advantages and unique features of the Chen-Yao array, the Kung-Hwang array, the Gaston-Irwin array and the new proposed array are summarized in Table 1.

Modified Square-Root Information Filter. The least squares formulation for the Kalman Filter is made possible by the use of a whitening operation on the system equations of (1). Denote the Cholesky factorization of the inverses of the noise covariance matrices as $R_W(K+1)^{-1} = W(K+1)^T W(K+1)$ and $R_V(K)^{-1} = V(K)^T V(K)$ where the matrices $W(K+1)$ and $V(K)$ are upper triangular. Multiplying the two equations of (1) by the whitening operators $W(K+1)$ and $V(K)$, respectively results in white noise processes, for which the minimum variance solution can be obtained by a least-squares computation. At each Kalman filter recursion, the QR triangularization of a $(2N+M) \times (2N+1)$ matrix is needed as follows

$$
T\begin{bmatrix} R(K) & O & b(k) \\ \tilde{F}(K) & W(K+1) & O \\ \tilde{C}(K) & O & \tilde{y}(K) \end{bmatrix} = \begin{bmatrix} R_{K,K} & Z(K+1) & a(K) \\ O & R(K+1) & b(K+1) \\ O & O & r_K \end{bmatrix}
\tag{2}
$$

where T is the orthogonal transformation matrix for the triangularization. The upper triangular form of $W(k)$ is the submatrix structure used to reduce the computations. $R_{k,k}$ and $R(K+1)$ are NxN upper triangular matrices. The coefficent matrices after whitening, $\tilde{F}(K)$ and $\tilde{C}(K)$, are $\tilde{F}(K) = -W(K+1)F(K)$ and $\tilde{C}(K) = V(K)C(K)$. Similarly, the observation vector after whitening is $\tilde{y}(K) = V(K)y(K)$. The minimum variance predictor of the state vector, $x(K+1)$, is $\hat{x}(K+1) = R(K+1)^{-1} b(K+1)$. A similar QR triangularization problem can be posed for the minimum variance estimator.

By performing the QR decomposition nullification in (2) in a particular order, the W matrix remains triangular [10]-[11]. Specifically, the matrix $\tilde{C}$ is first zeroed, and then the zeroing of the matrix $\tilde{F}$ starts in the lower left component and proceeds up each column before starting at the bottom of the next column. A two stage single assignment algorithm for the QR decomposition is given in this paper and dependence graph mapping is then applied to it to derive systolic arrays for Kalman filtering. First, a brief overview of the dependence graph mapping design methodology to be used will be given.

Dependence Graph Mapping. In an effort to make systolic array design a systematic process, a number of systolic array design methodologies have been proposed. In dependence graph mapping methodologies, an algorithm defined over a multi-dimensional dependence graph is mapped onto a processor array of lower dimension [12],[14]-[16]. In particular, the formulation of Rao [12], will be used in this paper.

The algorithm is first formulated as a single assignment algorithm in which every variable has a unique value during the evaluation of the algorithm. A single assignment algorithm can be represented as a dependence graph, where the algorithm computations are represented as nodes and the dependencies between the variables are represented by directed arcs. The nodes of the graph are denoted by integer-valued index vectors whose range is called the index space. Designing a systolic array for an algorithm requires the assignment of each node or algorithm computation, to a processor and to a time slot.

The dependence graph of an algorithm is mapped onto a processor space or processor array of lower dimension by the processor allocation function. The simplest form of allocation function is based on a linear projection in the direction of a projection vector, denoted here as U. The allocation function is specified for a S-dimensional index space as $A(I)=P^T I$, where I is an index point in the index space. $A(I)$ is an integer (S-1)-dimensional location in the processor space representing a processor location, and P is a $S \times (S-1)$ dimensional integer matrix whose columns form a basis for the processor space. Since the projection direction is orthogonal to the processor space, the relationship $P^T U = 0$ holds.

The scheduling function assigns a time slot for each calculation. An affine scheduling function is defined as $S(I)=\Lambda^T I+\gamma$, where Λ is a S-dimensional integer vector called the schedule vector and γ is an integer additive constant. $S(I)$ specifies the time slot during which the calculations for index point I are done. The scheduling function must satisfy the variable precedences of the algorithm, i.e., if variable $x(I+D)$ depends on the value of variable $Y(I)$, then Λ must satisfy $\Lambda^T D \geq 1$. The vectors D are called the index displacement vectors. To insure that no processor is assigned more than one calculation for a particular time slot, the projection vector and schedule vector must satisfy $|\Lambda^T U| \geq 1$.

Thus, the systolic array is completely defined by specifying the scheduling function and the allocation function. The index displacement vectors are transformed to processor array communication links by $D_p = P^T D$ and $\tau = A^T D$ where D_p gives the communication link processor displacement in the array and τ specifies the time difference between performance of the calculations.

QR Triangularization Single Assignment Algorithm. The QR triangularization of equation (2) to be performed at each filter iteration can be divided into two stages. In the first stage, the matrices $\tilde{C}$ and $\tilde{F}$ are nullified, while in the second stage the Givens rotations generated in stage one are applied to the W and Z matrices. The application of the rotations to the vectors b and $\tilde{y}$ are included in the first stage. For each of the stages, a single assignment algorithm can be given to which dependence graph mapping is applied.

The single assignment algorithm for stage one consists of three types of variables defined over the index space (i,j,k) with $1 \le i \le N+M$, $1 \le k \le N$, and $k \le j \le N+1$. The computation precedences in the algorithm provide for the nullification of the matrices' components in a bottom-to-top, left-to-right order. The variable definitions at each index point are:

$$r(i-1,j,k) = \begin{cases} r_{kj} & i=N+M+1, \ j \le N \ , \\ b_k & i=N+M+1, \ j=N+1, \\ r(i,j,k)\cos\phi(i,j,k) + f(i,j,k)\sin\phi(i,j,k) & i \le N+M \ . \end{cases}$$

$$f(i,j,k+1) = \begin{cases} \tilde{f}_{1,j} & k=0, \ i \le N \ , \\ \tilde{C}_{i-N,j} & k=0, \ i>N, \ j \le N \ , \\ \tilde{y}_{i-N} & k=0, \ j=N+1 \ , \\ -r(i,j,k)\sin\phi(i,j,k) + f(i,j,k)\cos\phi(i,j,k) & k \ge 1 \ . \end{cases}$$

$$\phi(i,j+1,k) = \begin{cases} \tan^{-1}(f(i,j,k)/r(i,j,k)) & j=k \ , \\ \phi(i,j,k) & j>k \ . \end{cases}$$

The single assignment algorithm for stage two also has three variable types defined over the index space (i,m,k) with $1 \le i \le N$, $i \le m \le N$, and $1 \le k \le N$. The variable definitions are:

$$w(i,m,k+1) = \begin{cases} w_{i,m} & k=0 \ , \\ -z(i,m,k)\sin\beta(i,m,k) + w(i,m,k)\cos\beta(i,m,k) & k \ge 1 \ . \end{cases}$$

$$z(i-1,m,k) = \begin{cases} 0 & i=m+1 \ , \\ z(i,m,k)\cos\beta(i,m,k) + w(i,m,k)\sin\beta(i,m,k) & i \le m \ . \end{cases}$$

$$\beta(i,m+1,k) = \begin{cases} \phi(i,k,k) & m=i-1 \ , \\ \beta(i,m,k) & m \ge i \ . \end{cases}$$

The rotational parameters, $\phi(i,k,k)$, calculated in the stage one algorithm are needed as initialization for the stage two algorithm. For simplicity, the single assignment algorithms above are based on using the Givens algorithm for the QR triangularization, but other algorithms such as the Modified Fast Givens can be used [12]. The dependence graph mapping can be applied to those QR triangularization algorithms and the resulting systolic array designs are similar since the algorithms have similar structure.

Kalman Filter Systolic Array Derivation. Dependence graph mapping can be applied to the single assignment algorithms for each stage separately. Since a single systolic array is used to perform both stages, the scheduling functions and allocation functions for both stages can not be selected independent of each other. However, there remains a large amount of flexibility in the design choices for the scheduling and allocation function components. In this section, the particular dependence graph mapping choices for the Kung-Hwang and Gaston-Irwin arrays are specified. In addition, a new efficient Kalman filter processor array is derived.

First, consider the stage one single assignment algorithm. There are an infinite number of directions in which the dependence graph can be projected. The selection of projection vector $U^T=[1, 0, 0]$ defines a projection requiring the fewest number of processors. The corresponding projection matrix is not unique, but the choices represent different processor space bases. For the projection matrix

$$P^T = \begin{bmatrix} 0 & 1 & 0 \\ 0 & 0 & 1 \end{bmatrix} , \qquad (3)$$

the range of the processor space is $P^T[i, j, k]^T=[j, k]$ with $1 \leq k \leq N$ and $k \leq j \leq N+1$. Thus, the array is a triangular array with an additional column of processors. The index displacement vectors from the stage one single assignment algorithm, $\{ [-1, 0, 0]^T, [0, 0, 1]^T, [0, 1, 0]^T \}$, impose the following constraints on the schedule vector, $\Lambda^T=[\lambda_1, \lambda_2, \lambda_3]$: $\lambda_1 \leq -1, \lambda_2 \geq 1$, and $\lambda_3 \geq 1$. These are satisfied by $\Lambda^T=[-1, 1, 1]$. The stage one scheduling function additive constant is only important in relation to the scheduling function for the second stage. For the choice of $N+M-1$, the stage one processing starts at time slot 1 with the computation at index $(N+M,1,1)$. The scheduling function for stage one, denoted as $S_1(i,j,k)$, is then

$$S_1(i,j,k) = -i + j + k + (N+M-1) . \qquad (4)$$

The index displacement vectors for the r, f, and ϕ variables are transformed to communication links with processor space displacement $[0\ 0]^T$, $[0\ 1]^T$,

$[1 \ 0]^T$, respectively, by P^TD. The r variables thus remain in place in the array, while the f and ϕ variables move in orthogonal directions. Since for all the index displacement vectors, $\Delta^TD=1$, no delay registers are required on the communication links. As given in the single assignment algorithm, the external input of the matrices, $\tilde{F}(K)$ and $\tilde{C}(K)$, and the observation vector, $\tilde{y}(K)$, is through the variables f(i,j,0) with $1 \le i \le N+M$ and $1 \le j \le N+1$. These indices are mapped to processors by $P^T[i, \ j, \ 0]^T = [j, \ 0]$ and time slots by $S_1(i,j,0)=-i+j+N+M-1$. Hence, the components of a column of $\tilde{F}$ and $\tilde{C}$ are loaded into the array through the same processor, starting with the bottom component since the coefficient of the scheduling function for the i dimension of the index space is negative. Each component of a matrix row enters the array one time slot prior to the component to the right since the j dimension scheduling function coefficient is one. Therefore, the input is loaded into the array in a skewed manner with the bottom rows of the matrices loaded first. Because the $\tilde{C}$ matrix initializes the f variables in the single assignment algorithm with the larger i dimension values, it is input to the array before the $\tilde{F}$ matrix.

The updated W matrix is used as the initial R matrix for the next recursion of the Kalman filter, i.e., the initial r_{kj} for the next iteration equals the final w_{1m} for the current iteration. To simplify the transition to the next recursion, it is desirable for the final processor space location of the w variables to be the same as the initial location of the r variables. The R matrix initializes the r variables at the stage one index space locations [N+M,j,k], which are mapped to processor space locations by $P^T[N+M, \ j, \ k]^T=[j, \ k]^T$. Let Q^T denote the projection matrix for stage two. The final w variables are at the stage two indices (i,m,N), hence, the final processor space location for the w variables is $Q^T[i, \ m, \ N]^T$. The selection of Q^T should, therefore, be such that $Q^T[i, \ m, \ N]^T=P^T[N+M, \ j, \ k]$ for i=k and m=j which is satisfied for

$$Q^T = \begin{bmatrix} 0 & 1 & 0 \\ 1 & 0 & 0 \end{bmatrix}. \tag{5}$$

The corresponding projection vector is $U^T=[0, \ 0, \ 1]$ and the range of the processor space is $[m, \ i]^T$ where $1 \le i \le N$ and $i \le m \le N$. This processor space is identical to that specified earlier for stage one, excluding the stage one extra column of processors. Normally, for dependence graph mapping the projection vector is selected first and then a projection matrix is determined. The processor spaces from the different projection matrices are rotations of one another corresponding to different bases. For the stage two processor space to overlay the stage one processor space so that the appropriate r and w variables are mapped to the same locations, the stage two processor space basis must be chosen accordingly. The constraints on the scheduling vector for the second stage imposed by the set of displacement vectors { $[0, 0, 1]^T$, $[-1, 0, 0]^T$, $[0, 1, 0]^T$ } and the projection vector are satisfied by $\Delta^T = [-1, 1, 1]$. The communication links for the w, z, and β variables are $[0\ 0]^T$, $[0\ -1]^T$, $[1\ 0]^T$ respectively. The w variables remain in place, although since they are external inputs, they will have to be first loaded into the array. The z variables move in the opposite direction as the f variables of stage one.

Up to this point, the two stages have been treated as two separate algorithms with regards to the scheduling functions. They are, however, linked through the rotational parameters, ϕ_{ik}. These are calculated in the first stage at the index points (i,k,k), while for the second stage, they are external inputs needed in initializing the β variables at index points (i,i,k). This initialization must not be scheduled prior to the calculation of the corresponding rotational parameter. The scheduling functions for the stages, $S_1(i,j,k)$ and $S_2(i,m,k)$, must therefore satisfy $S_2(i,i,k) > S_1(i,k,k)$ for $1 \le k \le N$ and $1 \le i \le N$. This places a condition on the additive constant γ for S_2. It must satisfy the inequality for $1 \le i \le N$ and $1 \le k \le N$ given by

$$[-1, 1, 1][i, i, k]^T + \gamma > [-1, 1, 1][i, k, k]^T + (N+M-1) . \tag{6}$$

Although stage one computes rotational parameters at indices with i>N, i.e., those used to nullify the $\tilde{C}$ matrix, these are not used to initialize any stage two variables. Thus, γ can be chosen as $2N+M-1$ and the resulting stage two scheduling function is given as

$$S_2(i,m,k) = -i + m + k + (2N+M-1) . \tag{7}$$

The rotational parameter, ϕ_{ik}, is calculated at processor location $[k, k]^T$ and is needed for stage two initialization at processor location $[i, i]^T$ after a delay of i-k+N time slots. These processors are all boundary processors and a complex data buffer is required to transport the rotational parameters to the appropriate processor with the specified delay.

By design, the projection matrices for the two stages were selected so that the two index spaces were mapped to overlaying processor spaces. This could result in a conflict, in which an index point, i.e., computation, from each stage is mapped to the same processor and the same time slot. For a realizable design, a particular processor must not be assigned for any time slot, the calculations associated with more than one index point. The stage one allocation function assigns to processor (a,b) the calculation of the stage one index points, I, such that $P^T I = [a, b]^T$. The stage one scheduling function assigns this set of index points, i.e., j=a, k=b, $1 \le i \le N+M$, time slots a+b-1, a+b, ... , a+b+N+M-2. Similarly, processor (a,b) is assigned the stage two index points, such that $Q^T I = [a, b]^T$, which are scheduled for calculation at time slots a-b+2N+M, a-b+2N+M+1, ... , a-b+3N+M-1. Therefore, there is a conflict between the stage one and stage two assignments at processor (a,b) if a+b+N+M-2 ≥ a-b+2N+M, which simplifies to b ≥ N/2 + 1. Such a conflict occurs at the processors of the lower portion of the array specified by N/2+1≤b≤N and b≤a≤N.

There are two options to resolve this conflict:
Option 1) Increase the additive constant for the stage two scheduling function,
Option 2) Provide additional processors as needed to remove the conflict.
For Option 1, the additive constant is increased by an offset to delay the stage two processing relative to the stage one processing. The offset must satisfy a+b+N+M-2 < a-b+2N+M+offset for all (a,b) with 1≤b≤N and b≤a≤N. Modifying the stage two scheduling function of equation (7) to include the additional offset results in a new scheduling function of

$$S_2(i,m,k) = -i + m + k + (3N+M-1). \qquad (8)$$

The resulting systolic array is the array derived by Kung and Hwang, and is shown in Figure 1. For Option 2, the duplicate processors are needed at processor locations (a,b) such that N/2+1≤b≤N and b≤a≤N, which is approximately one quarter of the processor locations as shown in Figure 2. For the array of Figure 1, the W matrix can be input after the $\tilde{F}$ matrix. Since the stage two processing starts earlier in the array of Figure 2, the input of the W matrix begins before the $\tilde{F}$ matrix is complete. Thus, additional communication links are necessary.

For recursive Kalman filter updating using these arrays, the next recursion can start once the required updates from the current recursion are available. The updated W matrix is used to initialize the next recursion's R matrix. The w variable w(i,m,N) is the initial R matrix component $r_{k,j}$ for the next filter recursion. In the stage one single assignment algorithm, $r_{k,j}$ is used to initialize the r variable at index (N+M+1,j,k). Therefore, the minimum time between Kalman filter recursions, called the recursion step, must satisfy the following equation for i=k and m=j:

$$S_1(N+M+1,j,k)+(\text{recursion step}) > S_2(i,m,N). \qquad (9)$$

For Option 1, this results in a recursion step of 4N+M, and for Option 2, the recursion step is 3N+M, as shown in Figures 1 and 2.

These two systolic array Kalman Filters were derived under the criteria of minimizing the number of processors required. That is, the projection vectors for each stage were chosen so that the projected index space resulted in the minimal number of indices in the resulting processor space. The processor allocation function for stage two was chosen so that the processor space location of the final w variables were identical to the location of the r variables. This simplified the transition from one Kalman Filter recursion to the next, since the r variables are initialized by the w variables of the previous filter recursion. Other selections of iteration vectors for the two stages will result in arrays with different characteristics.

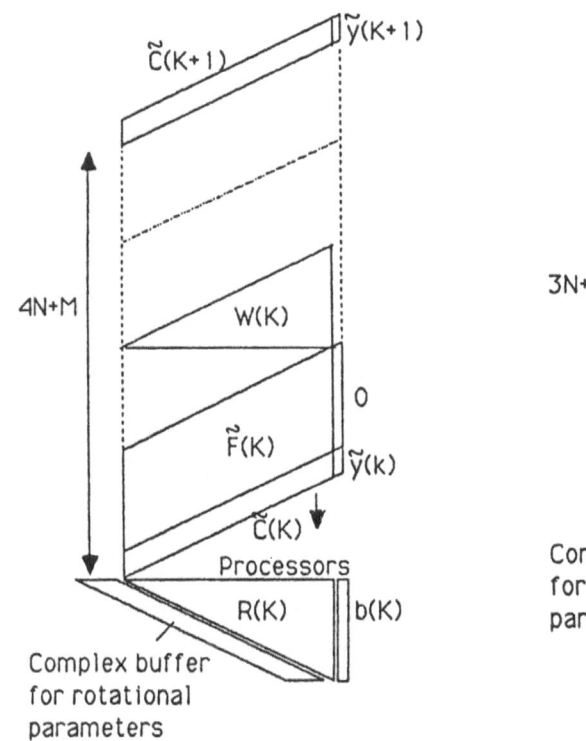

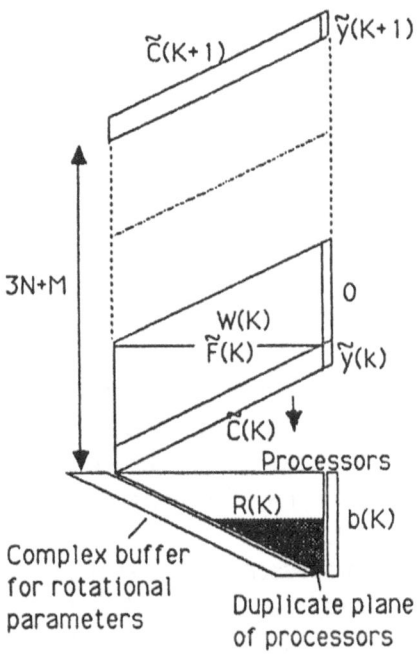

Figure 1. Kung-Hwang Kalman
filter processor array.

Figure 2. New Kalman filter
processor array.

One alternative is to select the stage two projection as $U^T=[\ 1,\ 0,\ 0\ ]$, instead of $U^T=[\ 0,\ 0,\ 1\ ]$. The systolic array proposed by Gaston and Irwin [13] can be derived by starting with this selection for the stage two projection direction.

For the choice of $U^T=[1,\ 0,\ 0]$, the projection matrix can be selected as:

$$Q^T = \begin{bmatrix} 0 & 1 & 0 \\ 0 & 0 & 1 \end{bmatrix}. \tag{10}$$

The processor space resulting from this choice is $Q^T[i,\ m,\ k]=[m\ k]$ with $1 \le m \le N$ and $1 \le k \le N$, which is a NxN square of processing elements overlaying the stage one triangular processor space. The three stage two index displacement vectors are mapped to communication links of [0, 1], [0, 0], and [1, 0] for the w variables, z variables, and β variables respectively. Note that in contrast to the previously derived arrays, the z variables remain in a particular processing element, while both the w variables and β variables flow through the array. Since the final w variables are the components of the matrix R(K+1), having them exit the array is advantageous in the calculation of $\hat{X}(K+1)=R(K+1)^{-1}b(K+1)$.

The scheduling functions given in equations (4) and (7) satisfy the constraints on the scheduling functions. The rotational parameters are still calculated in the diagonal processors (k,k), but the stage two index points (i,i,k), where they are used for initialization is mapped to processor (i,k). In addition, the diagonal processors are no longer on the boundary of the array, since the new stage two projection direction results in a rectangular array. However, since the rotational parameters are propagated in the stage one single assignment algorithm without being modified, the initialization of the stage two β variables need not use ϕ at index (i,k,k), but can use the index (i,N,k) for example. Making such changes to the single assignment algorithms, results in the Gaston-Irwin sytolic array.

The processor (a,b) is assigned stage one calculations during the time slots: a+b-1, a+b,..., a+b+N+M-2. The processor is needed for the stage two processing in time slots: b+2N+M-1, b+2N+M, ..., a+b+2N+M-2, hence there is no conflict between the requirements for the processors for the two stages. Since the scheduling functions for the two stages are the same as for Option 2, the minimum recursion time step is 3N+M.

<u>Kalman Filter Systolic Array Comparisons</u>. The processing power available with an array is the number of processors times the number of time slots needed for completion of the algorithm. Thus, the processing power available for each systolic array is the product of the number of processors and the recursion step. The processing power used for a Kalman filter recursion is determined from the size of the index spaces for the two stages. The index space of the first stage consists of the index points (i,j,k) with $1 \le i \le N+M$, $1 \le k \le N$, and $k \le j \le N+1$. Thus, the total number of computations or index points in the first stage is $(N+M)[(N(N+1)/2)+N]$. The term computation is used to denote the set of operations needed to evaluate the three variables at an index point. Similarly, the second stage index space of $1 \le i \le N$, $1 \le k \le N$, and $i \le m \le N$ requires $N(N/2)(N+1)$ computations. The two stages together require a total of $N^3 + N^2M/2 + 2N^2 + 3NM/2$ computations. A measure of an array's efficient use of its processing resources is the ratio of the processing power used to the

processing power available. This ratio, called the average processor utilization, gives the percentage of the time that an average processor is performing algorithm computations. Processor utilization is not affected by the use of a processor to transport values since the numerator contains only the number of computations given in the single assignment algorithm. The processor utilization for a particular processor can also be determined using the dependence graph mapping specification of the array. The allocation function specifies the index space points assigned to a particular processor (a,b). The ratio of the number of these indices to the recursion step gives the particular processor utilization for processor (a,b). For the Gaston-Irwin array and the new array, some processors are used only for computations of one of the stages. For example, since the processor space for the second stage of the Gaston-Irwin array is square, while that for the first stage is triangular, some processors are only assigned stage two values.

Table 1 compares the Kalman Filter systolic arrays regarding the number of processors, the recursion step, the average processor (PE) utilization for the cases N>>M and N=M>>1, and the particular processor utilization for N>>M. The Chen-Yao array uses a different algorithm than the single assignment algorithm given, i.e., the nullification of the $\tilde{F}$ and $\tilde{C}$ matrix elements is performed in a different order, which destroys the upper triangular form of W. Additional computations are then needed to retriangularize W. When comparing arrays, the processor utilization for an array should not appear higher just because its algorithm is inherently inferior or less efficient. Thus, for the Chen-Yao array average processor utilization given in Table 1, the same numerator for the ratio, i.e., the processing power used, is employed as for the other arrays. Because the algorithm of the Chen-Yao array requires more computations spread among the processors, meaningful comparison of the particular processor utilization is not possible.

Table 1. Comparison of Kalman Filter Systolic Arrays

Array	Processors	Recursion Step	Average PE Utilization		Particular PE Utilization
			N>>M	N=M>>1	N>>M
Chen-Yao	(N/2)(3N+5)	2N+M	1/3	1/3	not comparable
Kung-Hwang	(N/2)(N+3)	4N+M	1/2	3/5	1/2
Gaston-Irwin	N(N+1)	3N+M	1/3	3/8	1/3 to 2/3
(PEs for stage two only)					0 to 1/3
New Array	(N/8)(5N+13)	3N+M	8/15	3/5	2/3
(PEs for stage one or two only)					1/4

Acknowledgement. This work is partially supported by NASA/AMES Research Contract NAG-2-304 and the UC MICRO program. The authors are with the Electrical Engineering Department, UCLA, Los Angeles, CA, 90024-1594.

References

[1] Kalman, R. E., "A New Approach to Linear Filtering and Prediction Problems," J. Basic Engineering, vol. 82, pp. 35-45, March 1960.

[2] Dyer, P., and S. McReynolds, "Extension of Square-Root Filtering to Include Process Noise," J. Optimiz. Theory Appl., vol. 3, pp. 445-459, 1969.

[3] Kaminiski, P. G., et al, "Discrete Square-Root Filtering: A Survey of Current Techniques," IEEE Tran. on Auto Control, vol 6, pp. 727-736, Dec. 1971.

[4] Paige, C. C, and M. A. Saunders, "Least squares estimation of discrete linear dynamic systems using orthogonal transformation," SIAM J. Numer. Anal., vol. 14, pp. 180-193, 1977.

[5] Jover, J. M., and T. Kailath, "A Parallel Architecture for Kalman Filter Measurement Update and Parameter Estimation," Automatica, vol. 22, pp. 43-57, 1986.

[6] Chen, M. J., and K. Yao, Systolic Array, edited by W. Moore, etc., Adam Hilger, 1986, pp. 161-170.

[7] Chen, M. J., and K. Yao, "On realization and implementation of Kalman filtering by systolic array," in Proc. of the 1987 Conf. on Information Sciences and Systems, John Hopkins Univ., March 1987.

[8] Chen, M. J., and K. Yao, "Systolic Kalman Filtering based on QR Decomposition," in Proc. SPIE, vol. 826, Aug. 1987, pp. 25-32.

[9] Chen, M. J., On Realizations and Performances of Least-Squares Estimation and Kalman Filtering by Systolic Arrays, Ph.D. thesis, University of California, Los Angeles, April 1987.

[10] Kung, S. Y., and J. N. Hwang, "Systolic Designs for State Space Models: Kalman Filtering and Neural Network," in Proc. of the 26th Conf. on Decision and Control, Dec 1987, pp. 1461-1467.

[11] Kung, S. Y., and J. N. Hwang, "An Efficient Triarray Systolic Design for Real-Time Kalman Filtering," in Proc. ICASSP, April 1988, pp. 2045-2048.

[12] Rao, S. K., Regular Iterative Algorithms and Their Implementations on Processor Arrays, Ph.D. thesis, Stanford University, Oct. 1985.

[13] Gaston, F. M. F., and G. W. Irwin, "A Systolic Square Root Information Kalman Filter," in Proc. of International Conference on Systolic Arrays, 1988, pp. 643-652.

[14] Quinton, P., "The systematic design of systolic arrays," IRASA International Publication 193, April 1983.

[15] Moldovan, D. I., "On the design of algorithms for VLSI systolic arrays," Proc. IEEE, vol. 71, pp. 113-120, Jan. 1983.

[16] Kung, S. Y., "On supercomputing with systolic/wavefront array processors," Proc. IEEE, vol. 72, pp. 867-884, July 1984.

Generalized Distributive Memory Arrays

William A. Porter
Jorge L. Aravena

Louisiana State University

1.0   Introduction.   In recent years neuroanatomical models of brain functioning have given an impetus to the development of structures with associative (often called distributive) memory.   Such structures were referred to generically as neural networks.   In the initial studies of neural networks the processing nodes emulated neurons while the connecting linkages emulated synaptic channels.

In previous studies the neuron is considered to be a bistable device, typically assuming values ± 1. The neural structure is densely interconnected with linear channels of gain α_{ij} for transmissions from the ith to jth neuron.   Other assumptions; symmetry, zero diagonal, etc. are often also invoked.

Each neuron performs an arithmetic computation.   In the bistable case the state of the ith neuron is related to the other neuron states by a threshold function such as

$$x_i = \text{sgn} \left\{ \sum_{j=1}^{n} \alpha_{ij} x_j \right\}, \quad i = 1,\ldots n. \tag{1}$$

For synchronous operation the above equation is apparently a state transition equation. An input to a neural network is usually modeled as an initial condition on the state equation.   The equilibrium points of the state equation represent recognition.

While the above remarks are simplistic they do underscore certain characteristics of neural computations.   In particular the map $x(t) \rightarrow x(t+1)$, implemented in equation (1) consists of a linear map (the matrix α), followed by nonlinear operations (the sgn function) on each component of $\alpha x(t)$.   In addition, the neural computation proceeds by iteration. Finally we note that the 'memory' of the neural net is generally attributed to the scalars α_{ij}.

In a recent study [1], the sgn function of equation (1) was generalized to an arbitrary polynomic function. Surprisingly it then follows that the computation no longer needs to iterate.   In addition, previous limits on memory capacity are eliminated, computational throughput is greatly enhanced.   The network design has an analytic basis and the results have a convenient compatability with VLSI technology.

The results of [1], summarized above, suggest a new approach to distributive memory devices.   This approach focuses on a triple $\{\Delta, F, \Phi\}$ of maps and the composition $\pi = \Delta F \Phi$. The composition π is referred to as a distributive memory module, DMM. Either directly or in series, parallel and feedback combinations the DMM provides powerful pattern recognition capabilities.

In the present study we consider the DMM in a generality not here to fore attempted. We consider arbitrary nonlinearities (as contrasted to the sgn function of [2] [3] and the polynomic functions of [1]) and identify the minimal properties that such functions must satisfy. The allocation of memory among $\{\Delta,F,\Phi\}$ is considered, and the effect of training or memory expansion is analyzed. We consider also questions of robustness, and zones of attraction, and computational throughput. Remarks concerning possible architectures for VLSI synthesis are also included.

2.0 <u>The DMM Model</u>. In this section we develop the structure and some basic properties of the DMM. We turn first to the specification of the domains and ranges of each of the functions of the triple $\{\Delta,F,\Phi\}$.

Let S denote the signal space on which π is to operate. Let $\{x_1,...,x_\ell\}\subset$ S denote an arbitrary set of apriori signals to be recognized. The function Φ maps S into the space C, where C is a tuplet space embeding the characteristics of the set $\{x_1,...,x_\ell\}$; when S,C are linear spaces we shall restrict Φ to be a linear map. As a continuing example we shall use S = R^n and C = R^p and Φ is then identified with a pxn matrix.

The function F is nonlinear. More specifically let $\{f_1,...,f_p\}$ denote scalar functions. For $\xi=(\xi_1,...,\xi_p)\epsilon C$ the function F is computed by

$$F(\xi) = (f_1(\xi_1),...,f_p(\xi_p)). \qquad (2)$$

In most examples the domain and range of F are identical. It is convenient, however, to use R to denote the range of F and refer to it as the recognition space.

The map, Δ, is a generalized format/decode operation. The range of Δ, denoted by Q, may be chosen at the convenience of the application. The map Δ is always 1:1 and may be linear. In some cases, however, Δ provides a code/decode capability, and is by no means linear. We refer to Q as the output space.

For the purpose of brevity selected abuses of notation will also be tolerated. The symbol X will denote both the sets $\{x_1,...,x_\ell\} \subset$ S and the matrix formed by using these vectors as columns. The symbol Φ will denote the set $\{\phi_1,...,\phi_p\}$ and the matrix formed using these vectors as rows. The symbol F will denote the set $\{f_1,...,f_p\}$ and the function computed in equation (2). Finally π will denote the triple $\{\Delta,F,\Phi\}$ as well as the composition $\Delta F\Phi$.

2.1 <u>Some Basic Properties</u>. To minimize the abstractness of our development we specialize S to R^n and C to R^p. The symbol $\langle x,y \rangle$ denotes the usual dot product on these respective Euclidian spaces.

Our first development concerns the separability of the set X using a single linear functional. We start with the following result.

<u>Lemma 1</u>. Given a set $\{x_1,...x_\ell\}\subset R^n$ such that $x_i\neq 0$ all i, then there exists a $c\epsilon R^n$ such that

$$\langle c, x_i \rangle \neq 0 \qquad \text{all } i.$$

Proof. Our proof is by induction. Let c_1 denote a vector such that

$$\langle c_1, x_i \rangle \neq 0 \quad i = 1, \ldots k$$
$$\langle c_1, x_{k+1} \rangle = 0.$$

For example with $c_1 = x_1$ we have $k \geq 1$. Construct c_2 by the formula

$$c_2 = c_1 + \lambda \, x_{k+1}$$

Hence

$$\langle c_2, x_i \rangle = \langle c_1, x_i \rangle + \lambda \langle x_i, x_{k+1} \rangle \quad i = 1, 2, \ldots, k,$$

$$\langle c_2, x_{k+1} \rangle = \lambda \langle x_{k+1}, x_{k+1} \rangle.$$

Since $\langle c_1, x_i \rangle \neq 0 \quad i = 1, 2, \ldots, k$; it follows that

$$m_k = \min_i |\langle c_1, x_i \rangle|$$

$$M_k = \max_i \|x_i\|$$

are strictly positive. Choosing now

$$\lambda = \frac{1}{2} \frac{m_k}{\|x_{k+1}\| M_k}$$

it follows that

$$|\lambda \langle x_i, x_{k+1} \rangle| \leq \frac{1}{2} |\langle c_1, x_i \rangle|, \quad i = 1, 2, \ldots, k.$$

Therefore

$$\langle c_2, x_i \rangle \neq 0, \quad i = 1, 2, \ldots, k+1.$$

Clearly, the process can be repeated until the collection is exhausted.

Corollary. For $\{x_1, \ldots, x_\ell\} \subset R^n$ there exists a $c \in R^n$ such that

$$\langle c_1, x_i \rangle \neq \langle c, x_j \rangle, \quad i \neq j \text{ all } i, j$$

iff the set is distinct (i.e., $x_i \neq x_j$ all $i \neq j$).

Proof. Consider the derived set $\{x_i - x_j : i \neq j, \text{ all } i, j\}$. If X is distinct this set does not contain zero, hence Lemma 1 applies. The resultant c meets the conditions of the corollary. If $x_i = x_j$ then the strengthened implication follows trivially.

A vector $c \in R^n$ satisfying the conditions of the corollary will be called a distinguishing vector for the collection $\{x_1, \ldots, x_\ell\}$.

The result of Lemma 1 and its corollary has a very simple interpretation. If X is distinct then there is a family of parallel hyperplanes with common normal c such that the points of X are isolated by the hyperplanes. We shall return to this interpretation in section 4.0 which is concerned with equivalence classes and robustness.

Our corollary provides also a rudimentary solution to the recognition problem.   For this let X be distinct and c distinguishing on X.   Let $\{g_1,...,g_\ell\}$ be arbitrary scalar signatures, desired for the elements of X.   Let f be any function interpolating the points

$$f(<c,x_1>) = g_1, \quad i = 1,...,\ell.$$

Then $\pi(.) = f(<c,.>)$ is the rudimentary DMM.   We refer to this solution as a unisensor case.

The DMM, multisensor case, is natural extension of the unisensor case.   For this recall our definitions of Φ and F.   Using these we introduce the matrix $G(F,\Phi,X)$ which is given by the formula

$$G(F,\Phi,X) = \begin{bmatrix} f_1(<\phi_1,x_1>) \cdots \cdot f_1(<\phi_1,x_\ell>) \\ \vdots \qquad\qquad \vdots \\ f_p(<\phi_p,x_1>) \qquad f_p(<\phi_p,x_\ell>) \end{bmatrix}.$$

It is easy to verify that
$$G(F,\Phi,X) = F\Phi X.$$
Consistent with these conventions we also write
$$F\Phi\, x = \mathrm{col}(f_1(<\phi_1,x>),...,f_p(<\phi_p,x>)).$$

From a visual inspection of $G(F,\Phi,X)$ it is apparent that the columns of this matrix can be made distinct whenever X is distinct.   Indeed if any ϕ_i is distinguishing on X and if the associated $f_i = I$, then the columns of G are distinct. Similarily one can proceed to develop a choice of F,Φ such that $G(F,\Phi,X)$ has full rank.   We shall demonstrate this for the $p=\ell$ case.

Lemma 2.   X is distinct iff there exists F and Φ such that $G(F,\Phi,X)$ is invertible.

Proof.   Suppose $x_1 = x_2$, then by inspection the first two columns of ΦX are identical and $G(F,\Phi,X)$ is singular.   Suppose X is distinct, and choose each ϕ_i to be distinguishing on X.   The functions f_i may then be chosen to assign the entries of $G(F,\Phi,X)$ arbitrarily.   Choosing a nonsingular assignment the lemma follows.

When $G(F,\Phi,X)$ has full rank we define the function $\pi(F,\Phi,X)$ by the formula

$$\pi(F,\Phi,X)x = G^{-1}(F,\Phi,X)F\Phi x, \qquad x\epsilon R^n. \tag{3}$$

Lemma 3.   If $G(F,\Phi,X)$ has full rank
$$\pi(x_i) = e_i \quad i = 1,...,\ell.$$

Here, e_i is the ith coordinate vector of R^ℓ.   To prove the lemma we note that

$$F\Phi^* x_i \quad i = 1,...,\ell$$

are exactly the columns of G(F,Φ,X). The rows of G^{-1}(F,Φ,X) are duals to these columns from which the result is immediate.

Corollary. If G(F,Φ,X) is invertable then the DMM, π = ΔFΦ, where Δ = G^{-1}(F,Φ,X), recognizes the set $\{x_1,...,x_\ell\}$, moreover T = Xπ has the properties

$$Tx_i = x_i \quad i = 1,...,\ell.$$

The corollary follows from the Lemma by inspection.

2.2 Variations on the DMM. It is of interest to compare the DMM to the Hopfield model, HM. For this let S denote the vector sign function of equation 1. The HM recursion is then apparently

$$x(j+1) = S A x(j).$$

The DMM may also be used recursively yielding

$$x(j+1) = Δ F Φx(j)$$

as its recursion. By a trival identification; Φ = A, F = S, Δ = I, we see that the HM is a special case of the DMM. The DMM, however, is not limited to bistable applications. Moreover, choosing Δ = XG^{-1} as in the corollary to Lemma 3 we have one step convergence with arbitrarily assignable fixed points. Finally we note that DMM does not have the memory limitations, $\ell \leq n$ normally attributed to the HM.

For a second variation on the DMM, reference is made to Lemma 2. The proof of this lemma notes that when the rows of Φ are distinguishing on X, the functions, f_i, may be chosen to assign the entries of G(F,Φ,X) arbitrarily. Indeed, the constraint

$$G(F,Φ,X) = I$$

can be met, thereby eliminating the calculation Δ which is then the identity. This constraint however, may require much more sophistication in the choice of F and carry with it substantial additional difficulties with regard to memory expansion. Section 3.1 gives additional detail on these issues.

We have passed lightly over the construction of functions $f_i \epsilon F$ which provide the requisite interpolations of the development. In theory there are indeed many alternatives ranging from interpolating polynomials to staircase functions. Each such choice, however carries with it properties of the function π off the set X. We consider these matters more fully in section 4.0.

3.0 The DMM Architecture. We consider now the synthesis of the prototype DMM. For this the cylindrical systolic architectures developed in [4], [5], are most convenient. We review selected aspects of these architectures as a prelude to the synthesis.

In the present discussion the term node will refer to the computing primitive depicted in Figure 1. The node has a local register for data storage. In Figure

1, λ is a scalar in the local register. The node acts on two scalar valued data streams, repeating one data stream and performing a simple multiplication/addition to update the other.

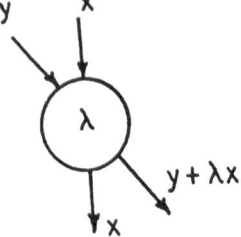

Figure 1. The Computing
Primitive.

To implement a linear operation on R^ℓ we form an $\ell \times \ell$ array of nodes. The array of Figure 2 illustrates this for the $\ell = 4$ case. The entries t_{ij} of the matrix T are placed in local registers. In figure 2 the subscripts ij indicate one possible allocation. The vector $\lambda = (\lambda_1, \lambda_2, \lambda_3, \lambda_4)$ enters at the top of the array.

It may be verified by inspection that Tλ is produced at the bottom of the array.

Turning now to the DMM we focus attention on the multisensor case, $\pi = \Delta F\Phi$, with p=ℓ. The other cases will be obvious simplifications of the architecture presented here. In keeping with the example of the above summary we let n=2, ℓ=4. Then Φ is a 4x2 matrix while Δ is 4x4.

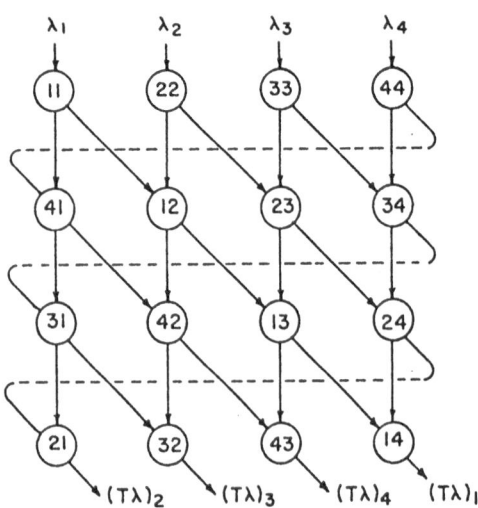

Figure 2. Cylindrical
Processing

The architecture of the prototype DMM namely $\pi(G,F,\Phi)$, is now almost transparent. Two arrays which implement Δ^{-1} and Φ respectively are interconnected by a set of nonlinear nodes, implementing F. To illustrate this let n=2 and ℓ=4, then Φ is a 4x2 matrix while Δ^{-1} is 4x4.

In Figure 3, the architecture of $\pi(\Delta,F,\Phi)$ is illustrated the notation x = (u,v) is used for convenience. The subscripts on the nodes of the upper array

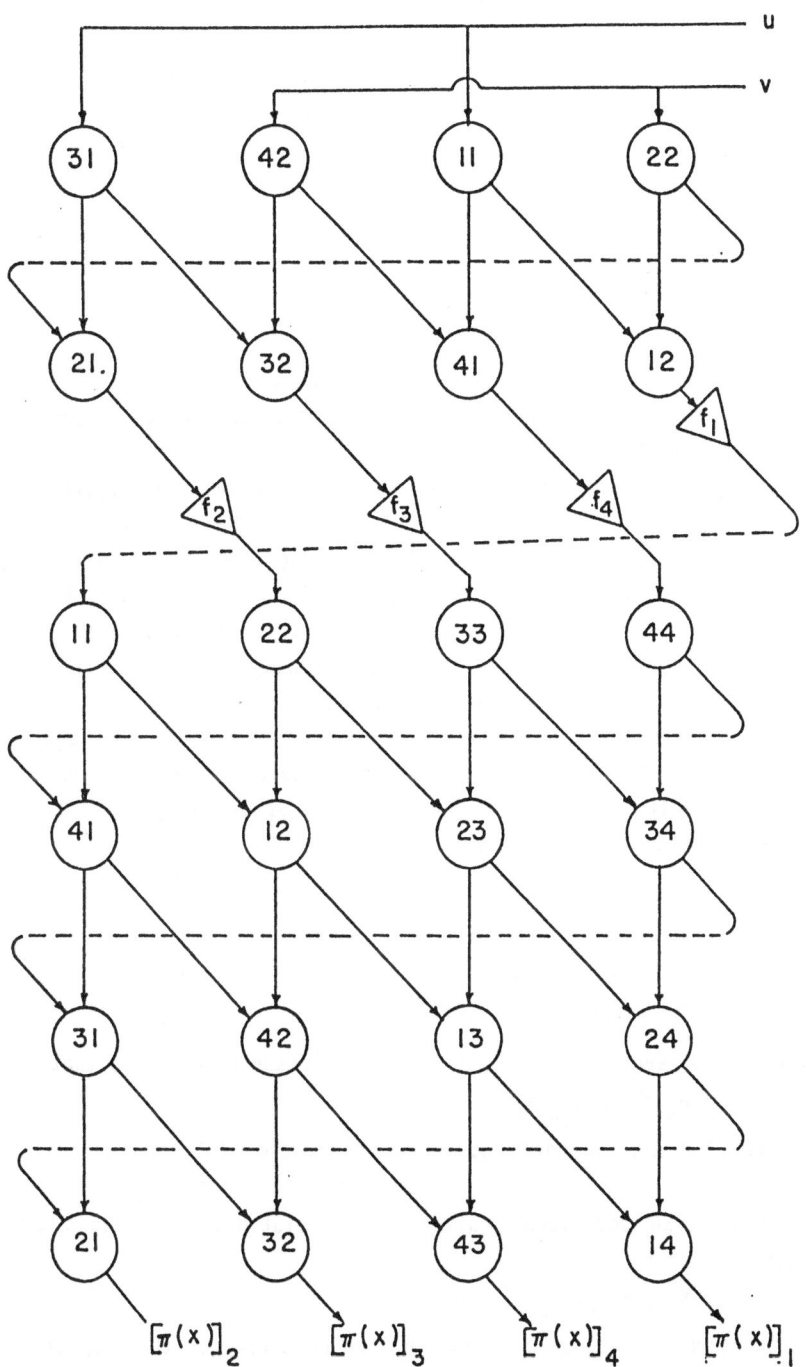

Figure 3. The DMM Architecture

denote the allocation of the entries of Φ. The triangle shaped boxes denote the functions $\{f_1,\ldots,f_4\}$. The lower array is identical with Figure 2 wherein T=Δ.

Remark. We note that the array is fully pipelined. It can operate on line with overlapping computations. Indeed once x(j) = (u(j),v(j)) clears the top row the data x(j+1) = (u(j+1),v(j+1)) may be entered. The sequential computations proceed down the array as non overlapping wavefronts. We note also that all computations are scalar valued.

As an example of the flexibility of the DMM architecture reconsider the corollary to lemma 3 and the map

$$Tx_i = x_i, \quad i = 1,\ldots,\ell$$

described therein. Apparently we need only choosé Δ = XG^{-1} to adapt the architecture of Figure 3. With this adjustment the array implements the computation of the map T.

3.1 Memory Expansion. The choice of functions F,Φ and the resultant G(F,Φ,X) tacitly developes memory of the characteristics of the set X. This memory information is explicitly allocated to the local registers of the DMM array. In addition it can be implicitly present in the specification of F and/or Φ.

To more clearly delineate the memory concept it is instructive to analyze the changes that occur in the DMM when the data set X changes. Such changes can occur during a training phase or when expanding the set of vectors to be recognized. In our study we consider the expansion of memory for the DMM. Our analysis has implicit hardware modification requirement. While we do not study these here they do appear to be tractable within the confines of VLSI technology. The following assumptions facilitate this development.

Let

$$X_\ell = \{x_1,\ldots,x_\ell\}$$
$$\Phi_\ell = \{\phi_1,\ldots,\phi_\ell\}$$
$$F_\ell = \text{diag}\{f_1,\ldots,f_\ell\}$$

denote pattern sets, vector sets and function sets respectively such that

$$G_\ell = G(F_\ell,\Phi_\ell,X_\ell)$$

is invertable. The affiliated function π_ℓ of equation (3) then describes a DMM on the set X_ℓ.

Consider then an additional pattern $x_{\ell+1}$, distinct from the elements of X_ℓ. The augmented pattern set

$$X_{\ell+1} = X_\ell \cup \{x_{\ell+1}\}$$

is to be recognized. A procedure (algorithm) which preserves the structure of π_ℓ, while evolving to $\pi_{\ell+1}$, a DMM on $X_{\ell+1}$, is said to be a <u>training procedure</u> for the recognition function.

The training procedure will, in general, require the specification of

$$\Phi_{\ell+1} = \{\phi_1', \ldots, \phi_\ell', \phi_{\ell+1}'\}$$

$$F_{\ell+1} = \{f_1', \ldots, f_\ell', f_{\ell+1}'\}$$

such that

$$G_{\ell+1} = G(F_{\ell+1}, \Phi_{\ell+1}, X_{\ell+1})$$

is nonsingular. The inversion of $G_{\ell+1}$ is a further consideration.

It is apparent that a variety of special cases can arise which may constrain the training process. One special case is to retain the previous Φ_k and F_k sets. That is

$$\phi_j' = \phi_j, \quad j = 1, \ldots, \ell$$

$$f_j' = f_j, \quad j = 1, \ldots, \ell$$

and focus attention on the choice of $\{\phi_{\ell+1}, f_{\ell+1}\}$ such that $G_{\ell+1}$ is invertable.

In some designs however (see [1]) the functions f_j are all identical. This might be preserved by including the constraint $f_{\ell+1} = f$. The training procedure then has available only the choice of vectors $\{\phi_j'\}$ to meet its requirements, that is $G_{\ell+1}$ invertable.

As a third special case we take note of the case where G_ℓ has been constrained to a particular structure. For example an easily invertable form, or a fast computational form. The training procedure may then be required to maintain that particular structure.

While each choice of constraints influences the solution of the training problem substantial commonality exists. Thus it suffices to consider here one specific example. This we do in the following section.

3.2 Training with Retention of Φ_k, and F_k.

We consider here the transition of $\pi_\ell \rightarrow \pi_{\ell+1}$ while retaining Φ_k and F_k. To be specific $\phi_j' = \phi_j$, and $f_j' = f_j$ for $j = 1, \ldots, \ell$. Our attention is focused on $G_{\ell+1}$ and its inversion.

For convenience we define

$$\psi_{\ell+1} = \text{col}[f_1(\langle\phi_1, x_{\ell+1}\rangle), \ldots, f_\ell(\langle\phi_\ell, x_{\ell+1}\rangle)]$$

$$\lambda_{\ell+1} = \text{row}[f_{\ell+1}(\langle\phi_{\ell+1}, x_1\rangle), \ldots, f_{\ell+1}(\langle\phi_{\ell+1}, x_\ell\rangle)] \tag{4}$$

$$r_{\ell+1} = f_{\ell+1}(\langle\phi_{\ell+1}, x_{\ell+1}\rangle).$$

In short; $\psi_{\ell+1}$ summarizes the original functions on the new pattern, $\lambda_{\ell+1}$ summarizes the new function on the original patterns, $r_{\ell+1}$ represents this new function on the new pattern. Using these definitions the matrix $G_{\ell+1}$ has the block form

$$
G_{\ell+1} = \begin{bmatrix} G_\ell & \psi_{\ell+1} \\ \lambda_{\ell+1} & r_{\ell+1} \end{bmatrix} .
$$

<u>Lemma 4</u>. If G_ℓ is invertible then $G_{\ell+1}$ is invertible iff

$$
\mu_{\ell+1} = r_{\ell+1} - \lambda_{\ell+1} \Delta_\ell \psi_{\ell+1} \neq 0 \tag{5}
$$

where $\Delta_\ell = G_\ell^{-1}$.

The above lemma is an immediate consequence of a well known matrix property. In fact it can be verified by inspection that, when equation 5 is satisfied, $\Delta_{\ell+1}$ has the block form

$$
\Delta_{\ell+1} = \mu_{\ell+1}^{-1} \begin{bmatrix} \mu_{\ell+1}\Delta_\ell + \Delta_\ell \psi_{\ell+1}\lambda_{\ell+1}\Delta_\ell & -\Delta_\ell \psi_{\ell+1} \\ -\lambda_{\ell+1}\Delta_\ell & 1 \end{bmatrix} . \tag{6}
$$

Lemma 4 supplies the key to most issues concerning the training in question. For instance we have the following corollary in which we presume that $X_{\ell+1}$ is distinct.

<u>Corollary</u>. Training of the DMM, from $X_\ell \rightarrow X_{\ell+1}$ preserving F_ℓ and Φ_ℓ is always possible.

<u>Proof</u>. Select $\phi_{\ell+1}$ such that the set $X_{\ell+1}$ is distinguished. Then

choose $f_{\ell+1}$ such that $\lambda_{\ell+1} = 0$ while $r_{\ell+1} = 0$. Thus equation 4 is satisfied.

We note in this case that Δ_ℓ remains intact at the same local

registers. Since $\Delta_\ell \psi_{\ell+1}$ is just $\pi_\ell(x_{\ell+1})$ the register data for the

requisite additional array row and column is also readily available.

Another special case discussed above is the situation where $G_\ell = I$. Since $\psi_{\ell+1} = \pi_\ell(x_{\ell+1})$ we have $\psi_{\ell+1} \neq 0$ in general. Thus $G_{\ell+1} = I$ is not an available result under the constraints that F_k, and Φ_k are retained. Therefore, the capability for recursive training is lost in this architecture. This is a potential drawback for real time training. We note, however, that upper triangularity can be preserved while retaining F_k and Φ_ℓ. We leave to the reader the simple verification of this assertion.

In the next section we examine the case $G(F,\phi,X) = I$. As illustration of the available capabilities we solve a problem in recognition sets.

4.0 **Recognition Sets and Robustness.** The concept of a recognition set is familiar to the neural network literature. To illustrate this concept let $\delta \epsilon R^n$ denote a fixed point of the right hand side of equation (1). An initial condition of equation (1) is said to be attracted to δ if the ensuing iteration limits to δ. The set of all initial conditions attracted to δ is called a recognition set.

It is apparent that the recognition set, affiliated with fixed point δ is an equivalence class under the function implicitly computed by equation (1). In the DMM format the system functions without iteration and hence a direct examination of equivalence classes is possible. In this regard we note that Δ is always invertible. Thus, without loss of generality, we tacitly set $\Delta = I$ and focus attention on $F\Phi$.

In the general case the specific form of the functions, F, will have a prominent effect on the equivalence classes inquestion. Indeed for arbitrary $\eta \epsilon \text{Range}(F\Phi)$ the equivalence class $[[\eta]]$ may be characterized by $[[\eta]] = \Phi^{-1}F^{-1}\eta$, where set inverses of Φ, F are intended. In this study, however, we are content to focus attention on a restriction of the DMM format. We note that the sgn function of equation (1) is a step function. In the following we restrict the functions F to be step functions. However, to retain the capability summarized in Lemma 1 we shall allow the DMM step functions to taken on (at most) $\ell+1$ distinct values.

Consider now the composite map $F\Phi$. Recall that F is diagonal and hence $F\Phi x$ is a vector, each of whose components is of the form $f(\langle\phi,x\rangle)$. That is the unisensor case with ϕ a row of Φ. We assume throughout that ϕ distinguishes X, that is the scalars

$$a_j = \langle\phi,x_j\rangle, \qquad j=1,\ldots,\ell$$

are distinct.

The scalars a_j have a natural order inherited from R. As a notational convenience we assume $a_i < a_{i+1}$ for $i=1,\ldots,\ell-1$ and define the 'midpoints'

$$\beta_0 = -\infty$$
$$\beta_i = (a_i+a_{i+1})/2 \qquad i = 1,\ldots,\ell-1 \qquad\qquad (7)$$
$$\beta_\ell = \infty$$

between adjacent a_i.

The sets $\Gamma_k \subseteq R$ are given by

$$\Gamma_k = \{x: x = mx_\phi \oplus \{\phi\}^\perp; \ \beta_{k-1} < m \leq \beta_k\}$$

for $k = 1,\ldots,\ell$ where $x_\phi = \phi/\|\phi\|^2$. It is readily verified that $\{\Gamma_k\}$ is a disjoint cover of R^n.

Now let f be a step function on R namely

$$f(r) = g_k, \quad \beta_{k-1} < r \le \beta_k, \quad k = 1,\ldots,\ell.$$

Then it is easily verified that

$$f(<\phi,x>) = g_k \longleftrightarrow x\epsilon \Gamma_k, \quad k = 0,1,\ldots,\ell.$$

These results are summarized in the following lemma.

<u>Lemma 5</u>. The sets of the disjoint cover $\{\Gamma_k\}$ of R^n are equivalence classes of the function $f(<\phi,.>)$ constructed above. If the $\{g_k\}$ are distinct then the set X is distinguished by $f(<\phi,.>)$.

In the spirit of this lemma we shall say that f recognizes the sets $\{\Gamma_k\}$. Since $x_j\epsilon\Gamma_j$, $j=1,\ldots,\ell$ the terminology 'f recognizes $X = \{x_1,\ldots,x_\ell\}$ is also appropriate. Of course when the g_k are not distinct then the Γ_k merge forming a less refined cover of R^n.

We turn now to the general form of map FΦ. The above analysis applies to each row $\phi_i\epsilon\Phi$ and affiliated function $f_i\epsilon F$. The effects of joint recognition are several fold. First a refined disjoint cover for R^n is generated with at most $(p)^\ell$ subregions. Secondly the recognition of $x_i\epsilon X$ has redundancy which provides an opportunity for enhancing fault tolerance and robustness. In addition the DMM can now recognize not only the points of X and their covers but each of the cover subregions.

The modifications necessary for FΦ are intuitive and hence we shall keep our remarks brief.

Let $a_{ij} = <\phi_i,x_j>$, $i,j = 1,\ldots,\ell$ and define

$$\Gamma_{ik} = \{x:x = m_{\phi_i} \oplus \{\phi_i\}^\perp, \; \beta_{i,k-1} < , \; m \le \beta_{i,k}\}$$

where $x_{\phi_i} = \phi_i/\|\phi_i\|^2$ and $\beta_{i,k}$ denote the natural extension of the midpoint definitions, $\beta_{i,k} = (a_{i,k} + a_{i,k+1})/2$. We now form a disjoint cover for R^n by taking all possible intersections

$$\Gamma = \{\Gamma_{1,i_1} \cdots \Gamma_{p,i_p} : 1 \le i_1,\ldots,i_p \le \ell\}.$$

Each of the f_i is taken to be a step function with constant value on Γ_{ik}, that is

$$g_{ik} = f_i(<\phi_i,x>) \longleftrightarrow x\epsilon \Gamma_{ik} \quad i=1\ldots p, \; k=1,\ldots,\ell.$$

The collection is a disjoint cover for R^n. The cardinality of this cover, #Γ is dependant on the set ϕ_i and the scalars a_{ij}.

However, the bounds

$$(p)^m \le \#\Gamma \le (p)^\ell$$

hold were $m = $ dim span $\{\phi_i\}$. If the set $\{\phi_i\}$ was linear independant the upper bound would be met. Unless the ϕ_i are redundant the lower bound is pessimistic.

We summarize this in the following theorem.

<u>Theorem</u>   If the vectors $\phi_i \epsilon \Phi$ are distinguishing on X.   If the $f_i \epsilon F$ are step functions on intervals $\beta_{i,k} < r \leq \beta_{i,l+1}$.   If the g_{ik} are chosen such that $G(F,\Phi,X)$ is invertable.   Then the DMM recognizes each subset of the partition Γ and satisfies the constraint conditions of Lemma (3).

4.1   <u>Robustness and Fault Tolerance.</u>   In section 3.0 we have seen that the multisensor DMM, with staircase functions F, has tacit memory capacity equal to the cardinality of Γ.   The primary task, of recognizing the set X, utilizes only a small fraction of this capacity.   This additional memory capacity can be used directly, as suggested in section 3.0 or utilized for fault tolerance in the performance of the primary task.

In this section we highlight this latter usage.   For this we assume that each row, of Φ is distinguishing on X.   For convenience we assume that $\{\delta_1,...,\delta_\ell\}$ are common recognition symbols for the elements of X.   We assume that the staircase functions, F, all incorporate these recognition symbols.   In short

$$f_i(<c_i,x_j>) = \delta_j \quad \text{all } i, \text{ each } j$$

holds on X.

The map Δ is now given a different interpretation.   For example, Δ may take the form of a majority logic network.   Each row of $F\Phi$ votes on the input, relative to the possible elements of X.   The map Δ provides the comparative logic.

It is apparent that

$$\pi(x_i) = \delta_i, \quad i = 1,...\ell$$

are satisfied.   It is of interest to determine the regions of attraction when the majority logic is in effect.   This require a variation on the development of section 3.0, the details of which are briefly sketched in the following.

Each function, f_k, partitions the underlying space into ℓ disjoint sets. The sets Γ_{ki} are given by

$$\Gamma_{ki} = \{x: x = mx_\phi + \{\phi_k\}^\perp, \quad \beta_{ki} < m \leq \beta_{k(k+1)}\}.$$

The outputs of the functions f_k will agree only on the intersection

$$\Gamma_i = \,_k \Gamma_{ki} \quad i = 1,2,...,\ell.$$

The region Γ_i will be called the primary attraction region for x_i. Clearly,

$$\Gamma_i \quad \Gamma_j = \phi;$$

any vector in Γ_i will be mapped onto the pattern x_i.   Obviously, by construction one has $x_i \epsilon \Gamma_i$.   If a simple majority rule is established, the attraction region is correspondingly modified.   In the following it will be shown that the attraction regions remain disjoint in this case.

Assume that the set of points mapped on the pattern, x_i, is specified by the set $\Gamma_i^{(\alpha)}$ defined by the rule:

$x \epsilon \Gamma_i^{(a)} \leftrightarrow x \epsilon \Gamma_{ki}$ for at least a distinct values of k.

An alternative definition for the attraction region is given by

$$\Gamma_i^{(m)} = \bigcup_{k\epsilon(s_1,\ldots,s_a)} \bigcap \Gamma_{ki} \qquad 1 \leq s_i \neq s_j \leq \ell$$

The union is taken over all possible combinations of distinct a-tuples.

From the above definition it becomes clear that

$$x \epsilon \Gamma_i^{(a)} \quad \Gamma_j^{(a)}$$

if and only if there exist a-tuples $(s_1,\ldots s_a)$ and $(\sigma_1,\ldots,\sigma_a)$ such that

$$k\epsilon\{s_1,\ldots,s_m\} \quad \text{and} \quad k\epsilon\{\sigma_1,\ldots,\sigma_m\}.$$

Since, for every k, the collection Γ_{ki} partitions R^n, the condition written is possible if the set of indices $(s_1,\ldots,s_a)$ and $(\sigma_1,\ldots,\sigma_a)$ have an empty intersection. In particular, if $2a > \ell$ (i.e. majority rule) the two sets of indices must have at least one common value, k_0. In this case

$$x \epsilon \Gamma_i^{(a)} \bigcap \Gamma_j^{(a)} \rightarrow k_0: x \epsilon \Gamma_{k_{0i}} \bigcap \Gamma_{k_{0j}}.$$

This condition is impossible since it contradicts the partitioning property of the collection $\{\Gamma_{k_{0i}}\}$. The attraction regions are, therefore, disjoint.

5.0 Discussion. The concept of a Distributive Memory Modules appears to be a powerful tool with application to problems in several fields. Connections with neural networks and coding have been briefly touched upon. In order to highlight the basic ideas, the details have been kept to a minimum. As a principal example, this study has focused, primarily, on the signal recognition problem.

The results include a design procedure which realizes a prescribed memory capacity. A comparison of unisensor and multisensor cases is given. The use of tacit memory in the multisensor case for fault tolerance is also discussed. Furthermore, the design procedure is shown to be compatible with array implementation.

Several additional comments are worth nothing. Consider first the choice of Φ. In view of Lemma 1 this choice is not unique. We note, however, that if $\phi\epsilon$ [Span X]$^\perp$ then $\phi X = 0$. Thus the constraint span $\Phi \subset$ Span X can be imposed without loss of generality.

Recall now, from section 4.0, that a linearly dependant choice of ϕ_i gives less refinement than a linearly independent choice. This together with Span Φ Span X suggests that a multisensor multiplicity of order p = dim Span X = dim Span Φ is a good balance between efficiency, and complexity and provides good redundancy as well.

In reference [1] the case $\Phi = X^*$ is studied, together with a fixed apriori choice of F. The Span Φ = Span X condition is obviously met. In this study, however, $p = \ell > \dim \text{Span } X$, is necessary to guarantee that an apriori choice of polynomic F will suffice.

Another intuitive choice of Φ is constructed as follows. Let R_X denote the covariance of the set X, and let $\{h_1,\ldots,h_p\}$ denote the unit eigenvectors of R_X associated with nonzero eigenvalues. It follows by a conventional analysis that; ϕ_i, $i=1,\ldots,p$, produces a linearily independant spanning set for Span X.

6.0   <u>Acknowledgement</u>: The authors are faculty with the Department of Electrical and Computer Engineering, Louisiana State University. This research was sponsored in part under grant 24962-MA-SDI through the U.S. Army Research Office, Durham, and State of Louisiana Research Contract LEQSF-A-17.

References

[1] Porter, W.A. "Distributive Memory Recognition Functions", Trans. Int. Conf. Communications and Control, Oct. 19-21, 1988, Baton Rouge, LA.

[2] Hopfield, I.J. and D.W. Tank, "Neural Computation of Decisions in Optimization Problems", Bio Cybern, Vol. 52, pp. 141-152, 1985.

[3] McEliece, R.J. and E.C. Posner, E.R. Rodemoch, S.S. Venkatesh, "The Capacity of the Hopfield Associate Memory", IEEE Trans. Inf. Theory, Vol. IT-33, No. 4, pp. 461-482, July 1987.

[4] Porter, W.A. and J.L. Aravena, "Orbital Architectures with Dynamic Reconfiguration", IEEE Proc., Vol. 134, Pt. E.

[5] Aravena, J.L. and W.A. Porter, "Non Planar Switchable Arrays, Circuits, Systems and Signal Processing, Vol. 7, No. 2, 1988, pp. 213-234.

DESIGN AND IMPLEMENTATION OF A CLASS OF NEURAL NETWORKS

Donald L. Gray, Anthony N. Michel and Wolfgang Porod
University of Notre Dame

Introduction. The study of neural networks and their applications has received tremendous interest in recent years [1]. Various design approaches have been presented for determining the values of network bias and interconnection weights. One particular approach holds that the dynamics of a neural network can be viewed as optimizing an associated energy function [2]. Using an energy function approach, Tank and Hopfield [5] proposed a particular network design for an analog-to-digital converter. In implementing their design, however, a mismatch in the neural thresholds will occur from the combination of the high number of feedback paths in conjunction with the consequences of the energy function approach. This combination leads to the appearance of spurious states in the analog-to-digital converter. This phenomenon is also encountered in the design of conventional analog circuits and it reveals itself in [5] since neural networks have their roots in analog circuits.

Current design methodologies generally leave out the effects of implementation on their performance. In an effort to develop neural networks with improved performance levels we have examined two areas. The first is a different structural approach to neural networks design, namely a lower block triangular form for the interconnection matrix. Secondly, the problems associated with network implementation has led to our presentation of a digital architecture for neural networks.

In this paper, we present an alternate design philosophy which is based on viewing the neural network as an interconnection of several subsystems [3,4]. In particular, we investigate lower block triangular interconnection structures. We illustrate this design method by applying it to sorting problems. The design method in this paper is superior in performance to past energy function designs due to the four following qualities. First, the network is composed of small blocks of strong local interconnections with weak coupling between the blocks themselves. The strong and weak couplings themselves both take on a lower triangular form. Secondly, due to this removal of the conventional symmetric interconnection the network thresholds can be better coordinated through the triangular form. This also causes a significant reduction in the potential number of spurious state sites in the network. Thirdly, the potential for modular design exists due to the functional block design strategy utilized. Lastly, in the analog-to-digital converter problem the need for exponentially increasing resistor values is eliminated. We will then present two further examples to illustrate the structure and design methodology. A modular analog-to-digital converter will be designed by this approach along with a network design to perform the classification of an unknown resistor to its closest standard part value and the determination of the tightest resulting standard tolerance.

Secondly, this paper investigates the possibility of creating a digitally based neurocomputer to perform the equivalent function of an analog circuit based neurocomputer. A design architecture is presented along with the calculation algorithms that are used for the sequencing and updating of the neurons that are made possible with this architecture. The motivation for presenting this design stems from the following detrimental consequences of the large scale implementation of neural networks with analog feedback systems [6]. The first of these problems is that of spurious state generation in a neural network caused by component tolerances and non identical amplifier transfer characteristics [7]. Secondly, the present inability to modify the hardware resistor values representing the interconnection weights inhibits the use of learning algorithms. Further, if a learning algorithm could operate on a modifiable hardware, the randomness of the analog neural network could wreak havoc in the system when coupled with the learning algorithm. This possibility exists since the exact order of the firing of a system's neurons will affect the system's solution trajectory. Lastly, the obvious problem of routing the high complexity interconnections hampers the implementation of these networks in the analog world. This fact, when accompanied with the relatively low four digits of precision usually available with analog equipment suggests that a new implementation technique may be required for large scale neural networks.

The Neural Model. We assume the now standard Hopfield model [2] for our neural network. The resulting equations for a system consisting of N neurons are:

$$\dot{U}_i = -\frac{U_i}{R_i C_i} + \frac{1}{C_i} \sum_{j=1}^{N} T_{ij} V_j(U_j) + I_i(t) + B_i X_i, \tag{0.1}$$

$$V_i = TANH(\beta \, U_i). \tag{0.2}$$

As usual, the T_{ij}'s and the B_i represent conductance values, I_i is the bias current and X_i the external input. We further assume that, by scaling, $R_i = 1$ Ohm and $C_i = 1$ Farad and, as in the Hopfield model, β will be allowed to approach infinity. This last assumption allows, from the design point of view, the neuron to operate as a comparator with a sigmoidal characteristic. As previously mentioned, we do not impose the condition of symmetry on the T_{ij} interconnection matrix. It is this condition which introduces additional and undesired spurious state sites as, for example, in the design of an analog-to-digital converter in Ref. 5. Furthermore, we will allow self feedback (T_{ii} nonzero) for each neuron creating a summing amplifier.

Decomposition into Lower Block Triangular Form. The first step is to construct a block diagram of the overall system showing the input and desired output lines. (It may be helpful here to refer to one of the two design examples for clarification of each step.) This diagram is then decomposed into as many smaller functional subsystem blocks as seen appropriate. These subsystems can be thought of as operations which when paralleled or cascaded will perform the overall sorting process. In most cases it will be found that the interconnection of the decomposed blocks will be of the lower triangular nature. Ideally, it is here that redundancy and modularity can best be exploited, but the granularity of this step is a judgement call to be made by the design engineer. Illustration of convenient subsystems selected in relation to an overall design can be found in the examples.

Secondly, each block can now be individually designed but with greater ease stemming from the decomposition. We will now present a design procedure which can be used in the design of blocks requiring a sorting capability.

A Sorting Block Design Procedure. A starting diagram for each functional block can be obtained from the decomposed block diagram done previously. For each block, a complete list, comprised of continuous ranges of the incoming analog signal, that together span the input range, must be made. Additionally, for each range, the associated digital outputs should be specified. For example, with a four bit A to D converter spanning a 0 to 16 Volt input, we would like the analog input voltage range from 1.5 Volts to 2.5 Volts to be associated with the binary word 0010, etc.

The next step is to analyze the list made in step 2 one output bit at a time. Again, continuing with the 4 bit analog-to-digital converter mentioned previously, a sample list appears in Table 1.

INPUT VOLTAGE	OUTPUT BITS
(-1.0 , 0.5)	0000
(.5 , 1.5)	0001
(1.5 , 2.5)	0010
(2.5 , 3.5)	0011
(3.5 , 4.5)	0100
.	.
(14.5 ,16.0)	1111

TABLE 1

In the analysis, it is the initial state and the on $(0 \rightarrow 1)$ and off $(1 \rightarrow 0)$ transitions of each bit that are of importance. In particular, for the 8's bit , the inital state (analog input equal to - 1.0 Volts) is off (0), and the first and only on transition occurs at the analog input of 7.5 Volts. Making a list of on/off specifications for each of the four output bits, we arrive at the data in Table 2.

Once a list similar to that in Table 2 is complete, the determination of the I_i and T_{ij} values can begin. The purpose of the I_i term in our scheme is to establish the lowest analog threshold level for each neuron. The T_{ij} terms are used to turn the ith neuron off at the analog level determined by the on threshold of neuron j. The analog input coefficient term B_i to each neuron will be set to a value of 1.0 on all neurons unless stated otherwise. We start with N neurons where N is the number of output bits from the network and we will add hidden neurons as the design develops.

8'S BIT		4'S BIT		2'S BIT		1'S BIT	
OFF	ON	OFF	ON	OFF	ON	OFF	ON
-1.0		-1.0		-1.0		-1.0	
	7.5		3.5		1.5		0.5
		7.5		3.5		1.5	
			11.5		5.5		2.5
				7.5		3.5	
					9.5		4.5
				11.5		5.5	
					13.5	.	6.5
					.		.
					13.5	.	
							14.5

TABLE 2

From Table 2, each I_i is determined as the negative of the level establishing the first desired on transition. Accordingly, $I_1 = -0.5$, $I_2 = -1.5$, $I_3 = -3.5$, and $I_4 = -7.5$.

Now working with one output bit at a time, in this case starting with the highest order bit first and proceeding downward, we determine the T_{ij} values. The need for a particular T_{ij} value arises when a neuron needs to be turned off. In the 8's bit case, an off transition does not exist and consequently no feedback is required. The 4's bit, from Table 2 though, requires an off transition at 7.5 Volts. Therefore, feedback must exist between a neuron that toggles on at 7.5 Volts and remains on over a range compatible with the the neuron it is to inhibit. This condition will result in a triangularity of the T_{ij} matrix. The analog to digital converter possesses an extreme in the availability of these levels with useful ranges. Since the 8's bit neuron toggles at 7.5 Volts and is upwardly stable it can function as the source of the feedback to the 4's bit without the addition of a hidden neuron. The value of the T_{ij} is determined as follows. The T_{ij} is equal to the negative of the difference of the next desired on transition level and the current value of the neuron threshold. For the 4's bit $T_{34} = -(11.5 - 3.5) = -8.0$. Similarly, for the 2's bit which first toggles at 1.5 Volts, to turn off at 3.5 Volts requires a feedback which is connected to a neuron that has a threshold of 3.5 Volts. In this case, again, the 4's bit meets this requirement in the range of -1.0 to 7.5 Volts. Thus, $T_{23} = -(5.5 - 1.5) = -4.0$. The next off-transition of the 2's bit is at 7.5 Volts where the T_{23} connection will become ineffective due to its forced off state by T_{34}. Therefore, we similarly create $T_{24} = -(9.5 - 1.5) = -8.0$ where 1.5 is the current threshold of the 2's bit neuron due to the inactivity of the only other connection T_{23} at 7.5 to 11.5 Volts. The third required off transition of the 2's bit is at 11.5 Volts. Analysis will show that no further connections or hidden neurons will be required since the T_{23} connection will reactivate at 11.5 Volts and turn off the 2's bit relocating its threshold to 13.5 Volts exactly where we would like it. The determination of the T_{ij}'s for the 1's bit will show that the off transition at 1.5 Volts can be accomplished by $T_{12} = -2.0$, the off transition at 3.5 Volts by $T_{13} = -4.0$, this combination accomplishing the off transition at 5.5 Volts. Lastly, the addition of $T_{14} = -8.0$ will complete the design by causing all of the other 1's bit specifications to be met. Thus, we have obtained the following design of a 4 bit analog-to-digital converter shown in Figure 1. This design can be compared to other designs as shown in References 8 and 9.

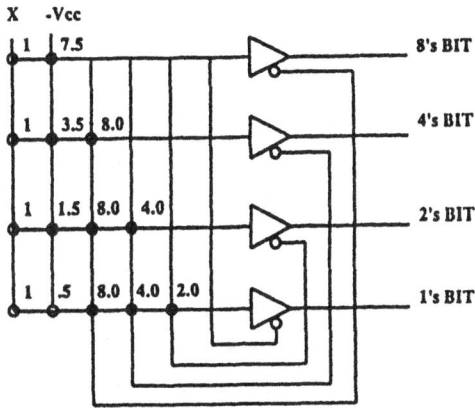

Figure 1. Design of a 4 bit analog-to-digital converter with lower block triangular interconnecting structure.

Further Examples: A Cascadable 2 Bit Analog-To-Digital Converter. In this section we proceed with an alternate design of the 4 bit analog-to-digital converter presented to illustrate a different block diagram decomposition. We will follow along the same design routine to implement a modular approach, as illustrated in Figure 2.

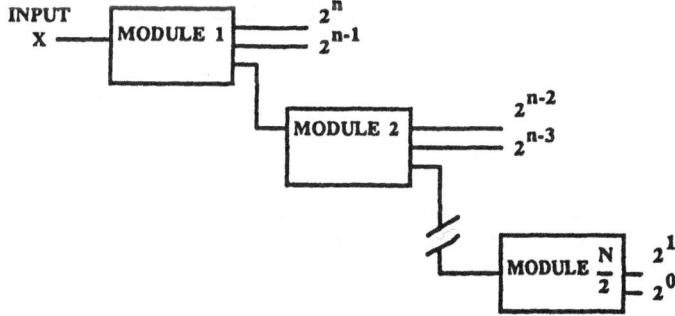

Figure 2. Block diagram of a modular analog-to-digital converter.

As before, we will assume that the converter is to operate on an incoming analog signal between 0 and 16.0 Volts. We wish each subsystem module to classify/sort the signal to 2 bits. Further, the output is to consist of the 2 bits and an error term to allow cascadability into further identical sections. If the 2 bit sorted output is subtracted from the incoming analog signal, X, and amplified by 2 squared, the following output equation is obtained: $E = 4X - 32V_2 - 16V_1$. Since the maximum error in the sort is $1/2^2 = 1/4$, the scaling of the internal error by 4 will allow identical sections to be used.

Now to generate the error term, $E = 4X - 32V_2 - 16V_1$, we add 1 neuron to the design, as shown in Figure 3 with two subsystem modules shown to illustrate cascadability. It is interesting to note here the appearance of $T_{33} = -1.0$. In order to produce the error signal, an analog summing cicuit was utilized resulting in the appearance of a self feedback term in the T_{ij} matrix. This concludes the second analog-to-digital converter design approach which by this particular decomposition has led to the design of a successive approximation A/D converter. This design is similar to designs discussed in [10].

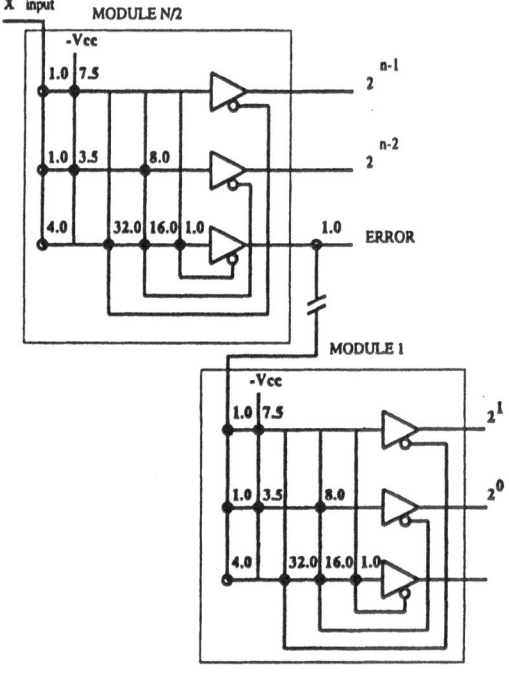

Figure 3. Modular design of an analog-to-digital converter.

Further Examples: A Resistor Sorter/Tolerancer. In this example, we design a network to classify an unknown resistor to the closest standard value and determine the tightest standard tolerance from that value. A block diagram for this system is shown in Figure 4. The unknown resistor is assumed to be in the decade 0 - 10 Ohms. Scaling can be used to transform this circuit to other ranges with reasonable part values for implementation.

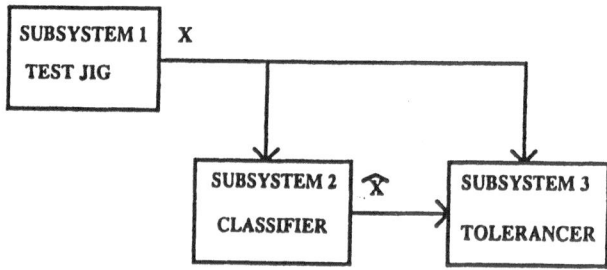

Figure 4. Block diagram for the resistor sorter/tolerancer.

The first subsystem is a test jig into which the unknown resistor R_x is placed. The output voltage will be directly proportional (in this case equal) to the unknown resistance. Subsystem 2

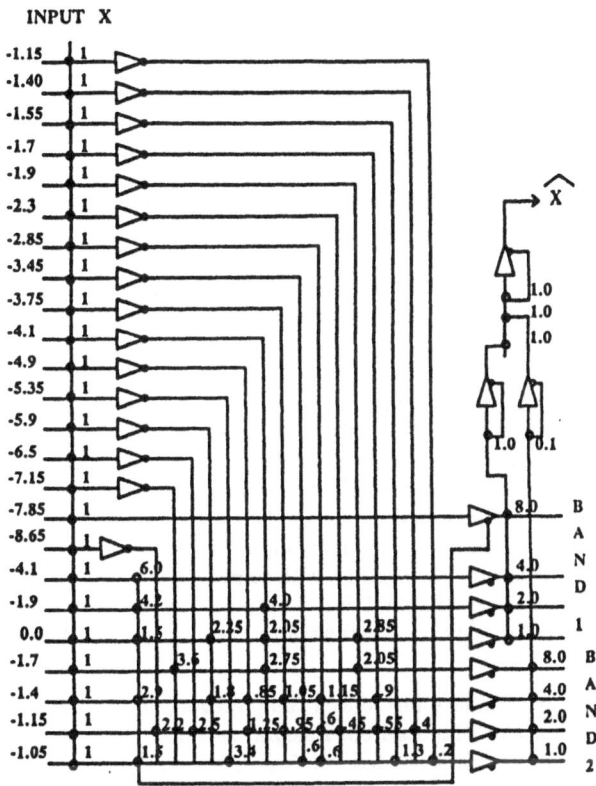

Figure 5. Resistor Sorter Classifier.

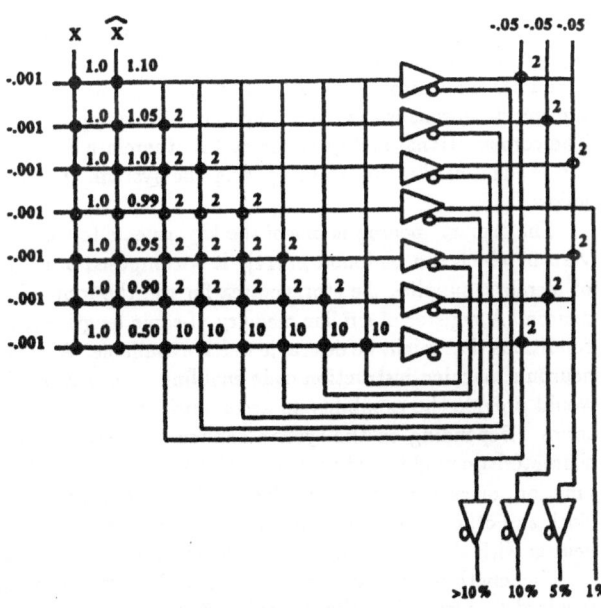

Figure 6. Resistor Sorter Tolerancer.

will associate the closest standard value to the input value and output three quantities. Two BCD digits will represent the closest standard resistor values while the third output will be an analog signal level representing the same value. This analog signal will allow subsystem 3 to establish the "error" between the unknown resistor and the classified value. This tolerance will be the output of subsystem 3.

Subsystem 2 is the classifier subsystem module and its design will be elaborated, a more detailed description can be found in [7].

Eight neurons are used to output the two BCD digits. We add additional hidden neurons to provide the other required cutoff values. To generate the analog output signal representing the classified part value we utilize 3 amplifiers to sum the output of the ones BCD digit, the one tenth BCD digit, and the two digits together. This results in the subsystem module shown in Figure 5.

The tolerancer subsystem is designed last. The two available inputs are analog voltages representing the the actual unknown resistor value and the classified value. We wish to determine if the part has a 1%, 5%, 10%, or > 10% tolerance from the classified value. A hidden layer of neurons is used to perform the function of determining if the unknown resistor is above a tolerance bound. Explicitly, the seven hidden neurons in Figure 6 test if the unknown resistor is greater than each of the following percentages of the classified part value; 50%, 90%, 95%, 99%, 101%, 105%, and 110% respectively. Each neuron in the hidden layer has the ability to inhibit all of the other neurons with lesser percentages when firing, thus allowing only one of the layer to remain on. An output layer of three additional neurons combine each of the two 5% outputs, the two 10% outputs, and the two > 10% outputs from the hidden layer. This subsystem module is shown in Figure 6.

In reviewing the design, it is interesting to note the contrast with respect to the analog-to-digital converter. The opportunity to exploit the linear combinations of outputs does not exist as in the A to D converter. It illustrates the need for additional hardware to cause an output state to be skipped.

A Digital Neurocomputer Architecture. We have introduced the preceding design approach to reduce the possibility of obtaining undesirable outputs from a neural network. The next step is to address the hardware that such a design is implemented on. The following design of a digital neurocomputer architecture addresses problems related to the randomness of an analog neural network and provides a flexible environment.

Hardware Implementation. From a macroscopic popint of view, the system is composed of a supervisory computer connected to N "smart" neurons, see Figure 7. The supervisory computer would serve numerous purposes in controlling and configuring the overall system. The addition of the smart neurons.is the main difference of this approach when compared to past implementations. The computational ability of the system therefore does not lie in only a few places but in each of the N identical "neurons" composing the system. A detailed description of the system will now follow.

The "smart" neuron is one of the key aspects to this implementation approach, a more detailed view of which can be found in [11]. It is composed of six major parts. A RAM, or another variety of memory, supplies the storage area for the the input, bias weight, and interconnections to each neuron. A register of similar memory of some length M, provides a storage area for special initial conditions that might be desirable to have available. A ROM memory would be provided to store the neuron's machine instruction code enabling it to perform its required operations. An accumulator would provide the capability to accomplish the summation of the input, bias, and interconnection terms. Depending on the system configuration/application, some special type of neural transfer characteristic might need to be available in the neuron. This could be generated by a trigonometric function, or custom made in a lookup table, etc. Finally, a control signal generation unit with a clock and other necessary components would be required to syncronize the internal workings of the neuron with the other neurons and the supervisory computer.

The length of the data word for the RAM memory is not specified since it can be of any length required to accomplish the desired system accuracy determined before the system is built. This

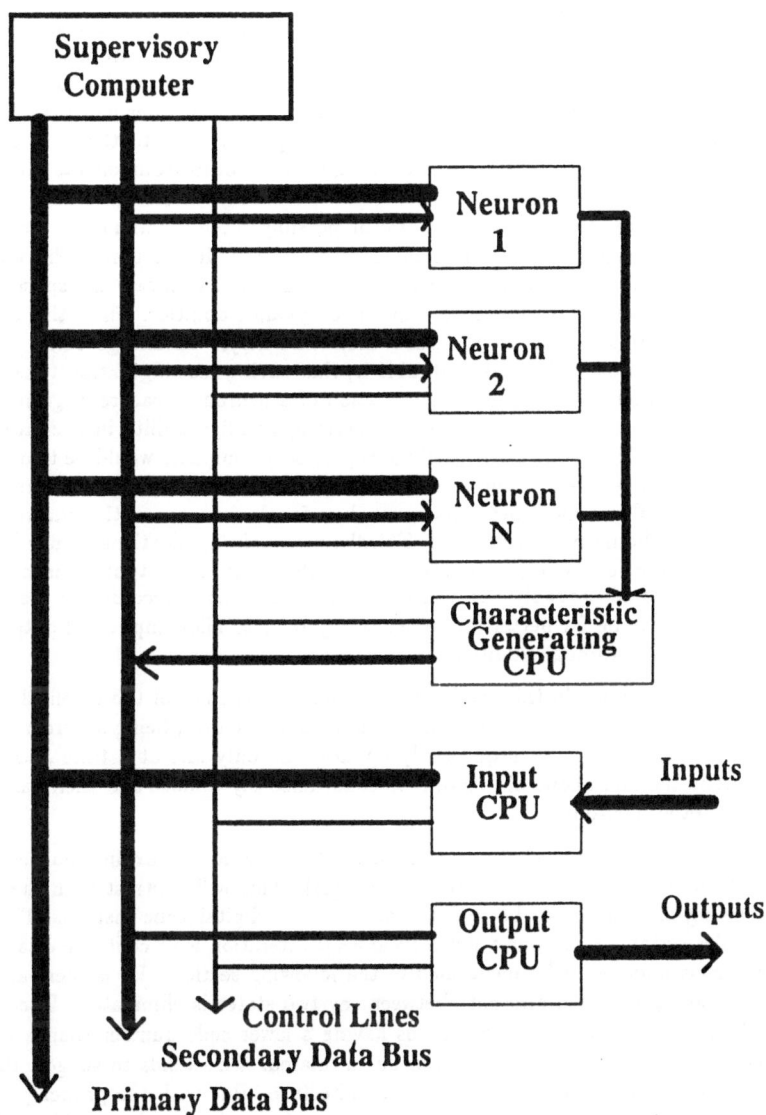

Figure 7. Digital Neurocomputer Architecture.

is one advantage of this approach over an analog system, a 128 bit value for the interconnection matrix, T_{ij}, can be accommodated and handled easily since the values never leave the internal storage and registers of the neuron. This fact will become clear when the overall system operation is described.

The length of the $1 \times N$ RAM register(s) is also variable, but related to the storage capacity of the RAM memory. Each bit of the register represents an interconnection from some other neuron. The length of the register and memory will determine the maximum number of interconnections allowed to a neuron, which will be less than or equal to the largest order network the system can handle. The size of the ROM memory will be directly proportional to the level of operational sophistication given to the neuron. More will be said on the other components after the system operation is covered.

The supervisory computer is to perform two or possibly three functions for the system. The first use would be to assign a neural code number to each individual neuron. This number would be different than the neurons system address code number, used to locate the specific neuron for communications purposes. This capability allows two dynamic abilities. First, this allows various network sizes to be implemented on the hardware. The major advantage being that it is not nessessary to load a neuron with zeros to disable it, thus saving reconfiguration time. Secondly, if the supervisory computer detects erroneous operation of a neuron, it can reassign its neural code number to another spare neuron. This gives the system reconfigurability in the face of faults or partial system failure. The second purpose of the supervisory computer would be to orchestrate the overall system's activities. This would also include operation of the system's learning algorithm. The system output could be monitored and new values for the interconnection matrix T_{ij} sent out over the data bus to the individual neurons. Thirdly, in certain applications it may be considered appropriate to have the supervisory computer conduct the input and output transactions with the outside world. This would probably result in the sacrifice of system speed due to the data transfer time. It would therefore be far better for large scale systems to allow input and output from each neuron either directly or through addition I/O CPUs.

Software and Algorithms. In this section two calculation schemes of the Hopfield model [5], as would be performed on this architecture, will be discussed. The first, being an ordered calculation where the neurons are allowed to sequentially update, i.e. only one at a time and in a specific order. The second method allows all neurons to simultaneously update and then pass along their new state to the other neurons.

Sequential Updating. The sequential calculation is presented for the explicit purpose of allowing a calculation to occur in a specific update sequence [11]. This will contrast with the digital time notion of allowing all neurons to be updated at the same digital time instant. The reason for this is that under certain conditions a finite cycle length iteration loop could be established when using simultaneous updates to be mentioned in the following section. By sequentially updating, the chances of having a system alternate between say two states is eliminated. The performance of this calculation scheme requires the neurons having a lower code number than a given neuron to update before the given neuron's state can be calculated. This tends to suggest that this is a serially calculated operation done on a parallel architecture. But with the proper phasing of the calculation, a serial updating can take place while maintaining the speed available with a parallel processing architecture. A physical description will now follow.

First, the given neural network bias values and interconnection weights are given to the supervisory computer. The supervisor then establishes the order of the system and assigns the appropriate number of neurons their respective neural code numbers.

The bias value, initial condition vector and external input weight along with each row of the interconnection matrix is transferred via the main data bus to each of the neuron's internal RAM memories and register. This completes the initialization of the system.

Next, the method of which depending on how the system is constructed, an input is applied to each neuron in one of the following four ways. These include direct presentation to each neuron of a 1 bit binary number or an N bit binary number from an external source, an N bit binary number passed from the supervisor, or an analog signal fed through an analog to digital converter which could exist within each neuron. With the completion of data input, the calculation can now proceed.

For the moment, consider an N neuron network whose parameters have been loaded into the architecture. Once the input data has been loaded an N+2 length calculation cycle can begin. This first calculation series performs the summation of the upper triangular terms of the interconnection matrix multiplied by the appropriate neural state value. The addition of the bias and weighted input terms based on the initial conditions vector stored in one of the RAM registers is also accomplished. Therefore, upon the completion of this series, the first neuron possesses a complete summation of the terms applied to its input. The second neuron would possess a sum comprised of all the terms applied to its input except from the neurons with lower code numbers, in this case only neuron one. The last neuron N, would posses the sum of only two terms the bias value and the weighted input. This step is completed utilizing only 50 percent of the available computing power of the system, and is only required when starting from non zero initial conditions. Once the first calculation series has been completed, the true power of the system can now be used. The output of neuron #1 can now be established. Thus the completed summation is presented to the neural characteristic and an output is established. This output could be as simple as a binary 0 or 1, or be an M bit binary number. The output value is then placed on the secondary data bus and simuleously presented to all $N-1$ other neurons. A write enable signal, or data ready signal is then sent out on a control line, followed by a signal to the supervisor over the release status bus line to signal completion of the update. Thus on the next cycle all N neurons can add one additional term to their sums, when given the signal to compute their next term from the supervisor over the control lines. At this point neuron #2 would have completed its sum and its updated output presented to all of the other neurons. Once the last neuron updates, the process can continue at the top with neuron one after two additional cycles. These two cycles are needed for the re-introduction of the input and bias terms to the current sums present in all of the neurons. Therefore, before the first neuron could complete its update the weighted input and bias terms would need to be added since only feedback information would have been entered up until that time. Once a neuron has established its updated output, the accumulator would be set to zero and the summation of new feedback terms begun. Complete downward passes could now continue to be made until the supervisory computer determined that a stable solution had been reached or an allotted solution stablization time had been used up.

This technique, when used for very long settling time problems, could approach 100 percent utilization of the computing power available with the N "smart" neurons used. At worst, the system would only utilize 50 percent of the computer power on a two sweep problem, the minimum possible. The speed increase over serial implementation of a nonsparsely interconnected system could range from approximately N/2 to N times faster.

Simultaneous Updating. This section will discuss the operation of a simultaneous updating technique for a Hopfield type network, when implemented on this architecture. Each complete update of the network state requires three groups of operations. The first is the simultaneous and independent calculation of all the terms applied to a neuron's input based on the previous state of the network. The second is the reflection of the neuron's input sum off the neural characteristic to determine an output value. Thirdly, the sequential passage of each neuron's output, over the secondary data bus, to all of the other neurons for use in the next calculation cycle. Under this scheme, the calculation speed can be no faster than N/2 times that of serial processing assumming equal calculation and data transfer times. Some possible variations to the hardware configurations of this architecture will be presented in the next section.

Design Variations. Deviations for this design are possible with respect to how certain tasks are performed. Specifically, system I/O and the neural transfer characteristic type along with their location within the system are discussed.

<u>Neural Characteristic Variations.</u> Since the purpose of this system is to simulate the operation of an analog neural network, time will be taken to examine the need for various neural characteristics to adequately describe the single analog device known as a comparator. Throughout this paper it has been assumed that an analog comparator would be used as the neural processing element in the analog implementation. When a comparator is utilized as a threshold device (T_{ii} is zero) an appropriate digital system characteristic would be the 'greater than' or 'equal to' function. This would allow selection of a 0 output when the applied input was less than zero, and 1 otherwise. But, when a negative self feedback is added to an analog comparator, a summing circuit results. This would require the use of a ramp or piecewise linear saturation function for the neural characteristic to provide adequate performance in the digital world. The use of positive self feedback generates a hysteresis characteristic for the neuron.

<u>Hardware Variations</u> As mentioned in a previous section various techniques can be used to provide input/output channels to the neurons. The type of neural transfer characteristic selected and data input form greatly influence the computational hardware and I/O selection for the system. Consider the effect of the type of input data applied to a neuron. If the data to be applied to a neuron using a threshold characteristic is simply a 1 bit binary number, the neuron need only to have the capability of addition. On the other hand, if a gray level is to be applied to a neuron having any type of characteristic, a multiplicative capability must exist within the neuron. This is also true of neurons which allow their outputs to take on real values and not just 0 or 1.

The actual mechanism for getting input data to the neuron has several possibilities. If an N vector of single digit binary numbers is to be the system's input it could be directed to the neuron through the supervisory computer or one bit directly applied to each "smart" neuron depending on system speed requirements. If a real valued N vector is to be applied to the system, numerous possibilities exist. Assuming the data is available as an M bit per element column vector, it could be applied directly to each neuron as M bits but this would not be very feasible. Depending on the load faced by the supervisory computer, it could perform the data distribution but operations would be slowed. A better method would be to add a data acquisition CPU which could perform the data distribution over the primary data bus while the updating neuron placed its output on the secondary data bus. Similarly, an output CPU could perform data collection and subsequently, if required, D/A conversion capabilities.

Finally, the placement of the neural characteristic is flexible. If only a threshold characteristic is required it could be acccomplished with a 'greater than' or 'equal to' operation within each neuron. Other more complicated neural characteristics can be implemented in one of two locations. The first, and obvious, is within the neuron itself. A trigonometric function generator, look - up table or other technique could be used. But, in a move that can save both hardware and at the same time better guarantee uniformity of all the neuron transfer characteristics would be to place the characteristic in the secondary data bus loop.

<u>Conclusions.</u> In this paper we have addressed issues relating to the design and the implementation of neural networks. The goal of these studies is to increase the level of performance that can be obtained.

The design procedures developed in this paper allow for arbitrary associations of one dimensional analog inputs and any desired digital outputs. The potential for spurious states existing is reduced due to the following reasons. First, the decomposition of the whole system into smaller subsystems in lower block triangular form reduces the number of feedback connections per neuron. Thus, cumulative resistor tolerance errors are reduced due to the localized interconnections. Secondly, the deliberate coordination of all neuron thresholds and triangularity of the system reduces spurious states arising from mismatches in the repositioned thresholds and the number of feedbacks generated by the energy function approach. The use of negative feedback around a neuron ($T_{ii} < 0$) allows the generation of error signals, etc., providing flexibility for the decomposition. Finally, the modular design approach allows simpler subsystems to be combined to form the composite system. A drawback of this scheme is the speed reduction caused by the cascading of the subsystem modules.

We also presented a discussion of design issues for a digital neural network architecture. The design addresses the problem of analog part tolerances, non congruent neural characteristics, and the inability to modify the hardware resistor interconnection weights. Further, an environment more suitable for a learning algorithm would be provided. The potential need to allow for custom neural characteristics which may be difficult to precisely generate and duplicate to each neuron has been addressed. Finally, methods were presented to interface this design to the outside world and maintain survivability through use of identical and reconfigurable neurons, and through a control and I/O scheme which can be made redundant.

Acknowledgments. The authors are with the Department of Electrical and Computer Engineering at the University of Notre Dame, Notre Dame, Indiana 46556. This work was supported in part by the National Science Foundation under grant ECS 84-19918 and the Office of Naval Research under grant N00014-86-K-0506. Helpful discussions with Dr. J. - H. Li and J. A. Farrell are gratefully acknowledged.

References.

[1] For a review, see, *Neural Networks for Computing*, J. S. Denker, Editor, American Institute of Physics Conference Proceedings **151**, Snowbird, Utah, 1986.

[2] J. J. Hopfield, *Proc. Natl. Acad. Sci. U.S.A.* **79**, 2554 (1982), and *ibid.* **81**, 3088 (1984).

[3] A. N. Michel and R. K. Miller, *Qualitative Analysis of Large Scale Dynamical Systems*, Academic Press, 1977.

[4] A. N. Michel, J. A. Farrell, and W. Porod, *Proceedings of the IEEE Conference on Neural Information Processing Systems*, Denver, Colorado, 1987.

[5] D. W. Tank and J. J. Hopfield, *IEEE Trans. Circuits and Systems* **CAS-33**, 533 (1986).

[6] H. P. Graf, L. D. Jackel , R. E. Howard, B. Straughn, J. S. Denker, W. Hubbard, D. M. Tennant, and D. Schwartz, in Reference [1], pp. 182 – 187.

[7] D. L. Gray, A. N. Michel, and W. Porod, *Proceedings of the Twenty - Second Annual Princeton Conference on Information Sciences and Systems*, Vol. I, pp. 276 – 281 (1988).

[8] D. F. Hoeschele, Jr., *Analog-to-Digital / Digital-to-Analog Conversion Techniques*, John Wiley and Sons, 1968.

[9] B. Loriferne, *Analog-Digital and Digital-Analog Conversion*, Heyden, 1982.

[10] H. Schmid, *Electronic Analog/Digital Conversions*, Van Nostrand, 1970.

[11] D. L. Gray, Masters Thesis, University of Notre Dame, August 1988.

NEURAL NETWORK MODELS FOR THE LEARNING CONTROL OF DYNAMICAL SYSTEMS WITH APPLICATION TO ROBOTICS

F. Pourboghrat and M.R. Sayeh

Southern Illinois University

Introduction

In classical systems theory, input-output descriptions are based on some assumed or predetermined mathematical structures, normally a set of linear differential equations. Replacement of these predetermined structures by learned stimulus-response associative memory mappings leads to more general, normally nonlinear, representations of the connections between inputs and outputs. Artificial neural networks are essentially a collection of highly interconnected simple parallel processors which can do fast computations [1]. Such networks may be utilized for the nonlinear input-output mappings and the controller design of dynamical systems [2-7].

In this paper we design a learning controller for a class of nonlinear dynamical systems based on the biological model of the cerebellum for motor function generation. Two neuromorphic controllers are developed as variants of [7]. One is a neural inverse dynamic controller, and the other is a neural servo state feedback controller. The first design, shown in Figure 1, has a neural network block in the feedforward part of the controller, together with a PD-type feedback block, as in [6], which realizes the model of the inverse-dynamics of the system (plant). However, the network here is a back propagation network in which each unit (neuron) has dynamics.

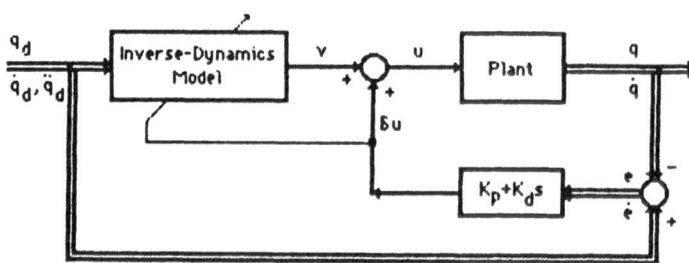

Figure 1

The second design, shown in Figure 2, is similar to the first one except that the feedback block is replaced by another neural network, which is to optimally compensate for unpredictable perturbations. The organization of the proposed controllers is inspired by the model of the cerebellum given by Kawato [2,3].

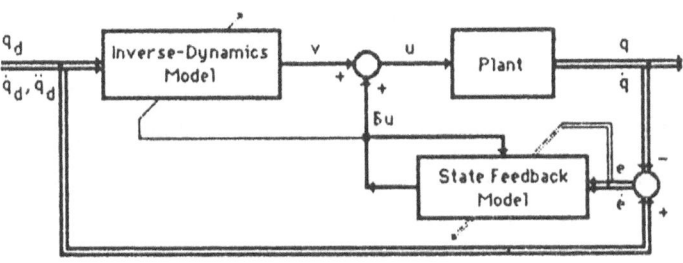

Figure 2

A complete design, in general, includes a trajectory planner which can be modeled by another neural network block. The proposed controllers enable the plant to track any desired smooth trajectory well, after a period of training, and to track other untrained trajectories satisfactorily. Also, these controllers will have adaptation capabilities so that they can compensate for unexpected disturbances.

Plant Dynamics

The dynamics of the systems (plants) under investigation, to be controlled here, are assumed to be given by

$$D(q)\ddot{q} + C(q,\dot{q})\dot{q} + G(q)q = T \tag{1}$$

where q, $\dot{q}$, and $\ddot{q}$ are n-dimensional vectors of the generalized positions, velocities, and accelerations, and the matrices $D(q)$, $C(q,\dot{q})$, and $G(q)$ are highly complex nonlinear functions of q and $\dot{q}$. This equation represents a 2n-dimensional nonlinear system, that is, a system with 2n degrees of freedom. Under mild conditions, many nonlinear systems, including robotic manipulators, can be represented by the above dynamics relationship.

Controller Design

In this section two controllers are designed. The first one is a neural inverse dynamic controller and the second one is a neural servo state feedback controller.

Neural Inverse Dynamic Controller: This controller has a single neural network in the feedforward block, together with a PD-type error feedback block which is used for the network's learning and also for the system's error compensation. The neural network block in this controller is intended to acquire a model of the inverse-dynamics of the plant. That is, the network adapts itself in such a way, that for a given desired trajectory, it will generate the input required by the plant to produce that trajectory.

The neural network block used in the proposed inverse dynamic controller is a generalized multi-layer error back propagation network, shown in Figure 3.

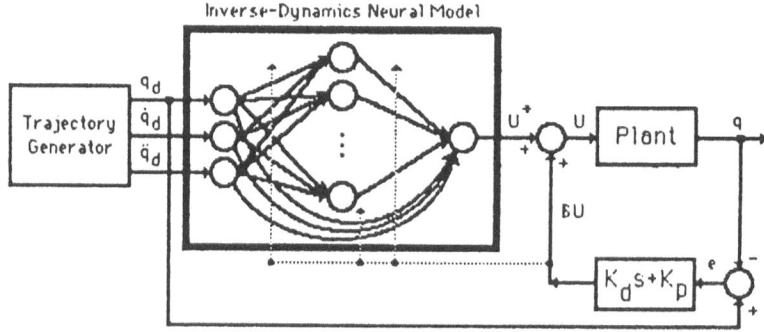

Figure 3

Unlike the usual error back propagation networks, each unit in the network has dynamics. Also, the network has a direct connection from the input layer to the output layer. The learning rule is an error back propagation algorithm (delta rule) [1], which is appropriately modified for the network.

Dynamics of the Network: In this controller, for a 2n degrees of freedom system, the inputs to its neural network block are the n desired angular positions q_d, the n desired angular velocities $\dot{q}_d$, and the n desired angular accelerations $\ddot{q}_d$. The outputs, v, of the network are the n input signals to the plant. The input-output relationship for each unit (neuron) in the network is given by

$$z_{i,1} = y_{i,0}$$

$$z_{i,2} = \Sigma w_{ij,1} y_{j,1} + \beta_{i,1}$$

$$z_{i,3} = \Sigma w_{ij,2} y_{j,2} + \Sigma w_{ij,3} y_{j,1} + \beta_{i,2}$$

$$x_{i,k} = (1 + \tau s)^{-1} z_{i,k}$$

$$y_{i,k} = f(x_{i,k})$$

$$f(x) = [1-\exp(-x)]/[1+\exp(-x)] = -1 + 2/[1+\exp(-x)]$$

where $y_{i,k}$ is the output level (firing rate) of the i-th unit in the k-th layer for k = 1,2,3 and $y_{i,3} = v_i$ is the output of the network, $y_{i,0} = q_d^{(i)}$ is the input to the network, f is a sigmoid function, $x_{i,k}$ is the inner state (membrane potential) of the i-th unit in the k=th layer, $z_{i,k}$ is the net input to the i-th unit in the k-th layer, $w_{ij,1}$ is the connecting weight from the j-th unit in the 1st layer to the i-th unit in the 2nd layer, $w_{ij,2}$ is the connecting weight from the j-th unit in the 2nd layer to the i-th unit in the 3rd layer, $w_{ij,3}$ is the connecting weight from the j-th unit in the 1st layer to the i-th unit in the 3rd layer, $\beta_{i,1}$ and $\beta_{i,2}$ are the threshold values of the i-th unit in the 2nd and 3rd layer, τ is the decay time-constant of the units, s is the differentiation operator, and $(1 + \tau s)^{-1}$ is the operator of the first-order time delay. That is, $x_{i,k} = (1 + \tau s)^{-1} z_{i,k}$ is given by the equation $\tau \dot{x}_{i,k} = -x_{i,k} + z_{i,k}$.

Error Back Propagation Learning Rule: The purpose of the learning here is to make the output v = $[v_1 ,..., v_n]^T$ of the neural network become equal to the input u = $[u_1 ,..., u_n]^T$ of the plant. That is, to indirectly force the feedback signal $\delta u = u-v$, which is a function of the output error $e = q_d-q$, to zero.

Let us denote the network's connection weights by $W = [W_1^T, W_2^T, W_3^T]^T$, where
$$W_1 = [w_{11,1} ,..., w_{km,1}]^T, \quad W_2 = [w_{11,2} ,..., w_{nk,2}]^T, \quad \text{and} \quad W_3 = [w_{11,3} ,..., w_{nm,3}]^T,$$
are respectively the weights connecting the network layers 1 to 2, 2 to 3, and 1 to 3, and m, k, and n are respectively the number of units in the first (input) layer, second (hidden) layer, and the third (output) layer of the network. In addition, let us define the algorithm for the adjustment of the connection weights of the network to be given by

$$\dot{w}_{ij,1} = -\gamma_{ij,1} \Sigma_{p=1,...n} (u_p -v_p)[(1-v_p^2)/2] \zeta_{pij,1} \tag{3a}$$

$$\dot{w}_{ij,2} = -\gamma_{ij,2}(u_1 -v_1)[(1-v_i^2)/2]\eta_{j,2}$$

$$\dot{w}_{ij,3} = -\gamma_{ij,3}(u_i -v_i)[(1-v_i^2)/2]\eta_{j,1}$$

and

$$\tau \dot{\eta}_{j,1} + \eta_{j,1} = y_{j,1}$$ (3b)

$$\tau \dot{\eta}_{j,2} + \eta_{j,2} = y_{j,2}$$

$$\tau \dot{\zeta}_{pij,1} + \zeta_{pij,1} = w_{pi,3}[(1-y_{i,2}^2)/2]\eta_{j,1}$$

with $w_{ij,k}(0)$, $k = 1,2,3$ taken as a random number between -0.2 to 0.2, and $\eta_{j,1}(0) = 0$, $\eta_{j,2}(0) = 0$ and $\zeta_{pij,1}(0) = 0$. Then we will have the following result.

Result 1: Consider a dynamic system given by the dynamic equation (1). Then the neural inverse dynamic controller (2), together with the learning rule (3), forces the plant's trajectory q and q̇ to follow the desired trajectory q_d and $\dot{q}_d$ after a sufficiently long time.

Proof: The proof is a modification of the convergence theory of gradient descent method given in [9]. A detailed proof can be found in [7].

Neural Servo State Feedback Controller: This controller contains both a feedforward and a feedback neural network block. The feedback neural block in this design has substituted for the PD-type error feedback block, as in Figure 4.

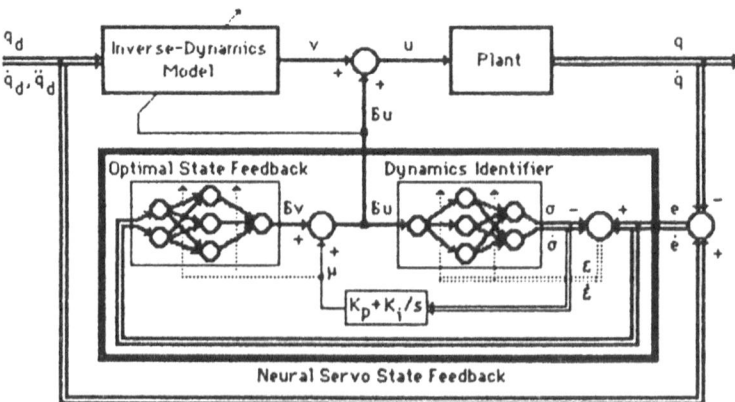

Figure 4

Dynamics of the Network: The inputs to this neural network block are the n angular position errors e, and the n angular velocity errors e. The outputs of the network are the n compensating torque signals δu. There are two sub-networks, dynamics identifier and optimal state feedback, each with three layers. The input-output relationship for each unit (neuron) in both the sub-networks are given by

$$z_{i,1} = y_{i,0} \tag{4}$$
$$z_{i,2} = \Sigma w_{ij,1} y_{j,1} + \beta_{i,1}$$
$$z_{i,3} = \Sigma w_{ij,2} y_{j,2} + \beta_{i,2}$$
$$x_{i,k} = (1 + \tau s)^{-1} z_{i,k}$$
$$y_{i,k} = f(x_{i,k})$$
$$f(x) = [1 - \exp(-x)]/[1 + \exp(-x)] = -1 + 2/[1 + \exp(-x)]$$

where for the dynamics identifier sub-network the inputs $y_{i,0}$ are δu_i, and the outputs $y_{i,3}$ are σ_i and $\dot{\sigma}_i$, and for the optimal state feedback sub-network the inputs $y_{i,0}$ are e_i and $\dot{e}_i$ and the outputs $y_{i,3}$ are δv_i, where $i = 1,...,n$.

Error Back Propagation Learning Rule: The purpose of the learning here is to make the output $\delta v = [\delta v_1,...,\delta v_n]^T$ of the optimal state feedback sub-network to become equal to the perturbation compensating torque $\delta u = [\delta u_1,...,\delta u_n]^T$ for the dynamic system. That is, to indirectly force the perturbation error $e = q_d - q$ equal to zero. For this, the dynamics identifier sub-network receives the perturbation torque δu as input and generates σ and $\dot{\sigma}$ as output, which is forced to equal e and $\dot{e}$. That is, through learning the error $\varepsilon = e - \sigma$ it is forced to become equal to zero. Also, the optimal state feedback sub-network receives the errors e and $\dot{e}$ as input and generates δv as output which is forced to equal δu. That is, through learning the error $u = \delta u - \delta v$, which is a function of e and $\dot{e}$, is forced to become equal to zero. The learning is complete when the quantities e, $\dot{e}$, ε, and $\dot{\varepsilon}$ are equal to zero. This, however, means that the optimal state feedback sub-network becomes the inverse of the dynamics identifier sub-network, and hence, produces the optimal compensating torque feedback corresponding to the trajectory error e, $\dot{e}$.

Let the connection weights for the dynamics identifier sub-network be denoted by $A = [A_1^T, A_2^T]^T$, where $A_1 = [a_{11,1}, ..., a_{km,1}]^T$ and $A_2 = [a_{11,2}, ..., a_{nk,2}]^T$ are respectively the weights connecting the network layers 1 to 2 and 2 to 3, and m, k, and n are respectively the numbers of units in the first (input) layer, second (hidden) layer, and the third (output) layer of the network. Also, let the connection weights for the optimal state feedback sub-network be denoted by $B = [B_1^T, B_2^T]^T$, where $B_1 = [b_{11,1}, ..., b_{kn,1}]^T$ and $B_2 = [b_{11,2}, ..., b_{mk,2}]^T$ are respectively the weights connecting the network layers 1 to 2 and 2 to 3, and n, k, and m are respectively the number of units in the first (input) layer, second (hidden) layer, and the third (output) layer of the network.

Let us define the learning rule for the dynamics identifier sub-network to be given by

$$\dot{a}_{ij,1} = -\gamma_{ij,1} \Sigma_{p=1,...n} (e_p - \sigma_p)[(1 - \sigma_p^2)/2]\zeta_{pij,1} \tag{5a}$$

$$\dot{a}_{ij,2} = -\gamma_{ij,2}(e_i - \sigma_i)[(1 - \sigma_i^2)/2]\eta_{j,2}$$

and

$$\tau\dot{\eta}_{j,1} + \eta_{j,1} = y_{j,1} \tag{5b}$$

$$\tau\dot{\eta}_{j,2} + \eta_{j,2} = y_{j,2}$$

$$\tau\dot{\zeta}_{pij,1} + \zeta_{pij,1} = a_{pi,3}[(1 - y_{i,2}^2)/2]\eta_{j,1}$$

with $a_{ij,k}(0)$, k = 1,2,3 taken as a random number between -0.2 to 0.2, and $\eta_{j,1}(0) = 0$, $\eta_{j,2}(0) = 0$, and $\zeta_{pij,1}(0) = 0$. Also, let us define the learning rule for the optimal state feedback sub-network as be given by

$$\dot{b}_{ij,1} = -\gamma_{ij,1} \Sigma_{p=1,...n} (\delta u_p - \delta v_p)[(1 - \delta v_p^2)/2]\zeta_{pij,1} \tag{6a}$$

$$\dot{b}_{ij,2} = -\gamma_{ij,2}(\delta u_i - \delta v_i)[(1 - \delta v_i^2)/2]\eta_{j,2}$$

and

$$\tau\dot{\eta}_{j,1} + \eta_{j,1} = y_{j,1} \tag{6b}$$

$$\tau\dot{\eta}_{j,2} + \eta_{j,2} = y_{j,2}$$

$$\tau\dot{\zeta}_{pij,1} + \zeta_{pij,1} = b_{pi,3}[(1 - y_{i,2}^2)/2]\eta_{j,1}$$

with $b_{ij,k}(0)$, $k = 1,2,3$ taken as a random number between -0.2 and 0.2, and $\eta_{j,1}(0) = 0$, $\eta_{j,2}(0) = 0$, and $\zeta_{pij,1}(0) = 0$. Then we will have the following result.

Result 2: Consider a dynamic system given by equation (1). Then the neural inverse dynamic controller (2) and its learning rule (3), together with the neural servo state feedback (4) and its learning rule (5-6), forces the dynamic system's trajectory q and $\dot{q}$ to follow the desired trajectory q_d and $\dot{q}_d$ after a sufficiently long time.

Proof: Again, the proof is a modification of the convergence theory of gradient descent method given in [9]. A detailed proof can be found in [7].

Example

In this section, we consider the learning control of a nonlinear system to demonstrate the feasibility of the proposed methodology. For that, a robotic manipulator is desired to track a known prespecified trajectory with high accuracy. The proposed neural controller is considered and the desired trajectory is applied repeatedly during the training (learning) period. Using the proposed learning algorithms, the connection weights of the neural networks in the controller are adjusted until the trajectory tracking error tends to zero.

For the sake of simplicity, a single-link robot with viscous friction and rotational spring is considered, which is shown in Figure 5. The dynamic equation of the robot is given by

$$J\ddot{q} = - B\dot{q} - Kq + T \tag{7}$$

where the values of the parameters in MKS units are given by $J = 0.25$, $B = 2.25$, and $K = 2.5 + 0.25\cos(q)$.

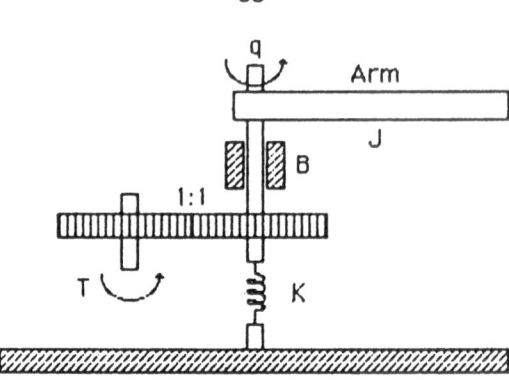

Figure 5

The arm is desired to track a prespecified trajectory over the interval $0 \le t \le 2$. A neural controller, as shown in Figure 4, is considered for the learning control of the arm. The trajectory generator is simply a double integrator with prespecified initial conditions. The desired acceleration is given by

$$
\begin{aligned}
\ddot{q}_d &= 2, \quad 0 \le t \le 0.5 \\
&= -6, \, 0.5 < t \le 1.0 \\
&= 6, \, 1.0 < t \le 1.5 \\
&= -2, \, 1.5 < t \le 2.0 \, .
\end{aligned}
\tag{8}
$$

The multi-layer neural network controller used in the design is intended to indirectly approximate the necessary functions in the feedforward inverse-dynamics block. That is, it approximately accounts for the linear combination of the unity gain, the first and second differentiators, and a nonlinear function $f = X\cos(X)$, which are the terms involved in the dynamic equation of the robot [6]. Therefore, the exact knowledge of the dynamics of the robot is not necessary for controller design. However, in this case, a large number of units in the hidden layers of the neural networks is usually required for a high degree of accuracy.

The closed-loop system was simulated on a digital computer. The system was allowed to repeat the movement with the same initial conditions for all the variables except for the values of the weights W_i, which were modified through learning during each trial. The weights converged to constant values after about 50 to 100 trials depending on the selected learning rate constants. For the case with 10 units in the

59

hidden layer of all three networks, the inverse dynamics network, the dynamics identifier, and the optimal state feedback, the angular positions are shown in Figure 6.

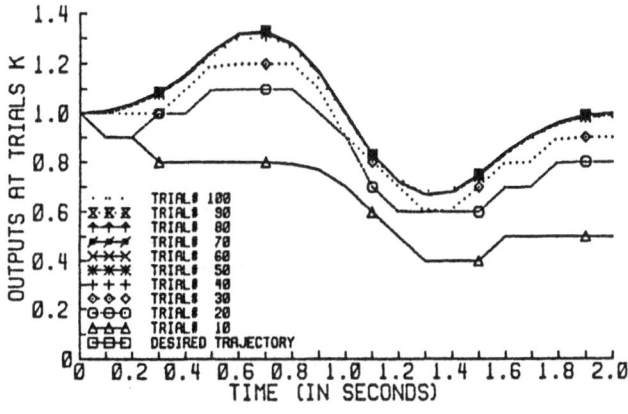

Figure 6

Conclusion

In this paper, two new brain-like control methodologies have been suggested. The significance of the designs rests in the novelty and simplicity of their architectures. The methodologies have several promising attributes that make them very feasible solutions to current problems in system control. First of all, the controllers mimic the functions of the cerebellum for the learning and control of voluntary movements. That is, these controllers learn the inverse-dynamics of the system by repeated trials of a given task, hence allowing the dynamic systems to track trained trajectories almost perfectly. But above that, with these controllers, systems can perform untrained trajectories quite satisfactorily. The schemes also have good adaptation capabilities which allow the controllers to respond to unexpected disturbances.

Another advantage of these methodologies is that the algorithms do not require the knowledge of the system parameters, and they are robust with respect to parameter variation and disturbances under a variety of tasks. Finally, they have parallel processing capabilities which make them fast and fault tolerant. Moreover, the parallel processing property of these architectures makes them highly suitable for the integration of a multitude of sensory information into the motion controller networks.

Acknowledgements: The authors are with the Department of Electrical Engineering at Southern Illinois University, Carbondale, Illinois 62901-6603.

References

[1] Rumelhart, D.E., McClelland, J.L, and the PDP Research Group, <u>Parallel Distributed Processing</u>, Volume 1, MIT Press, 1987.

[2] Kawato, M., Uno, Y., Isobe, M., and Suzuki, R., "A Hierarchical Model for Voluntary Movement and Its Application to Robotics," ICNN, 1987.

[3] Kawato, M., Furukawa, K., and Suzuki, R., "A Hierarchical Neural Network Model for Control and Learning of Voluntary Movement," Biol. Cybern., 57, 1987.

[4] Psaltis, D., Sideris, A., Yamamura, A., "Neural Controllers," ICNN, 1987.

[5] Eckmiller, R., "Neural Network Mechanisms for Generation and Learning of Motor Programs," ICNN, 1987.

[6] Pourboghrat, F., "Neuronal Controller for Robotic Manipulators," IASTED Int. Symp. Robotics Autom., 1988.

[7] Pourboghrat, F., "Neural Controls for Robots," Int. J. Robotics and Autom., 1988, (to appear).

[8] Parker, D.B., "Second Order Back Propagation: Implementing An O(n) Approximation to Newton's Method as An Artificial Neural Network," Neural Info. Proc. Conf., 1987.

[9] Fletcher, R., <u>Practical Methods of Optimization</u>, Second Edition, John Wiley and Sons, 1987.

COMPLEX MULTIPLICATION USING THE POLYNOMIAL RESIDUE NUMBER SYSTEM

Alexander Skavantzos
Louisiana State University

Thanos Stouraitis
The Ohio State University

I. INTRODUCTION

A number of important signal processing and communication algorithms including complex convolution, autocorrelation, cross-correlation, and computation of Fast Fourier Transforms (FFT) are complex-multiplication intensive. Such algorithms may become complex-multiplication bound in a high-speed real-time environment because of high multiplication data rate requirements. Array multipliers, being the high-speed design choice, require hardware investment proportional to n^2 for $(n \times n)$-bit multiplications, while their propagation delay is only proportional to n [Hwa79]. For traditional complex multiplication four multipliers and two adders are needed. For those cases where there does not exist a satisfactory "brute force" hardware solution, other alternatives must be sought. Many researchers, including Leung [Leu81], Krogmeier and Jenkins [Kro83], Soderstrand and Poe [Sod84], Taylor et al. [Tay85a, Tay85b, Tay86], and Krishnan et al. [Kri86] have published on the subjects of performing complex arithmetic using the Quadratic Residue Number System (QRNS), Quadratic-Like RNS (QLRNS), and the Modified QRNS (MQRNS). All of the above systems are interrelated. In fact, they represent special cases of a more general computer arithmetic system proposed and analyzed recently and are known as the Polynomial Residue Number system (PRNS) [Ska87].

The processing of complex numbers via PRNS requires that they be encoded in a manner allowing the parallel processing of the parts of their encoded versions. An attempt to find appropriate encoding schemes is described in this paper. More specifically, it examines the encoding of complex numbers as third-order polynomials so that their product can be computed via a fourth-order PRNS. This will be shown to require four $(\frac{n}{2}+1) \times (\frac{n}{2}+1)$ multipliers operating in parallel and no adders at all. This means that the multiplier hardware is only one fourth of the hardware required for traditional multiplying methods, while at the same time the speed is doubled.

The paper is organized as follows: The encoding of complex numbers is described in the second section, followed by an example of multiplication of their encoded versions. The third section reviews the basics of the PRNS, that are necessary for its employment in the multiplication of the encoded complex numbers. An example of PRNS processing is also given there. Finally, the fourth section concludes with a discussion of the new multiplier and assesses its potential for a wide range of applications.

II.   POLYNOMIAL REPRESENTATION OF COMPLEX NUMBERS

Encoding. Any complex number A can be written as $A = \Re(A) + j\Im(A)$, where $\Re(A)$ and $\Im(A)$ are the n-bit real and imaginary parts of A respectively. Without any loss of generality, n is considered to be an even number. If $\Re_H(A)$ and $\Re_L(A)$ are the $n/2$-bit higher and lower parts of $\Re(A)$ respectively, whereas $\Im_H(A)$ and $\Im_L(A)$ are the higher and lower parts of $\Im(A)$ respectively, then

$$\Re(A) = \Re_H(A)2^{n/2} + \Re_L(A)$$
$$\Im(A) = \Im_H(A)2^{n/2} + \Im_L(A). \tag{1}$$

An approximation of A may be given by a third-order polynomial representation of the form $(\hat{a}_0, \hat{a}_1, \hat{a}_2, \hat{a}_3)$, defined as

$$A \approx \hat{a}_0 2^{n/2} x^0 + \hat{a}_1 x^1 + \hat{a}_2 2^{n/2} x^2 + \hat{a}_3 x^3 \tag{2}$$

where $x = e^{j\pi/4}$ and, therefore, $x = \frac{\sqrt{2}}{2}(j+1)$, $x^2 = j$, and $x^3 = \frac{\sqrt{2}}{2}(j-1)$. In other words, the complex number A is now represented by its $\frac{n}{2}$-bit projections, $\hat{a}_i$s on vectors having magnitude equal to unity.

By equating the forms of A in Eq. (1) and Eq. (2) one finds that

$$\hat{a}_0 2^{n/2} + \hat{a}_1 \frac{\sqrt{2}}{2} - \hat{a}_3 \frac{\sqrt{2}}{2} + j\left[\hat{a}_1\frac{\sqrt{2}}{2} + \hat{a}_2 2^{n/2} + \hat{a}_3\frac{\sqrt{2}}{2}\right] =$$

$$\Re_H(A)2^{n/2} + \Re_L(A) + j[\Im_H(A)2^{n/2} + \Im_L(A)].$$

If $\hat{a}_0$ and $\hat{a}_3$ are defined as $\hat{a}_0 \equiv \Re_H(A)$ and $\hat{a}_2 \equiv \Im_H(A)$, then $\hat{a}_1$, and $\hat{a}_3$ are calculated to be

$$\hat{a}_1 = \frac{\Re_L(A) + \Im_L(A)}{\sqrt{2}} \quad \text{and} \quad \hat{a}_3 = \frac{\Im_L(A) - \Re_L(A)}{\sqrt{2}}. \tag{3}$$

The projections $\hat{a}_0$, and $\hat{a}_2$ are $n/2$-bit numbers and the two entities $\hat{a}_1$ and $\hat{a}_3$ are also taken to be rounded as $n/2$-bit integers. By forcing $\hat{a}_1$ and $\hat{a}_3$ to consist of only $n/2$ bits, an approximation error is introduced for the lower-weight parts of A.

The selection of powers of $x = e^{j\frac{\pi}{4}}$ as projection vectors for the complex numbers was motivated by the fact that $x^4 = -1$. This means that $x^i = -x^{<i>_4}$, where $<>_m$ denotes a mod m operation. Therefore, the product of two numbers approximated via Eq. (2) can always be given in a format similar to the same equation; a fact that is highly desirable for the cyclic polynomial multiplication of PRNS processing, as it will become apparent in the next section. The approximation error introduced by the rounding of the coefficients of Eq. (2) can be eliminated by slightly modifying the encoding scheme described above. By allowing the magnitude of some of the projection vectors to be different than unity, a new representation of A is proposed now, based on the relationship

$$2A = a_0 2^{n/2} + a_1(j+1) + a_2 2^{n/2}j + a_3(j-1), \tag{4}$$

where a_0 and a_2 are defined as

$$a_0 = 2\Re_H(A) \quad \text{and} \quad a_2 = 2\Im_H(A). \tag{5}$$

The other two projections, a_1 and a_3, are computed to be

$$a_1 = \Re_L(A) + \Im_L(A), \quad \text{and} \quad a_3 = \Im_L(A) - \Re_L(A). \tag{6}$$

By allowing the magnitude of the projection vectors to be different from each other, the representation of $2A$ through the use of the $(\frac{n}{2} + 1)$-bit integers (a_0, a_1, a_2, and a_3) is now exact.

Multiplication. In a similar manner, a complex number B may be represented by the four-tuple (b_0, b_1, b_2, b_3). The product of the two complex numbers A and B may assume two different representations, based on the way that the complex numbers are represented. If they are not decomposed, then their product is

$$P = AB = \Re(P) + j\Im(P)$$

where

$$\Re(P) = \Re(A)\Re(B) - \Im(A)\Im(B), \tag{7}$$

$$\Im(P) = \Re(A)\Im(B) + \Im(A)\Re(B).$$

The above form requires four real multiplications and two real additions. On the other hand, if they are decomposed according to Eq. (4), then

$$4P = 4AB = \left[a_0 2^{n/2} + a_1(j+1) + a_2 2^{n/2}j + a_3(j-1)\right] \cdot \tag{8}$$
$$\left[b_0 2^{n/2} + b_1(j+1) + b_2 2^{n/2}j + b_3(j-1)\right]$$

or

$$\begin{aligned}
4P = 4AB = \ &a_0 b_0 2^n + a_0 b_1 2^{n/2}(j+1) + a_0 b_2 2^n j + a_0 b_3 2^{n/2}(j-1) \\
&-a_1 b_3 2 + a_1 b_0 2^{n/2}(j+1) + a_1 b_1 2j + a_1 b_2 2^{n/2}(j-1) \\
&-a_2 b_2 2^n - a_2 b_3 2^{n/2}(j+1) + a_2 b_0 2^n j + a_2 b_1 2^{n/2}(j-1) \\
&-a_3 b_1 2 - a_3 b_2 2^{n/2}(j+1) - a_3 b_3 2j + a_3 b_0 2^{n/2}(j-1)
\end{aligned} \tag{9}$$

Then $4P$ can be written as

$$\begin{aligned}
4P = 4AB = \ &2(a_0 b_0 - a_1 b_3 - a_2 b_2 - a_3 b_1) + 2(2^{n-1} - 1)(a_0 b_0 - a_2 b_2) \\
&+2^{n/2}(a_0 b_1 + a_1 b_0 - a_2 b_3 - a_3 b_2)(j+1) \\
&+(a_0 b_2 + a_1 b_1 + a_2 b_0 - a_3 b_3)2j + 2(2^{n-1} - 1)(a_0 b_2 + a_2 b_0)2j \\
&+2^{n/2}(a_0 b_3 + a_1 b_2 + a_2 b_1 + a_3 b_0)(j-1)
\end{aligned} \tag{10}$$

Considering arithmetic in the modular ring $Z_{2^{n-1}-1}$—a ring that is closed with respect to addition and multiplication mod $(2^{n-1} - 1)$ [Her75a]—then $2^{n-1} \equiv 1$ and Eq. (10) becomes

$$4P = p_0 2 + p_1 2^{n/2}(j+1) + p_2 2j + p_3 2^{n/2}(j-1), \tag{11}$$

or

$$2P = p_0 + p_1 2^{\frac{n}{2}-1}(j+1) + p_2 j + p_3 2^{\frac{n}{2}-1}(j-1), \tag{12}$$

where

$$
\begin{aligned}
p_0 &= a_0 b_0 - a_1 b_3 - a_2 b_2 - a_3 b_1, \\
p_1 &= a_0 b_1 + a_1 b_0 - a_2 b_3 - a_3 b_2, \\
p_2 &= a_0 b_2 + a_1 b_1 + a_2 b_0 - a_3 b_3, \quad \text{and} \\
p_3 &= a_0 b_3 + a_1 b_2 + a_2 b_1 + a_3 b_0.
\end{aligned}
\tag{13}
$$

It can be seen that Eq. (12) and Eq. (4) will be of the same format if all of their coefficients are scaled by appropriate factors (powers of 2). Therefore, upon computing and scaling the coefficients p_0, p_1, p_2, and p_3 the value of $2P$ is obtained in the format of Eq. (4) and can be used in consecutive multiplications without any need for re-encoding. A technique for the fast computation of all p_is is necessary for an efficient multiplier. Such a technique is described next.

Suppose that the product, $P(x)$, of the two polynomials $A(x)$ and $B(x)$ is to be computed mod $(x^4 + 1)$. Then

$$
\begin{aligned}
P(x) &= \; < A(x)B(x) >_{(x^4+1)} \\
&= \; < (a_0 + a_1 x + a_2 x^2 + a_3 x^3)(b_0 + b_1 x + b_2 x^2 + b_3 x^3) >_{(x^4+1)} \\
&= \; p_0 x^0 + p_1 x^1 + p_2 x^2 + p_3 x^3.
\end{aligned}
\tag{14}
$$

The coefficients p_0, p_1, p_2, and p_3 which are necessary for the evaluation of the product of the two complex numbers A and B, given by Eq. (12) and Eq. (13) can be obtained from the polynomial product $< A(x)B(x) >_{(x^4+1)}$ described by Eq. (14).

The traditional computation of the tuple (p_0, p_1, p_2, p_3) using Eq. (13) requires 16 multiplications and 12 additions, offering thus no savings when compared to conventional multiplication techniques realizing Eq. (7). Recent advances in computer arithmetic have made possible the computation of the same tuple with only four $(\frac{n}{2} + 1) \times (\frac{n}{2} + 1)$-bit multiplications performed in parallel and no additions at all. The amount of hardware of an $(n \times n)$-bit array multiplier is proportional to n^2 and its propagation delay is proportional to n. This means that these new advances reduce the multiplier hardware down to one fourth and the propagation delay down to less than one half compared to the traditional complex multiplication described by Eq. (7) which uses four $(n \times n)$-bit multipliers and two adders.

These substantial savings are possible if the tuple (p_0, p_1, p_2, p_3), which is necessary for the evaluation of the polynomial product, is computed by using the Polynomial Residue Number System (PRNS) described in the next section.

In an Nth-order PRNS, or PRNS(N), the coefficients of $(N-1)$st-order polynomials belong to an appropriately specified modular ring Z_m. The PRNS(N) is fast when computing products of polynomials mod $(x^N \pm 1)$. This form of cyclic polynomial multiplication is guaranteed by the adopted representation of complex numbers in Eq. (4). In fact, this is a case of multiplication modulo $(x^4 + 1)$.

An example illustrating the use of the proposed encoding scheme for complex multiplication follows.

Example of Multiplication of Encoded Numbers. Suppose that $A = 8 - j11$, $B = -3 + j9$, and $n = 6$. Performing arithmetic mod $(2^{n-1} - 1)$, or mod 31 we find

$$\Re(A) = \; <8>_{31} \qquad\qquad \equiv (001000)_2 = 1 \times 2^3 + 0$$
$$\Im(A) = \; <-11>_{31} \equiv 31 - 11 = 20 \quad \equiv (010100)_2 = 2 \times 2^3 + 4$$
$$\Re(B) = \; <-3>_{31} \equiv 31 - 3 = 28 \quad \equiv (011100)_2 = 3 \times 2^3 + 4$$
$$\Im(B) = \; <9>_{31} \qquad\qquad \equiv (001001)_2 = 1 \times 2^3 + 1.$$

Following the encoding procedure described by Eq. (4)

$$2A = a_0 2^{n/2} + a_1(j+1) + a_2 2^{n/2} j + a_3(j-1)$$

and

$$2B = b_0 2^{n/2} + b_1(j+1) + b_2 2^{n/2} j + b_3(j-1)$$

with

$a_0 = 2 \times 1 = 2$	$b_0 = 2 \times 3 = 6$
$a_1 = 0 + 4 = 4$	$b_1 = 4 + 1 = 5$
$a_2 = 2 \times 2 = 4$	$b_2 = 2 \times 1 = 2$
$a_3 = 4 - 0 = 4$	$b_3 = 1 - 4 = -3$

Application of Eq. (13) gives

$$p_0 = 12 + 12 - 8 - 20 \quad = -4$$
$$p_1 = 10 + 24 + 12 - 8 \quad = 38$$
$$p_2 = 4 + 20 + 24 + 12 \quad = 60$$
$$p_3 = -6 + 8 + 20 + 24 \quad = 46$$

and $2P$ is given by Eq. (12) as

$$2P \; = -4 + 38 \times 4(j+1) + 60j + 46 \times 4(j-1)$$
$$= -4 + 152j + 152 + 60j + 184j - 184$$
$$= -36 + 396j = \; <26 + j24>_{31} \; .$$

Computing the product AB in the conventional way,

$$AB \; = 75 + j105 \quad \text{or}$$
$$2AB \; = 150 + j210 \quad \text{or}$$
$$<2AB>_{31} \; = 26 + j24$$

which is the same as $< 2P >_{31}$. □

Some important aspects of PRNS arithmetic that are of interest to the multiplication of complex numbers are reviewed next.

III.   THE POLYNOMIAL RESIDUE NUMBER SYSTEM (PRNS)

It has been demonstrated that the PRNS achieves Winograd's lower bound for multiplication complexity [Ska87]. To obtain the lowest possible multiplication count, the polynomials $x^N \pm 1$ must be factorized in N distinct factors in a modular ring, Z_m, as

$$x^N \pm 1 \equiv (x - r_0)(x - r_1) \ldots (x - r_{N-1}) \tag{15}$$

with $r_0, r_1, \ldots, r_{N-1} \in Z_m$.

Suppose $P(m)$ is a finite structure containing $(N-1)$st-order polynomials with coefficients in Z_m or

$$P(m) = \left\{ A(x)/A(x) = \sum_{i=0}^{N-1} a_i x^i \right\} \tag{16}$$

where $a_i \in Z_m$ for $i = 0 \ldots, N-1$. Then, there exists an isomorphic mapping, f_N, of $P(m)$ onto $Z_m^N \triangleq \overbrace{Z_m \times Z_m \times \ldots \times Z_m}^{N}$, called the PRNS($N$) isomorphic mapping, which is given by

$$f_N : A(x) = a_0 + a_1 x + \cdots + a_{N-1} x^{N-1} \xrightarrow{f_N} A^*(x) = (a_0^*, a_1^*, \ldots, a_{N-1}^*) \tag{17}$$

with

$$a_i^* = << A(x) >_{(x-r_i)} >_m = < A(r_i) >_m, \quad i = 0, 1, \ldots, N-1 \tag{18}$$

where r_i are distinct roots of $x^N \pm 1 \equiv 0$ in Z_m.

The inverse mapping is defined as

$$f_N^{-1} : A^*(x) = (a_0^*, a_1^*, \ldots, a_{N-1}^*) \xrightarrow{f_N^{-1}} A(x) = a_0 + a_1 x + \cdots + a_{N-1} x^{N-1} \tag{19}$$

where $A(x)$ is given as

$$A(x) = \sum_{i=0}^{N-1} a_i^* Q_i(x) \tag{20}$$

and

$$Q_i(x) = N^{-1} \left(1 + r_i^{-1} x + r_i^{-2} x^2 + \cdots + r_i^{-(N-2)} x^{N-2} + r_i^{-(N-1)} x^{N-1} \right). \tag{21}$$

It is reminded that N^{-1} and r_i^{-j} are the multiplicative inverses of N and r_i^j in Z_m respectively, which means that $< N^{-1} N >_m \equiv 1$ and $< r_i^{-j} r_i^j >_m \equiv 1$.

Operations in PRNS(N) are defined as follows:

Addition:

$$(a_0^*, a_1^*, \ldots, a_{N-1}^*) + (b_0^*, b_1^*, \ldots, b_{N-1}^*)$$

$$= (\ < (a_0^* + b_0^*) >_m, \ < (a_1^* + b_1^*) >_m, \ \ldots, < (a_{N-1}^* + b_{N-1}^*) >_m). \tag{22}$$

Multiplication:

$$(a_0^*, a_1^*, \ldots, a_{N-1}^*)(b_0^*, b_1^*, \ldots, b_{N-1}^*)$$

$$= (\ < (a_0^* b_0^*) >_m, \ < (a_1^* b_1^*) >_m, \ldots, < (a_{N-1}^* b_{N-1}^*) >_m). \tag{23}$$

Eq. (23) makes clear that if $A(x) = \sum_{i=0}^{N-1} a_i x^i$ and $B(x) = \sum_{i=0}^{N-1} b_i x^i$, then a polynomial product $< A(x) B(x) >_{(x^N \pm 1)}$ requires N multiplications mod m and no additions. The same polynomial product needs N^2 multiplications and $N(N-1)$ additions mod m, if performed in $P(m)$. In addition, the computation of the polynomial product in Z_m^N requires only one level of operations instead of the multiple levels that are required for operations in $P(m)$.

To utilize the full potential of the PRNS(N), the polynomials $x^N \pm 1$ must be factored into N distinct first-degree factors in Z_m. The facts that are stated below provide the necessary theoretical background for the appropriate choice of modulus m so that the factorization in N distinct factors is feasible, and describe some useful relationships between the roots [Ska87]. They are just stated here without proof.

Fact 1: If N, m are positive integers with a prime decomposition of m given in terms of powers e_i of its prime factors p_i, as $m = p_1^{e_1} \cdot p_2^{e_2} \cdots p_L^{e_L}$, with $N < p_i$ for $i = 1, \ldots, L$, then $x^N + 1$ can be factorized in N distinct factors in Z_m as $x^N + 1 \equiv < (x - r_0)(x - r_1) \ldots (x - r_{N-1}) >_m$ if and only if $N | (p_i - 1)/2$, $i = 1, \ldots, L$, where $a | b$ reads "a divides b". The necessary and sufficient condition for the factorization of the polynomial $x^N - 1$ as $< (x - r_0)(x - r_1) \ldots (x - r_{N-1}) >_m$ becomes $N | p_i - 1$, $i = 1, \ldots, L$. There are $(N!)^{L-1}$ different ways to factor $x^N \pm 1$ in N distinct factors.

Fact 2: If $x^N + 1$ can be factorized in N distinct factors in Z_m as

$$x^N + 1 \equiv < (x - r_0)(x - r_1) \ldots (x - r_{N-1}) >_m,$$

then the jth powers of the multiplicative inverses of the roots, r_i^{-j}, exist for every $i = 0, \ldots, N-1$ and $j = 1, \ldots, N-1$ and are given by $r_i^{-j} \equiv < -r_i^{N-j} >_m$. In the case of the polynomial $x^N - 1$ the inverses of the roots are given by $r_i^{-j} = < r_i^{N-j} >_m$.

Fact 3: For both congruences $x^N \pm 1 \equiv < 0 >_m$, the multiplicative inverses of their roots are also roots of the congruences, while their additive inverses are roots of the congruences only when N is even.

Example of PRNS(4) Multiplication. Consider the modular ring Z_{17} for arithmetic operations. Then $x^4 + 1$ can be decomposed by using the roots of $x^4 + 1 \equiv < 0 >_{17}$. The roots and the powers of their multiplicative inverses are

$$
\begin{array}{llll}
r_0 = 2 & r_0^{-1} = 9 & r_0^{-2} = 13 & r_0^{-3} = 15 \\
r_1 = 15 & r_1^{-1} = 8 & r_1^{-2} = 13 & r_1^{-3} = 2 \\
r_2 = 9 & r_2^{-1} = 2 & r_2^{-2} = 4 & r_2^{-3} = 8 \\
r_3 = 8 & r_3^{-1} = 15 & r_3^{-2} = 4 & r_3^{-3} = 9
\end{array}
$$

and $N^{-1} = 4^{-1} = 13$. Then $x^4 + 1$ can be written as

$$
x^4 + 1 \equiv < (x - 2)(x - 15)(x - 9)(x - 8) >_{17} .
$$

Suppose that

$$
A(x) = 5 + 6x + 8x^2 + 13x^3 \quad \text{and}
$$

$$
B(x) = 9 + 14x + 10x^2 + 12x^3.
$$

From Eq. (18) we find

$$
\begin{array}{ll}
a_0^* = 0 & b_0^* = 3 \\
a_1^* = 6 & b_1^* = 10 \\
a_2^* = 1 & b_2^* = 3 \\
a_3^* = 13 & b_3^* = 3
\end{array}
$$

and the product of the two polynomials is given in PRNS(4) as $(0, 6, 1, 13) \cdot (3, 10, 3, 3)$, or

$$
\begin{aligned}
p_0^* &= < 0 \cdot 3 >_{17} = 0 \\
p_1^* &= < 6 \cdot 10 >_{17} = 9 \\
p_2^* &= < 1 \cdot 3 >_{17} = 3 \\
p_3^* &= < 13 \cdot 3 >_{17} = 5.
\end{aligned}
$$

Based on Eq. (21) we find

$$
\begin{aligned}
Q_0(x) &= 4^{-1}(1 + 9x + 13x^2 + 15x^3) \\
Q_1(x) &= 4^{-1}(1 + 8x + 13x^2 + 2x^3) \\
Q_2(x) &= 4^{-1}(1 + 2x + 4x^2 + 8x^3) \\
Q_3(x) &= 4^{-1}(1 + 15x + 4x^2 + 9x^3).
\end{aligned}
$$

Therefore, from Eq. (20) we compute .

$$p_0 = \; < 4^{-1}(p_0^{\bullet} + p_1^{\bullet} + p_2^{\bullet} + p_3^{\bullet}) >_{17} \qquad = \; < 13(17) >_{17} = 0$$

$$p_1 = \; < 4^{-1}(9p_0^{\bullet} + 8p_1^{\bullet} + 2p_2^{\bullet} + 15p_3^{\bullet}) >_{17} \; = \; < 13(0 + 4 + 6 + 7) >_{17} = 0$$

$$p_2 = \; < 4^{-1}(13p_0^{\bullet} + 13p_1^{\bullet} + 4p_2^{\bullet} + 4p_3^{\bullet}) >_{17} \; = \; < 13(0 + 15 + 12 + 3) >_{17} = 16$$

$$p_3 = \; < 4^{-1}(15p_0^{\bullet} + 2p_1^{\bullet} + 8p_2^{\bullet} + 9p_3^{\bullet}) >_{17} \; = \; < 13(0 + 1 + 7 + 11) >_{17} = 9$$

which means that in Z_{17}

$$P(x) = \; < A(x)B(x) >_{(x^4+1)} = 16x^2 + 9x^3.$$

This product is the same with the one computed traditionally mod $(x^4 + 1)$ which is given as

$$
\begin{aligned}
< A(x)B(x) >_{(x^4+1)} &= \; < (5 + 6x + 8x^2 + 13x^3) \cdot \\
&\qquad \cdot (9 + 14x + 10x^2 + 12x^3) >_{(x^4+1)} \\
&= \; 45 + 70x + 50x^2 + 60x^3 \\
&= \; -72 + 54x + 84x^2 + 60x^3 \\
&= \; -80 - 96x + 72x^2 + 112x^3 \\
&= \; -182 - 130x - 156x^2 + 117x^3 \\
&= \; 16x^2 + 9x^3.
\end{aligned}
$$

$\square$

IV. CONCLUSIONS

The introduction of a model to represent complex numbers as 3rd-order polynomials allows the use of a 4th-order PRNS for performing a complex multiplication. Since the n-bit numbers have been decomposed in $(\frac{n}{2} + 1)$-bit entities, the multiplications can be performed in a shorter time. This multiplication technique presents advantages for Digital Signal Processing tasks like the FFT, convolution, correlation, and others. Instead of the two n-bit channels for the real and imaginary parts of the operands, the $(\frac{n}{2}+1)$-bit channels $a_0, a_1, a_2,$ and a_3 can be used operating in parallel.

Some of the advantages of the PRNS complex multiplication, when array multipliers are used, have already been discussed. The same kind of favorable comparison can be established when lookup-table-based RNS implementations are considered. In this case, the conventional implementation of the complex product $[\Re(A) + j\Im(A)] \cdot [\Re(B) + j\Im(B)]$ requires four memory tables with $2n$ input address lines. Since the size of a memory table increases exponentially with the table address space, any address space reduction is much desired. The decomposition of complex numbers in four channels results in an address space—for each of the four tables—equal to $2(\frac{n}{2} + 1) = n + 2$; almost half than without modeling. The reduced address space results in higher throughputs as well. Furthermore, since the polynomial product of the complex numbers is a product mod $(x^4 + 1)$, it can be mechanized through the PRNS decomposition with only four real multiplications performed in parallel. The lack of an addition stage in the proposed complex multiplication scheme allows the operation to be completed in the time required for a $(\frac{n}{2} + 1) \times (\frac{n}{2} + 1)$-bit multiplication.

In conclusion, complex multiplication is accelerated and requires less hardware when performed using the PRNS. This allows its employment in a rich variety of application areas, especially multidimensional ones. Efficient coding is necessary for the use of PRNS in higher-order problems. The research described in this paper sets a precedent and guidelines for applying the PRNS in applications of higher order with potential advantages.

REFERENCES

[Her75] Herstein, I.N., *Topics in Algebra*, 2nd ed., Xerox College Publishing, Lexington, Mass., 1975.

[Hwa79] Hwang K., *"Computer Arithmetic,"* John Wiley & sons, New York, 1979.

[Kri86] Krishnan, R., G.A. Jullien, and W.C. Miller, "The Modified Quadratic Residue Number System (MQRNS) for Complex High Speed Signal Processing," *IEEE Trans. on Circuits and Systems*, vol. CAS-33(3), pp. 325-327, March 1986.

[Kro83] Krogmeier, J.V, and W.K. Jenkins, "Error Detection and Correction in Quadratic Residue Number Systems," *Proceedings of 26th Midwest Symposium on Circuits and Systems*, pp. 408-411, August 1983.

[Leu81] Leung, S.H., "Application of Residue Number Systems to Complex Digital Filters," *Proceedings of Fifteenth Asilomar Conference on Circuits, Systems, and Computers*, pp. 70-74, November 1981.

[Ska87] Skavantzos, A., "The Polynomial Residue Number System and its Applications," Ph.D. Dissertation, University of Florida, August 1987.

[Sod84] Soderstrand, M.A., and G.D. Poe, "Applications of Quadratic-Like Complex Residue Number System Arithmetic to Ultrasonics," *Proceedings, IEEE International Conference on Acoustics, Speech, and Signal Processing*, pp. 28A.5.1 – 28A.5.4, March 1984.

[Tay85a] Taylor, F.J., "A Single Modulus Complex ALU for Signal Processing," *IEEE Trans. on Acoustics, Speech, and Signal Processing*, vol. ASSP-33(5), pp. 1302-1315, October 1985.

[Tay85b] Taylor, F.J., G. Papadourakis, A. Skavantzos, and A. Stouraitis, "A Radix-4 FFT Using Complex RNS Arithmetic," *IEEE Trans. on Computers*, vol. C-34(6), pp. 573-576, June 1985.

[Tay86] Taylor, F.J., "On the Complex Residue Arithmetic System (CRNS)," *IEEE Trans. on Acoustics, Speech, and Signal Processing*, vol. ASSP-34(6), pp. 1675–1677, December 1986.

On Routing and Performance Analysis of the Doubly Connected Networks for MANs & LANs

Tein Y. Chung
North Carolina State University

Suresh Rai
Louisiana State University

Dharma P. Agrawal
North Carolina State University

I. Introduction

As local area network (LAN) has become popular in the office automation application, the development of an integration network for a metropolitan area has received much attention in the past years. A metropolitan area network (MAN) offers various commercial applications for the city. Note, the development cost of a MAN is reduced if the existing LAN topologies and protocols are adopted for it. Among the topologies for a LAN, a doubly-connected directed graph (digraph) is popular as it suits point-to-point behavior of an optical fibre which will be the main communication medium in the future network development. Moreover, the existing LAN protocols for loop topology considers a directed loop topology.

Some of popular doubly-connected topologies include distributed data loop computer network (DDLCN), daisy chain, forward loop backward hop (FLBH), 2-dimension 3-jump (2-D 3-J), 3-ary 2-cube, and Manhattan Street Network [1–4]. This paper discusses only three topologies namely 3-ary 2-cube, 2-D 3-J, and MSN. It, then, introduces the concept of a macro-graph. Definitions and examples illustrate the macro-graph notation. Later, two routing schemes, named conservative and refined-conservative routing algorithm, are proposed for these topologies. Schemes use the macro-graph representation of a topology. A simulation program generates performance results for the proposed routing schemes. This paper critically reviews five medium access protocols too. These topologies include Store and Forward model (SF), Cut Through model (CT), Quasi-Cut Through model (QCT), Bux and Schlatter model (BS) and Hilal et. al model (HL) [5–9]. Finally, we have also discussed the implementation scheme of a node in a loop network.

The layout of the paper is as follows: Section II outlines the preliminaries of 3 topologies and develops the notation of macro-graph too. Routing issues are described in section III. Section IV considers protocols, and discusses the hardware implementation of a node. Finally, section V concludes the paper.

II. Preliminary

2.1 Two-dimensional topology

Generally, a computer network is modeled as a digraph G(n,d), where n represents the number of processors in the network and d refers the maximum number of adjacent processors. Note, the vertices (nodes) V(G) in the digraph stands for the processors while edges (links) E(G) specify the communication links between the processors.

For two dimensional topology group, n is factored as n_2*n_1. The node address is, then, represented by two-tuples $X_2 X_1 \mid n_2, n_1$. A generalized link equation for these networks is expressed as:

Given node (X_2, X_1),

Obtain the adjacent nodes $(X_2, b_1) = \{X_2, [(X_1 \ op1 \ 1) \bmod n_1]\}$,

and $(b_2, X_1) = \{[(X_2 \ op2 \ 1) \bmod n_2], X_1\}$

where '*op1*' and '*op2*' are arithmetic operator '+' or '-'. Three networks that satisfy different operations *op1* and *op2* are obtained as 3-ary 2-cube, 2-D 3-J, and MSN. For 3-ary 2-cube, the '*op1*' and '*op2*' are always '+', while for MSN, '*op1*' is '+' ('-') when X_2 is even (odd) and '*op2*' is '+' ('-') when X_1 is even (odd). In a 2-D 3-J network, '*op1*' is always '+' while '*op2*' is '+' ('-') when X_1 is even (odd). Figure 1 illustrates all these three topologies.

2.2 Macro-graph for 2-D topologies

This subsection describes the concept of macro-graph, and various other terms related to its development. To define a mapping of a two-dimensional network uniquely, the macro-graph should be able to reflect the information of (1) the size of each dimension, and (2) the connection pattern of the topology. In what follows, a macro-graph is considered as one which is composed of two macro-cycles. A macro-cycle includes macro-nodes and macro-edges, where a macro-node denotes a cycle in each dimension (hereafter referred as dimensional cycle) of the original network G(n,d), and a macro-edge denotes the group of the edges interconnecting two different dimensional cycles. A dimensional cycle is defined as the cycle in which all nodes have the same address in a specific address field. Associated with each macro-node is a direction bit indicating the connection pattern (discussed later) of the corresponding dimensional cycle. Hence, a two dimensional network is well defined by its macro-graph since the number of macro-nodes in each macro-cycle represents the size of the dimension one and two respectively, while the macro-edges and the associated direction bit for each macro-node may uniquely represent the network's connection pattern.

For a two dimensional network of size $n=n_2*n_1$, the i-th address field of node X is denoted as $A_i(X)=X_i$ for i=1,2.

Definition 1. A dimensional cycle at the i-th dimension, denoted as $DC_{X_j}(i)$; for i≠j, consists of vertices

$V\{DC_{X_j}(i)\}=\{v \mid A_j(v)= X_j\}$; where $0 \leq X_j < n_j$,

and links

$E\{DC_{X_j}(i)\}=\{(uv) \mid (uv) \in E(G); u \text{ and } v \in V\{DC_{X_j}(i)\}\}$;

For example, for a dimensional cycle $DC_{X_j}(i)$ in a MSN network of size 16=4*4 with i=1 and $X_2=0$, we have the addresses of the nodes in $DC_0(1)$ as $V\{DC_0(1)\}=\{0, 1, 2, 3\}$ and edges $E\{DC_0(1)\}=\{(0,1),(1,2),(2,3),(3,0)\}$.

Definition 2. An edge (uv) is defined as $e_{+(-)}$ link if any two nodes

$(u)_{n2,n1} = u_2 \ u_1$, and

$(v)_{n2,n1} = v_2 \ v_1$

satisfy

$$v_j = (u_j + (-)\ 1)\ \text{mod}\ n_j$$
$$v_i = u_i;\ i \neq j$$

For instance, in a MSN of size $16 = 4*4$, link (1,2) connecting node $(1)_{4,4} = (0,1)$ to node $(2)_{4,4} = (0,2)$ belongs to e_+ link, on the other hand, link (1,13) is an e_- link.

Definition 3. Union of $e_+(e_-)$ type links is called a well-formed (wf) link, e_{wf}, and is denoted as

$$\alpha \xrightarrow{\hspace{2cm}} \beta$$

$$e_{wf}\ \text{link}$$

Mathematically, it is represented as follows:

$$e_{wf}\ (\alpha,\beta) = U(\varepsilon)$$
$$\text{where, } \varepsilon = \{(uv) \mid (uv) \in e_+\},\ \text{ or } \varepsilon = \{(uv) \mid (uv) \in e_-\}$$
$$\alpha = \{\ u \mid u \text{ is the source vertex of } (uv) \in \varepsilon\}$$
$$\beta = \{\ u \mid u \text{ is the terminal vertex of } (uv) \in \varepsilon\}$$

Alternatively, α (β) can be taken as a vertex that identifies all the source (terminal) vertices of the edges belong to ε.

Definition 4. Union of e_+ and e_- types links is called ill-formed (if), e_{if} and is denoted as

$$\alpha \xleftarrow{\hspace{2cm}} \beta$$

$$e_{if}\ \text{link}$$

Mathematically, it is expressed as follows:

$$e_{if}\ (\alpha,\beta) = U(\varepsilon)$$
$$\text{where, } \varepsilon = \varepsilon_1 \cup \varepsilon_2$$
$$\varepsilon_1 = \{(uv) \mid (uv) \in e_+\},\ \text{and}\ \varepsilon_2 = \{(uv) \mid (uv) \in e_-\}$$

and

$$\alpha = \{\ u \mid u \text{ is the source vertex of } (uv) \in \varepsilon_1 \text{ or}$$
$$\text{the terminal vertex of } (wu) \in \varepsilon_2\ \}$$
$$\beta = \{\ u \mid u \text{ is the terminal vertex of } (vu) \in \varepsilon_1 \text{ or}$$
$$\text{the source vertex of } (uw) \in \varepsilon_2\ \}$$

An edge $(uv) \in e_{if}\ (\alpha,\beta)$ iff $u \in \alpha$, $v \in \beta$ and $(uv) \in E(G)$.

A node induced from a dimensional cycle $DC_{xi}(j)$ is called the macro-node $M_{xi}(j)$. In other words, a dimensional cycle $DC_{xi}(j)$ is deduced or mapped into the macro-node $M_{xi}(j)$. Edge $e_{wf}(\alpha,\beta)$ or $e_{if}\ (\alpha,\beta)$ is called a macro-edge and is denoted as $E_m(M_{xi}(j), M_{xk}(j))$, where $\alpha = V(DC_{xi}(j))$ and $\beta = V(DC_{xk}(j))$.

Definition 5. A macro-cycle $MC(j)$ is a cycle with

$$V(MC(j)) = \{M_{xi}(j) \mid M_{xi}(j), \forall\ 0 \leq x_i < n_i\ \}$$
$$E(MC(j)) = \{E_m(M_{xi}(j), M_{xk}(j)) \mid M_{xi}(j) \text{ and } M_{xk}(j) \in V(MC(j))\}$$

Therefore, a macro-graph is a disconnected graph consisting of two macro-cycles MC(1) and MC(2) by the definition above. Since $e_{if}(\alpha,\beta)$ is not well defined (the same node sets of α and β, and different edge sets of ε_1 and ε_2 can be mapped to the same $e_{if}(\alpha,\beta)$ link). For example, for $\alpha=\{1,2,3\}$ and $\beta=\{4,5,6\}$, then $U(\varepsilon)= U\{\varepsilon_1=\{(1,4),(2,5)\}, \varepsilon_2=\{(6,3)\}\}$ is the same as $U\{\varepsilon_1=\{(1,4)\}, \varepsilon_2=\{(6,3),(5,2)\}\}$. Hence, a direction bit is needed to be associated with each edge (uv) $\in$ $e_{if}(\alpha,\beta)$ so that $e_{if}(\alpha,\beta)$ and the associated direction vectors for the edges (uv)$\in$ ε can uniquely denote ε. However, we can reduce the size of direction vectors dramatically by the fact that every dimensional cycle is a directed cycle, and hence the edges in a dimensional cycle must belong to the same edge type. Therefore, instead of associating every edge a direction bit, we can denote the connection pattern by associating a direction vector $R_k=(r_0,r_1,...,r_{n_k-1})$, k=1 or 2, with each macro-cycle, where r_i, $0\leq i < n_k$, is 0 (1) if edge (uv) $\in$ $E(DC_i(j))$, j$\neq$k, is of type e_+ (e_-). We can then deduce a two-dimensional network by its macro-graph and the direction vector associated with each macro-cycle {(MC(1), R_2), (MC(2), R_1)}. A simple example, given in Figure 2, illustrates the various notations and concepts of macro-graph and direction vectors. The diameter, the longest path of all node pairs in the network, of a macro-cycle is define as follows:

Definition 6. the diameter of a macro-cycle $D_{MC(j)}|_{n_i}$, j$\neq$i, is as follows

$$D_{MC(j)}|_{n_i} = n_i- 1 \quad \text{if } \forall\ E_m\ (M_{xi}(j), M_{xk}(j)) \in e_{wf}, 0\leq x_i,x_k< n_i;$$
$$\lfloor n_i/2 \rfloor \quad \text{if } \forall\ E_m\ (M_{xi}(j), M_{xk}(j)) \in e_{if}, 0\leq x_i,x_k< n_i;$$

For example, the diameter $D_{MC(1)}|_3$ of MC(1) shown in Figure 2 is equal to $\lfloor 3/2 \rfloor=1$ and $D_{MC(2)}|_4$ = 4 − 1 = 3.

2.3 Performance parameter for the routing comparison.

The parameters that are considered important for the routing performance comparison include (i) the average routing distance and its increased percentage over average path length, the shortest path hit ratio, and (ii) the percentage of bifurcated traffic over the network traffic. Note, average path length of the network is the average of path lengths of all node pairs and, therefore, symbolizes the average message delay in the network. By considering the average routing distance, we note which network performs better with the same simple routing scheme. On the other hand, the increased percentage of average routing distance over the average path length and the shortest path hit ratio reflect the efficiency of the routing algorithm for the network. The percentage of bifurcated traffic over the network traffic means the ratio of traffic that can be routed along either macro-cycle randomly selected. Boorstyn and Livne [10] claims that if the bifurcated traffic for a node with two outgoing links is over 15%, then the node behaves like a M/M/2 system instead of two parallel M/M/1 system. Hence, the ratio of bifurcated traffic does affect the performance of the network.

Assume that a node has equal probability to transmit a messages to any other nodes and the network traffic is balanced; every link is equally utilized. Then, the average routing distance and the ratio of bifurcated traffic is defined as follows:

Definition 7. The average routing distance for a routing scheme applied to a network topology is defined as

$$\text{Average routing distance} = \frac{1}{n(n-1)} \sum_{\substack{0 \le i < n}} \sum_{\substack{0 \le j < n \\ j \ne i}} \text{routing distance } (i,j)$$

where, routing distance is obtained by computer simulation over the routing scheme.

Definition 8. The ratio of bifurcated traffic is defined as

$$\% \text{ of bifurcated traffic} = \frac{1}{n(n-1)} \sum_{\substack{0 \le i < n}} \sum_{\substack{0 \le j < n \\ j \ne i}} \frac{\text{\# of random out-link selection in routing path } (i,j)}{\text{routing distance } (i,j)}$$

III. Adaptive Routing Schemes

By mapping a two-dimensional network to its macro-graph, the message routing can be considered separately in each macro-cycle. Since a node is mapped into a macro-node in each macro-cycle, the relative position between nodes in one of the dimensions can be reflected by the corresponding macro-cycle, and hence, the relative position between nodes in the network can be reflected by the macro-graph of the network. If a current node is mapped into the same macro-node in a macro-cycle as the destination but to a different macro-node in the other macro-cycle, then the current node is in the same dimensional cycle as the destination. When the current node is mapped to the same macro-node in both macro-cycles as the destination, then the destination is reached. Using this concept, we obtain a simple routing scheme for routing the messages along macro-cycle MC(1) or MC(2) independently.

Before the routing scheme is described, we first define the distance function, $D_{ni}((X_i)_s, (X_i)_d)$, for i=1 or 2 as:

$$D_{ni}((X_i)_s, (X_i)_d) = [(X_i)_d - (X_i)_s] \bmod n_i - D_{MC(j)} \mid n_i ; j \ne i$$

The proposed conservative-, and refined conservative-routing algorithms are based on the macro-graph and its associated direction vectors. The difference between these algorithms lies basically in their routing approaches used when the source and destination node are matched in one of the macro-cycle. In conservative-routing algorithm, the messages are always routed along the unmatched macro-cycle while refined conservative-algorithm routs messages along the unmatched macro-cycle only when the distance between the macro-nodes of the source and destination in the unmatched macro-cycle is smaller than its diameter. These two routing algorithms are described as follows:

3.1.1 conservative routing algorithm (Scheme 1)

Step 1. Compute the distance function for each macro-cycle.

Step 2. Return the preference direction DP1 and DP2 corresponding to MC(1) and MC(2).

The preference direction DP_i is given

$DP_i = 1$; if $D_{ni}((X_i)_s, (X_i)_d) > 0$;

      0 ; if $D_{ni}((X_i)_s, (X_i)_d) < 0 \ \& \neq - D_{MC(j)} \mid ni$;

      * (don't care) ; otherwise;

Step 3. Obtain matched preference routing status bit $P_i = \overline{A_{(X_i)_s}(R_j) \oplus DP_i}$, where $i \neq j$, for MC(i) and

denotes the matched preference routing vector $P = (P_2, P_1)$. Note, $A_{(X_i)_s}(R_j) = r_{(X_i)_s}$ and

$\overline{A_{(X_i)_s}(R_j) \oplus *} = 1$.

Step 4. Traverse MC(i), if the source and destination node are matched in MC(j), $j \neq i$, else

routs the messages according to P as follows:

P=(1,0)   transmits the messages along MC(2);

P=(0,1)   transmits the messages along MC(1);

P=(0,0) or (1,1) transmits the messages along either macro-cycle randomly selected;

3.1.2 refined conservative-routing algorithm (Scheme 2)

Basically, this algorithm uses the same three steps 1 to 3 of conservative routing algorithm, however step 4 is modified as follows:

Step 4. If the source and destination node are matched in MC(j), then

    A. traverses MC(i), $i \neq j$, if $P_i = 1$;

    B. traverses MC(j), otherwise,

or else, routs the message according to P as described in conservative algorithm.

3.2 Simulation and comparison

Since the proposed unified algorithms are so simple that they may not use the shortest paths between node pairs, it is necessary to evaluate their performances when they are applied to the three different two-dimensional networks.

We have implemented a simulation on the proposed routing schemes applied to the two dimensional networks, assuming that a node has equal probability to transmit messages to any other nodes and the network traffic is balanced. Simulations for two different sizes networks are shown in Table I. When the network size is small, the three two-dimensional networks have very close diameter and average distance, and hence, the simulation results for a small network may not reflect the pros and cons of the routing schemes. However, it is interesting to compare the influence of network size over the efficiency of the routing schemes, so, we select a small network with size n=4*4 and a larger network of size n=8*8. Table I.a shows that both routing algorithms work well for both 3-ary 2-cube network and MSN of size n=4*4, but perform poorly for 2-D 3-J network. 3-ary 2-cube network has more than twice as large ratio of bifurcated traffic as MSN does. This is obviously true because 3-ary 2-cube network has a larger number of shortest paths existing between node pairs. When applied to 2-D 3-J network, conservative-algorithm is better than refined conservative-algorithm because the latter tends to hit smaller ratio of shortest paths. In Table I.b, the average routing distance shows that MSN performs the best although the routing schemes do not hit all the shortest paths between node pairs. It is also interesting to note that conservative-algorithm works better for 2-D 3-J network but worse for MSN while refined conservative-algorithm works in the other way. This is because in MSN most short distance paths between node pairs [located in the same dimensional cycle and with distance larger than the diameter of the macro-cycle] is obtained by traversing the unmatched macro-cycle first. But in 2-D 3-J network, a traverse along the unmatched macro-cycle means a non-shortest path used except when the size in each dimension has a large difference. From the analysis above, we can see that MSN is the best topology among these two-dimensional networks in terms of average routing distance. Also, the proposed routing schemes fit 3-ary 2-cube network the best as they always hit the shortest path in this network. These two routing algorithms also offer more than 15 % bifurcated traffic when applied to the two dimensional networks and, hence, can reduce the average response time of the network significantly.

It may not be efficient to keep direction vectors in every node from practical point of view. An alternative way to obtain the needed macro-cycle direction information is to keep the addresses of a node's out neighbors; the terminals of a node's out-going links. With the addresses of a node's out neighbors, a node can detect the type of link it has according to the definition given previously. This may reduce the information that a node need to keep for the execution of the proposed routine schemes.

IV. Network protocol and performance modeling

In this section, we briefly review (1) some LAN's medium protocols that are useful for the two-dimensional networks, (2) the proposed interface hardware to the medium, and (3) the performance models that are suitable for the register insertion ring.

4.1 Medium access protocol

Medium access protocol plays a key role on the performance of a doubly linked network. Some of the existing protocols for a single ring network are still useful for a doubly connected network, but not all. The medium access protocols, for example, slotted ring and register insertion ring are used by Maxemchuk [11] and Raghavendra [1] as the access protocols. The token ring, however, is no longer appropriate for a doubly connected network because it would poorly utilize the additional links. Hence, we do not discuss token ring in this article.

In a register insertion protocol shown in Figure 3a, each node contains a shift register and an output buffer. The shift register has size large enough to store a maximum message. When a message arrives from an adjacent node, the message is shifted into the register, and the header of message is checked. If the message is not for this node, it is transmitted out right away when the output link is not busy, otherwise, it is buffered in shift register until the output link becomes free. On the other hand, when a station has a packet to transmit, it loads the packet into the output buffer and waits until the output link is idle. In this protocol, the messages can be of variable size and multiple messages can be transmitted in the network. Hence, it can be utilized in a doubly-linked network to take the benefit of the additional links.

In a slotted ring protocol, shown in Figure 3b, the round trip delays are slotted into a fixed time slot or packet slot with one associated bit indicating whether the slot is full or empty. When a station wishes to transmit, it waits for an empty slot to come around, marks it as full, and puts its data in the slot. The packets of data in this protocol are of fixed size and are transmitted synchronously. Since this protocol can operate with multiple packets at the same time, it can be extended to a doubly connected network and the added links can be well utilized to improve the network throughput.

4.2 Hardware structure

Hardware structures for a node in a doubly linked network, employing both register insertion ring and slotted ring protocols, have been proposed in past years. They are basically extended from the hardware for a single register insertion or slotted ring. Although the hardware for a node is simple in structure, the software becomes much more complicated than that of single loop network because of the added functionality requirements for buffer management, deadlock prevention, and traffic conflict management in the increased node degree environment.

Three similar hardware structures [1,9,11] for a node in a doubly linked register insertion network have been proposed. Here, only one hardware structure [6] is illustrated in Figure 4 and is detailed enough to conveniently explain its underneath operation philosophy. Generally, a doubly-linked network is considered as a two-loop network. Then, the function of buffers w11 and w22 of Figure 4 could be considered to be the same as that of the buffer in a single register insertion ring. The other buffers w12 and w21 can be considered as the buffers for the messages transferring from loop 1 to loop 2 and loop 2 to loop 1 respectively, which make w21 (w12) as an additional source for loop 1 (2), and an additional sink for loop 2 (1). If the scheduling policy for each loop is kept the same as a single register insertion ring, the incoming messages are always given priority over local source. When traffic conflict occurs among traffic from local loop, local source, and the other loop, then a message queue may build up in both w21 and w12. Since a queue may exist, the problems of buffer overflow and deadlock arise and make the network hardware and software much more complicated than a single register insertion ring. Likewise, queues may also be built up in the buffers associated with each output link in a slotted system shown in Figure 5 wherein only one queue is required for each link as the packet transmission is synchronized and traffic conflict can only occur between two incoming data packets.

4.3 Performance models

Since a slotted system transmits fixed size packet in each time slot, a message with size larger than a data packet must be dissembled before being transmitted. Hence, overhead incurs due to the header associated with each data packet and partial usage of the time slot when the packet is smaller than the fixed packet size. Register insertion ring protocol can asynchronously transmit variable message length and hence is more efficient than a slotted system. Therefore, we will only consider the performance models for a doubly linked register insertion ring. We study those performance models that are evolved from the simplest model to the most accurate model for a register insertion ring. These representative models are: SF [5], CT [6], QCT [7], BS [8], and HL [9].

Generally, for purpose of mathematical tractability, the following assumptions are made in analyzing the performance models:

(a) The message arrival processes are poisson.

(b) The message lengths are exponentially distributed.

(c) Propagation delay between nodes is neglected.

The accuracy of a performance model then depends on (1) how close is the model to the physical protocol implementation, and (2) how well the relationship between interarrival and service times of the messages is taken care of. The message interarrival time changes when a message arrives at a node while the preceding message is queued or being served in the node. Thus, there exists a statistical dependence between the message interarrival time and service time. Based on the above two considerations we analyze the representative models in the following text.

SF is the simplest model. It assumes that every message is always stored completely and takes exponential time when it is served. However, in a register insertion protocol, a message is served immediately by examining its header in a fixed service time if the link is free even though the message is not completely received. Hence, this model does not reflect the protocol closely. Also, independence assumption [5] is made in computing the average end-to-end transmission delay which forces to discard the interdependence between the interarrival time and message service time. Therefore this model substantially over-estimates the end-to-end transmission delay.

The CT [6] improves SF by assuming the service time for a message arriving at an idle output link as constant. This means that a message can fully cut-through a node when the output link selected is idle. However, when the outgoing link is busy while a message arrives, it assumes that the arriving message has to be received completely before it can be transmitted, which is not exactly what register insertion procedure really is. This constraint is eliminated in QCT [7]. Both of these two models assume independent message interarrival time and service time, and therefore they globally overestimate the average response time of the system.

BS [8] adopts the beauty of QCT without any of its defects. It overcomes the independence assumption by considering the message interarrival and service time relationship between two adjacent nodes, and considers a source node differently from an intermediate node. However, as Hilal et al [9] pointed out, this model is incapable of handling cases in which the traffic leaving a station has a significant volume, and hence its estimation for the average response time become relatively less accurate as the link utilization exceeds 50%. Hilal et al, then, proposes a model extended from the superposition model by Rubin [12] to remedy the defect of BS. HL, basically, improves the estimation for average response time when the link utilization is over 50% because separate formulas of average time delay is used for network load over 50%.

Among all these models, the former three models are originally proposed for a general network and arbitrary topology, and the latter two models are for register insertion ring. It is no wonder that the former three models give poor approximation for the register insertion protocol while the latter two models offer much more accurate performance estimation. The later two models, however, are for a single ring network, and may not be appropriate for a doubly linked network. For example, these two models are incapable of reflecting the adaptive routing for a doubly linked network in their models. The same defect occurs in the model proposed by Raghavendra and Silvester [13,14], which modifies the message delay for an intermediate node in Bux and Schlatter model to fit FLBH network using static routing scheme. As we know, two parallel M/M/1 queues with arrival rate λ and service rate μ respectively have larger average waiting time than an M/M/2 queue with arrival rate 2λ and service rate μ for each server. Therefore, these models will overestimate the average response time when they are applied to a doubly linked network utilizing an adaptive routing scheme. In Table II, we list the closed-form solutions for these studied models. Also an example for their application to an optimal FLBH with number of nodes $n=h^2$ is given in Table III, wherein a static routing algorithm and uniform traffic is assumed.

V. Summary

In this paper the concept of a macro-graph is developed and is used to reduce a group of two-dimensional networks to two simple one dimensional cycles. Two-dimensional graphs like 3-ary 2-cube, 2-D 3-J, and MSN topologies are uniquely defined using two macro-cycles and the associated direction vectors. Based on the macro-graph of the network, two unified simple routing algorithms, conservative and refined-conservative, applicable to all two-dimensional networks are proposed. The parameters including average routing distance, bifurcated traffic ratio, and shortest path hit ratio are adopted to evaluate the performance of the network and the applicability of the routing on the network. Computer simulation for the proposed routing scheme is also developed. The simulation results show that the proposed routing schemes works well for 3-ary 2-cube network. MSN behaves the best in terms of routing average distance when the same routing schemes are applied. The large ratio of bifurcated traffic reflects the traffic adaptivity of the routing schemes and hence may reduce the average response time significantly.

Moreover, we have described medium access methods namely slotted ring and register insertion ring, along with their hardware interfaces for a node in a doubly connected network. Five performance models such as SF, CT, QCT, BS & HL for register insertion ring are then compared. It is shown that these models do fail to reflect the bifurcated traffic effect on the network performance, and hence, a new performance model is to be investigated to reflect routing adaptivity.

Acknowledgements: Authors acknowledge the financial support provided by U.S. Army and NASA under contract numbers DAAG-29-85-k-0236 and NAG 2-449, respectively.

Reference

[1] C. S. Raghavendra, M. Gerla, A. Avizienis, "Reliable loop topologies for large local computer networks", IEEE Trans. on computers, vol. c-34, Jan. 1985, pp. 46-55.

[2] T. Y. Chung, S. Rai, and D. P. Agrawal, "Doubly connected Multi-dimensional networks for LANs and MANs", in Proc. 1988 INFOCOM, pp. 551–557.

[3] W. J. Dally and C. L. Seitz, "Deadlock-free message routing in multiprocessor interconnection networks", IEEE Trans. Computers, vol. c-36, May 1987, pp. 547-553.

[4] N. F. Maxemchuk, "The manhattan street network", Globecom '85, Dec. 2-5, 1985, New Orleans, Louisiana, pp. 252-261.

[5] R. R. Boorstyn and A. Livne, "A technique for adaptive routing in networks", IEEE Trans. on Communications, vol COM-29, April 1981, pp. 474-480.

[6] N. F. Maxemchuk, "Regular mesh topologies in local and metropolitan area networks", AT&T Technical Journal vol. 64, Sept. 1985, pp. 1659-1685.

[7] J. A. Silvester and C. S. Raghavendra, "Analysis and simulation of a class of double loop network architecture", in Proc. of 1984 INFOCOM, pp. 30-35.

[8] N. F. Maxemchuk, "Routing in the Manhattan Street Network", IEEE Trans. on Communications, vol. COM-35, May 1987, pp. 503-512.

[9] W. Hilal, J. F. Chiu, and M. T. Liu, "A study of register-insertion rings", in Proc. of 1987 INFOCOM, pp. 479-491.

[10] L. Kleinrock, Queueing Systems Vol. II: Computer Applications, Wiley Interscience, New York, 1976 (book).

[11] P. Kermani and L. Kleinrock, "Virtual Cut Through: A new computer communication switching technique", Computer Networks and ISDN system, vol. 3, 1979, pp. 267-286.

[12] M. Ilyas and H. T. Mouftah, "Quasi Cut Through: New hybrid switching technique for computer communication networks", IEE Proceedings, vol. 131, Pt. E, no. 1, Jan, 1984, pp. 1-9.

[13] W. Bux and M. Schlatter, "An approximate method for the performance analysis of buffer insertion rings", IEEE Trans. on Communications, vol. COM-31, Jan. 1983, pp. 50-55.

[14] I. Rubin, "An approximate time-delay analysis for packet-switching communication networks", IEEE Trans. Communications, vol. COM-24, Feb. 1976, pp. 210-222.

[15] C. S. Raghavendra and J. A. Silvester, "Doubly loop network architecture- A performance study," IEEE Trans. on Communications, vol. COM-33, Feb. 1985, pp 185-187.

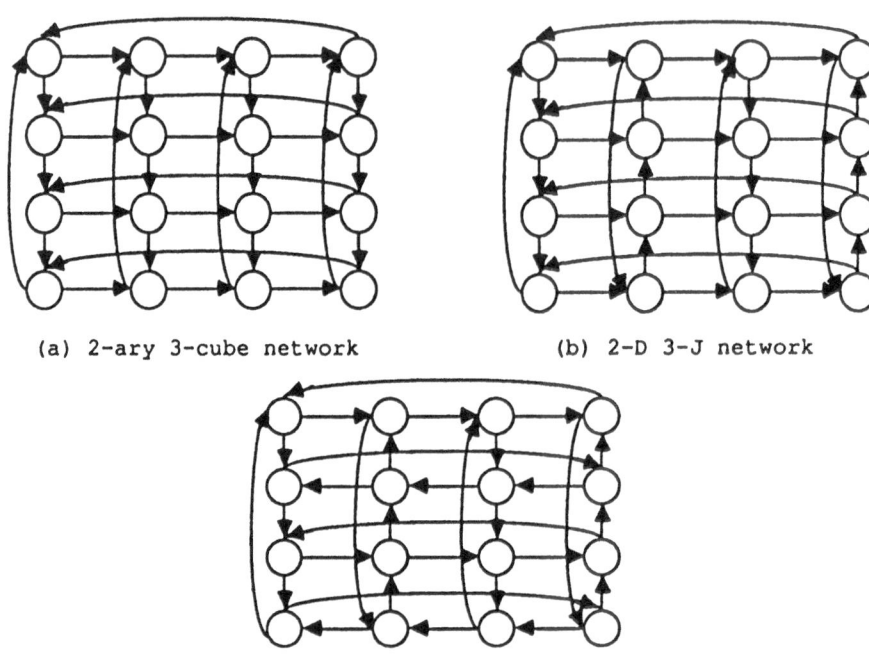

(a) 2-ary 3-cube network   (b) 2-D 3-J network

(c) Manhattan Street Network (MSN)

Figure 1. Two dimensional networks

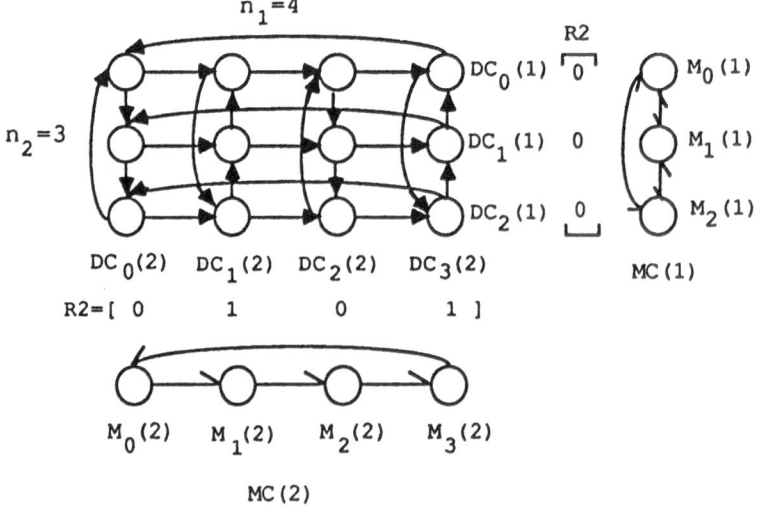

Figure 2. Macro-graph and various notations for a
two dimensional networks

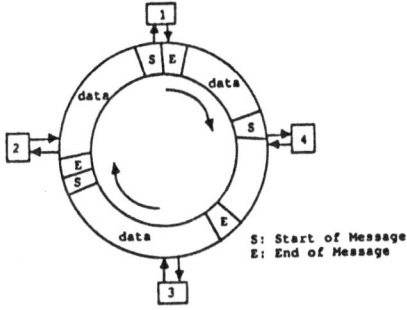

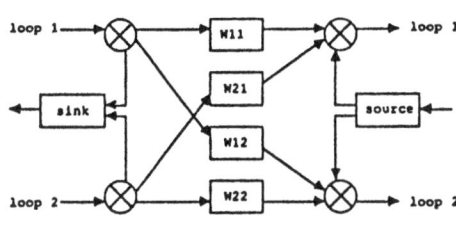

Figure 4. Hardware structure for a node in a
doubly-linked register insertion ring.

(a) Register insertion ring protocol

S: Start of Message
E: End of Message

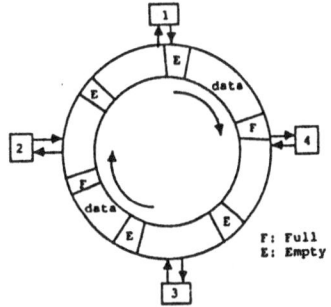

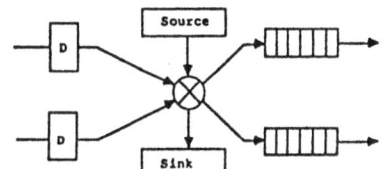

Figure 5. Hardware structure for a node in a
doubly-linked slotted ring.

F: Full
E: Empty

(b) Slotted ring protocol

Figure 3. Two basic ring access protocols that
are useful for a doubly linked network.

Table I, Performance of the two routing schemes for the two
dimensional networks with size n= 4*4 and 8*8.

I.a Network size n=4*4

	Routing scheme	Avg. routing distance	% increased on avg. distance	% bifurcated traffic	% shortest path hit ratio
3-ary 2-cube	c	3.2	0	34.47	100
	rc	3.2	0	34.84	100
2-D 3-J	c	3.04	8.57	20.31	88
	rc	3.47	23.8	23.24	84
MSN	c	2.93	0	15.44	100
	rc	2.93	0	15.44	100

I.b Network size n=8*8

	Routing scheme	Avg. routing distance	% increased on avg. distance	% bifurcated traffic	% shortest path hit ratio
3-ary 2-cube	c	7.11	0	43.92	100
	rc	7.11	0	43.92	100
2-D 3-J	c	6.51	12.74	26.24	83.17
	rc	7.86	37.14	29.20	75.55
MSN	c	6.03	20.25	17.06	74.60
	rc	5.62	12.15	22.21	84.76

Note: c — conservative
rc — refined-conservative

Table Ii. Close form formulas for five register insertion ring performance models

model	End-to-end delay	Comments
Store and Forward	$T_{sd} = \frac{1}{\gamma} \sum_i \frac{\rho_i}{1-\rho_i}$ $\rho_i = \lambda_i \cdot l_m$	
Cut through	$T_{sd} = \frac{1}{\gamma} \sum_i \frac{\rho_i}{1-\rho_i} - (\bar{k}-1)(1-\rho_i)(l_m - l_h)$	γ : total traffic input to the network λ_i : traffic rate in each link ρ_i : link utilization l_m : average message transmission time
Quasi cut through	$T_{sd} = \frac{1}{\gamma} \sum_i \frac{\rho_i}{1-\rho_i} - (\bar{k}-1)(1-\rho_i)(l_m - l_h)$ $-(1-\rho_i)\rho_i(\bar{k}-1)(l_m - l_h - \frac{\lambda_i \overline{l_m}^2}{2})$	l_h : header processing time $\bar{k}$: average path length T_m^2 : second moment of l_m λ_s : input arrival rate at each node
Bux and Schlatter	$T_{sd} = w_i + \sum_i w_i(i) + l_m$ $w_i(i) = \frac{\rho_s l_m}{(1-\rho_i(i))}$ $w_i = \frac{(\rho_i + \rho_s)l_m}{(1-\rho_s-\rho_i)(1-\rho_i)}$	ρ_s : source utilization w_i : waiting time of source node w_i : waiting time at an intermediate node $\bar{w}$: average waiting time at each node
Hilal et al	$T_{sd} = \bar{k}\bar{w} + l_m + l_h\bar{k}$ $(\rho_s + \rho_i - \rho_o) < 0.5$ $\bar{w} = \frac{\rho_s l_m}{(1-\rho_i+\rho_o)(1-\rho_i-\rho_i+\rho_o)} + \frac{\rho_i(1-\rho_o)l_m}{(1-\rho_s)(1-\rho_i-\rho_i+\rho_i\rho_o)}$ $- \frac{\rho_i l_m}{(1-\rho_i)}$ $(\rho_s + \rho_i - \rho_o) > 0.5$ $\bar{w} = \frac{\rho_s l_m}{(1-\rho_i+\rho_o)} + \frac{\rho_i(1-\rho_o)l_m}{(1-\rho_s)(1-\rho_i-\rho_i+\rho_i\rho_o)} - \frac{\rho_i l_m}{(1-\rho_i)}$	ρ_o : traffic departure rate at each node

Table III. Illustrating models for an FLBH network.

model	End-to-end delay	Assumptions and Network parameters
Store and Forward	$T_{sd} = \frac{2n\, l_m}{2(n+1) \cdot \gamma l_m}$	The network is an FLN with size $n=h^2$ and average path length $\bar{k}=n/(h+1)$. Assumptions: (1) Message processing time is exponential distributed with mean l_m.
Cut-through	$T_{sd} = \frac{2n\, l_m}{2(n+1) \cdot \gamma l_m} - (1 - \frac{\gamma l_m}{2(n+1)})\times(l_m - l_h)\times(n/(h+1)-1)$	(2) message arrival process is poisson with mean λ. (3) propagation time is neglected. (4) Traffic is uniform for every link.
Quasi-cut-through	$T_{sd} = T_{sd}(\text{cut through}) -$ $(1 - \frac{\gamma l_m}{2(n+1)})\frac{\gamma l_m}{2(n+1)}(\frac{n}{h+1}-1)(l_m - l_h - \frac{\gamma\, l_m^2}{2(h+1)})$	By the assumption (4), the following parameters are obtained: $\lambda_i = \frac{\gamma}{2(n+1)}$ $\rho_i = \lambda_i l_m$
Bux and Schlatter	$T_{sd} = l_m + (k-1)\frac{\gamma l_m^2}{2n(1-\rho_i)} + \frac{\rho_i + l_m\gamma/2n}{1-\rho_i} \cdot \frac{l_m}{1-\rho_i - l_m\gamma/2n}$	$\rho_s = \frac{\gamma}{2\lambda} l_m$ $\rho_o = \frac{\gamma}{2\lambda} l_m$
Hilal et al	$T_{sd} = \bar{k}\bar{w} + l_m + l_h\bar{k}$ $(\rho_s + \rho_i - \rho_o) < 0.5$ $\bar{w} = \frac{\rho_s l_m}{(1-\rho_i+\rho_o)(1-\rho_i-\rho_i+\rho_o)} + \frac{\rho_i(1-\rho_o)l_m}{(1-\rho_s)(1-\rho_i-\rho_i+\rho_i\rho_o)}$ $- \frac{\rho_i l_m}{(1-\rho_i)}$ $(\rho_s + \rho_i - \rho_o) > 0.5$ $\bar{w} = \frac{\rho_s l_m}{(1-\rho_i+\rho_o)} + \frac{\rho_i(1-\rho_o)l_m}{(1-\rho_s)(1-\rho_i-\rho_i+\rho_i\rho_o)} - \frac{\rho_i l_m}{(1-\rho_i)}$	Note ρ_i and ρ_o contain a factor "2" in their denominators because the traffic entering and leaving each link of a node are assumed to be equal. By substituting these formulas, the end-to-end delay for Hilal et al model could be obtained

BICAMERAL NEURAL COMPUTING

Michael C. Stinson
Central Michigan University

Subhash C. Kak
Louisiana State University

1. Introduction: Neural networks have been one of the most exciting areas of research in computer science in the past five years. They have been studied for applications to associative information storage, optimization, pattern recognition, semantic information processing, cognitive science, and robotics (for reviews see [1],[2]). In spite of the variety of applications that are being investigated, the fundamental properties of neural networks are not well understood. This is owing to the fact that a neural network is a highly non-linear structure and, therefore, it is extremely hard to obtain analytic results on its dynamics. Most applications are thus based on empirical studies.

A neural network may be represented as an interconnection of threshold nodes that is potentially fully connected (Figure 1). When a probe is fed into a network it moves along certain trajectories through attraction basins to stable points (Figure 2). These stable points generally represent learned memories.

That neural network behavior includes pathological modes is wellknown. For instance, when used as associative memory, a neural network may, upon the presentation of a pattern, converge to a wrong memory or a spurious memory. A wrong memory is one that the presented pattern is not closest to, whereas a spurious memory is a pattern that does not belong to the set that the network has been programmed to store. The next sections present new ways to improve the convergence of neural networks to the correct memory. The improvement is obtained by postulating the use of an asynchronous controller C that implies the following update of a probe vector x by the connection matrix T:

$$x' = sgn [C(x,Tx)]$$

For a VLSI-implemented neural network the controller block can be easily placed in the feedback loop as shown in Figure 3.

The introduction of the asynchronous controller may be viewed as a first step in the development of bicameral neural structures (Figure 4). Since the bicameral nature of the human brain appears to endow it with considerable capability [3], [4], it may be assumed that bicameral neural networks would offer computational advantage. (See also [3] for related discussion). The work has been discussed earlier in [6],[7], [8],[9].

One omission in the earlier neural network research is the question of how the stored memories may be indexed, ordered and interrelated. Evidently this is a major question because human memory is characterized by rich interrelationships between memory fragments. In other words, human memory is not merely associative, it is also ordered like the memory of a conventional computer where each fragment

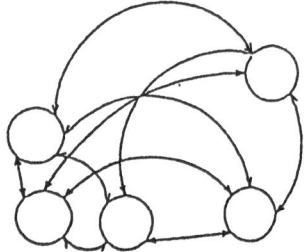

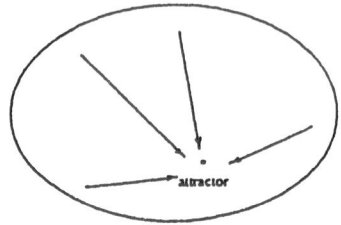

Figure 1. A Neural System.   Figure 2. An attractor in an attraction
                                                 basin.

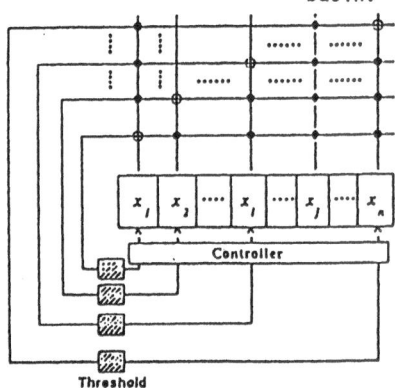

Figure 3. A neural net with control after threshold.

is given a unique label and the memory can also be retrieved fully by the use of
this label.   We present a solution to the problem of indexing information in a
neural network. This establishes that a neural-network associative memory can also
be ordered.   We further show that spurious states that degrade the recall
capabilities of a neural-network memory system can be fully eliminated by using
indexing keys of appropriate size. It is conceivable that a mechanism similar to
our model may be at the basis of human memory and recall.   We also show that a
memory can be recalled using its indexing key alone.

2. The Hopfield Model: In this neural network model [10] if x =
$(x_1, x_2, \dots, x_n)$ represents the current state of network, the new state x_i of
the ith neuron is determined by

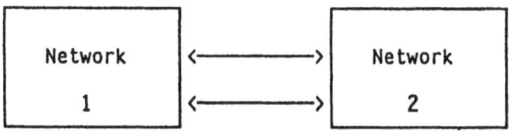

Figure 4. A bicameral neural network where each network operates at a
different level of abstraction.

$$x_i' = \text{sgn}\left(\sum_{j=1}^{n} T_{ij}x_j\right) = \begin{cases} 1 & \text{if } \sum T_{ij}x_j \geq 0 \\ -1, & \text{if } \sum T_{ij}x_j < 0 \end{cases}$$

The T_{ij} are the synaptic interconnection weights.

The network can operate in the synchronous or the asynchronous modes. In the
synchronous mode all the neurons are updated simultaneously. In the asynchronous
mode, only one neuron is chosen, randomly, for an update, the state of the network
is redefined, and the update process repeated. The network is guaranteed to end in
a stable state only in the asynchronous mode of operation. However, the synchronous behavior often produces results similar to that of the asynchronous mode.

The Hopfield model assumes that there is no self-feedback for the neurons, or
$T_{ii} = 0$, though the requirement that each computation end in a fixed point in an
asynchronous operation only requires that $T_{ii} \geq 0$.

Let $x(1)$, $x(2)$, ..., $x(m)$ be the memories to be stored in a n-neuron
network. The interconnection matrix T is given by

$$T_{ij} = \sum_{k=1}^{m} T_{ij}^{(k)}$$

where

$$T_{ij}^{(k)} = \begin{cases} x_i^{(k)} x_j^{(k)} & , i \neq j \\ 0 & , i = j \end{cases}$$

Example 1: Consider the following three memories:

$x(1) = -1 \quad 1 \quad -1 \quad 1 \quad 1 \quad 1 \quad 1$

$x(2) = 1 \quad 1 \quad 1 \quad -1 \quad 1 \quad -1 \quad 1$

$x(3) = -1 \quad -1 \quad -1 \quad -1 \quad -1 \quad 1 \quad 1$

To store these in a Hopfield net:

$$T = \begin{bmatrix} 0 & 1 & 3 & -1 & 1 & -3 & -1 \\ 1 & 0 & 1 & 1 & 3 & -1 & 1 \\ 3 & 1 & 0 & -1 & 1 & -3 & -1 \\ -1 & 1 & -1 & 0 & 1 & 1 & -1 \\ 1 & 3 & 1 & 1 & 0 & -1 & 1 \\ -3 & -1 & -3 & 1 & -1 & 0 & 1 \\ -1 & 1 & -1 & -1 & 1 & 1 & 0 \end{bmatrix}$$

We stress here that apart from the memories to be stored, also called the
fundamental memories (though not necessarily all will be stored), some of the

complements of these memories as well as spurious memories are stored in the network. All the complements of the stored fundamental memories will be stable states if the neuron updating rule were to be:

$$x_i' = \begin{cases} \text{sgn } (\sum T_{ij} \, x_j) & , \sum T_{ij} x_j \neq 0 \\ x_i & , \sum T_{ij} x_j = 0 . \end{cases}$$

Even in the more conventional asymmetric update where sgn $(\sum T_{ij} x_j) = +1$ if $\sum T_{ij} x_j \geq 0$, as n becomes large the probability of $\sum T_{ij} \, x_j = 0$ becomes small and, therefore, almost all stored memories have their complements as stored values as well. For small n, the assertion that all complement states are stored is valid as long as $\sum T_{ij} \, x_j \neq 0$ for any neuron.

Unfortunately the complement states have been ignored by many researchers. They are extremely important since a probe vector may be closer to the complement of a 'wrong' memory than to the 'correct' memory.

Furthermore, some of the spurious states of the asymmetric model are nothing but the complement states modified by the update rule asymmetry. To see this let us return to Example 1. One of the spurious states is

$$-1 \quad -1 \quad -1 \quad 1 \quad -1 \quad 1 \quad 1$$

which is complement of $x^{(2)}$ excepting for the last neuron owing to

$$\sum_j T_{7j} x_j^{(2)} = 0.$$

Now if the probe sequence is

$$x_1 = -1 \quad 1 \quad -1 \quad 1 \quad -1 \quad 1 \quad 1$$

which is at a Hamming distance of 1 from $x^{(1)}$, it has an equal probability of ending up in the above spurious state since it is at the same Hamming distance from that state. Indeed when using asynchronous updating one obtains

$$x_1' = \text{asyn. up. sgn } (-8, -4, -8, 2, 0, 8, 0)$$

One must choose between an update of the second or the fifth neurons. Updating the second takes one to the spurious memory, whereas updating the fifth takes one to the memory $x^{(1)}$. Also note that $x^{(3)}$ is not a stable state.

3. **When are Memories Fixed Points?** McEliece et al [11] have recently investigated the problem of memories being fixed points for the Hopfield model. The essential result is that for memories that are n-tuples of ± 1, most of $m = n/(2 \log n)$ random memories are exactly recoverable in an asymptotic sense. We follow a heuristic argument in [11] to see the behavior of fixed points for finite n.

Let us consider whether memory $x^{(1)}$ is a stored state. If the total number of stored memories is m, with an equal number of complements one may assume the $n(m-1)$ bits to be independent identically distributed ± 1 r.v.'s. Now

$$(x^{(1)})_i^{'} = \text{sgn} \left[\sum_j T_{ij}^{(1)} x_j^{(1)} + \sum_{k=2}^{m-1} \sum_j T_{ij}^{(k)} x_j^{(1)} \right]$$

Since

$$T_{ij}^{(1)} = \begin{cases} x_i^{(1)} x_j^{(1)} & , \ i \neq j \\ 0 & , \ i = j \end{cases}$$

$$\sum_j T_{ij}^{(1)} x_j^{(1)} = (n-1) x_i^{(1)}$$

We get an incorrect storage, therefore, if

$$Z = \sum_{k=2}^{m-1} \sum_j T_{ij}^{(k)} x_j^{(1)} < -(n-1) x_i^{(1)}$$

for a symmetric thresholding function and about the same result for an asymmetric thresholding function. Using Hoeffding inequality (relation (53) of [12]) and noting that as $x_i^{(1)} = \pm 1$ one may consider $Z \geq n$,

$$\text{Prob} \ (Z \geq n) \leq \exp\left[\frac{n}{2(m-1)}\right] .$$

If $m = n/\log n$, we obtain

$$\text{Prob} \ (x_i^{(1)} \text{in error}) \leq \exp \ (- \frac{1}{2} \log n) = n^{-1/2}$$

This shows how error probability for each neuron decreases rapidly as n increases, and considering m = n/log n is appropriate.

The motivation for this analysis was to get a result on fixed memories that is valid for small n, as the results of McEliece et al [11] are strictly valid only asymptotically.

4. <u>Neural Net with Controller</u>: Consider the new model of Figure 3. The controller can be placed in that model either before the thresholding or after it. The two cases are not equivalent since the actual magnitudes, before thresholding, may offer easier choices between alternatives. On the other hand, using a controller before threshold may lead to the domination by the strongest eigenvalue of the underlying system. This paper will, therefore, consider the controller to be placed after thresholding.

We have already indicated in the Introduction the motivation to introduce the controller from the analogy with the human brain. Let us now briefly present some other reasons. Note that all characteristics of memories are not captured by the outer product formulation of the interconnection matrix. Furthermore, subsets of memories may have clearly defined patterns so that if there is a large probability that the probe belongs to a subset, one should really use the corresponding correlation properties. Also, if the net converges to a state that is clearly either spurious, or wrong, or complement, one needs to backtrack so as to trace some other branches of the decision tree. A controller, with memory, can allow such a backtracking. Backtracking is not being mentioned as a realistic procedure but as a technique that could be used in principle.

A controller also allows the possiblity of modifying the size of the net. This may be useful in increasing the redundancy of the stored patterns.

The controller may be programmed to do different kinds of computing tasks. We briefly describe two of these:

    1. The algorithm O [Oracle]

    2. The method of handles

    5. The Oracle Approach: Basically one wishes to use side information to assist in the updating of a probe in the asynchronous mode. Assume that these probabilities are pre-computed for the entire set of memories, as well as selected subsets, and they are also made available to all users. To look at this in its barest essentials consider Example 1 again. Note that bit 2 has prob(+1) equal to 2/3 and bit 5 has prob(+1) also equal to the same value. Thus, if the probe is 1 and -1 at these locations and one must choose between the two for an update, changing the 5th bit first would be more reasonable. Incidently, this does lead to the correct solution.

The choice between different branches of the decision tree can be either probabilistic or follow the most likely alternative at each decision point. In the probabilistic approach one would use a random experiment to decide on the path to take at each fork.

Other kinds of oracles may be used. In a hash oracle one computes a hash function on each possible successor state and picks one that has a more likely hash function. In a cluster oracle one uses some clustering algorithm to see if a successor state is likely to be a valid state[15].

If all paths lead to an invalid or incorrect state one can perform backtracking. If that also does not converge to a valid solution, one can randomly change a few bits in what is analogous to simulated annealing and continue the previous algorithm.

    6. The Approach of Handles: This implies adding known bit sequences to valid memories to assist in convergence. To see why this idea works, consider the problem of rejecting the complement memories. Let an extra bit, + 1, be appended to the left of each valid memory. The complement memories will have -1 at that location. Therefore, if the neural associative memory ends up in a sequence where the first bit is -1, we know that sequence needs to be rejected. The probability of reaching a valid memory increases greatly by the use of this approach.

To increase redundancy still further, one may append several bits at the start. The disadvantage of this method is that it does not increase the Hamming distance between the modified fundamental memories. To overcome this disadvantage one stores the memories after cyclically shifting the appended memories through different values. This amounts to inserting shifted handles. The position of the handle can serve as an index to the stored memories. Of course, the handle pattern cannot be allowed to occur in the data which can be accomplished by bit stuffing.

The capability to index data is extremely important for establishing relationships amongst the memories themselves. Clearly this indexing cannot be performed in the old Hopfield model.

Example 2:

Consider the memories of Example 1 and append a single + 1 bit to the left. Generate the new transition matrix. It will be found that the spurious memories have now been eliminated. The complement memories can be easily identified by the first bit. In general not all spurious memories will be eliminated.

Table 1 presents results on random memories that are 7-bit long stored using the Hopfield model and by the new method of using a 1-bit additional handle to eliminate complement states. We observe that except for one case where 3 complement states are replaced by 4 spurious states there is a substantial improvement in performance. The results for different values of N(number of neurons) are summarized in Table 2.

7. Keys for Indexing Information: Our earlier investigations established that each memory should be tagged by a key that consists of two components: (i) one bit to distinguish a valid memory from its complement (ii) several additional bits to sequence the memory in such a fashion that the linear sums of stored memories can be identified and discarded. This lies at the basis of the superior performance of the bicameral network. Now we present a result that is valid for all learning schemes where spurious memories are linear combinations of the stored memories.

Theorem: A key where the sequencing is done by an appropriate m+1 bit binary pattern, where m is the total number of memories, that can be otherwise stored by the network, will identify all spurious memories.

Proof (Outline): Let the label sequences run

$$
\begin{array}{cccccc}
1 & 1 & -1 & -1 & . & -1 \\
1 & -1 & 1 & -1 & . & -1 \\
1 & -1 & -1 & 1 & . & -1 \\
. & . & . & . & . & . \\
1 & -1 & -1 & -1 & -1 & 1
\end{array}
$$

where each contains a single 1 at the left and another 1 whose location changes. This set may be viewed as an orthogonal basis set. Since the original memories are storable, so are the memories with the labels. Linear combinations of these label sequences can be identified now. In other words, our method identifies spurious attraction basins associated with the network.

8. Recall Mechanism: Since each memory is correlated strongly with a unique label, the label itself may be used to retrieve the memory. We propose the following algorithm for recall:

Step 1. Probe the network by the key followed by a random sequence. Update

Step 2. Update the probe but do not change the key bits at any stage.

A Sample Table of Results for One Added Bit Using 7-bit Information							
Hopfield Model				New Method			
f	c	s	% Correct	f	c^t	s	% Correct
1	1	0	50	1	0	0	100
1	1	0	50	0	0	2	0
3	2	0	60	2	0	0	100
2	1	0	66	1	0	1	50
2	2	1	40	3	0	0	100
3	3	0	50	3	0	4	43
3	3	0	50	3	0	0	100
2	2	0	50	2	0	0	100
2	0	0	100	2	0	0	100
4	2	0	67	4	0	0	100
1	1	0	50	3	0	0	100
1	1	2	25	1	0	0	100

where:

f	potential fundamental stable points
c	potential complement stable points .
s	potential spurious stable points
c^t	$c = 0$, no stable point complements allowed

Table 1: Results for One Added Bit Using 7-bit Information

A Sample Table of Observed Improvement for Certain Values of N			
N	Hopfield % Correct	New Method % Correct	Observed Improvement
7	51.4	86.4	35.0
8	49.8	89.5	39.7
9	48.6	88.6	40.0
10	46.4	68.0	21.6
11	44.7	78.8	32.1
12	38.2	61.6	23.4

Table 2: Observed Improvements for Certain Values of N.

Step 3.   If the network can update only the key bits, then add a random
pattern to the memory probe and return to Step 1.

This implies that if a choice in an asynchronous update of the neural network
includes key bits and other memory bits, any of the memory bits can be chosen for
change but the key bits can never be allowed to change. In Step 1 one may use the
all 1 or all minus 1 sequences since these may be viewed as random sequences. In
other words, the use of a structured sequence does not affect the performance of
the algorithm.

We have used this retrieval procedure on several examples with Hebbian
learning and it has worked with complete success. (Our experiments suggest that
for Hebbian learning the spurious memories are indeed the linear combinations of
fundamental memories and their complements.)

Example 3

There are n=10 neurons in the system and three memories of 6 neurons each:

$$x^{(1)} = (+1 \quad +1 \quad -1 \quad +1 \quad +1) \quad -1)$$
$$x^{(2)} = (-1 \quad -1 \quad -1 \quad +1 \quad +1 \quad -1)$$
$$x^{(3)} = (+1 \quad -1 \quad +1 \quad +1 \quad +1 \quad -1)$$

To the memories we append three keys, each of length 4:

$$k^{(1)} = (+1 \quad -1 \quad -1 \quad +1)$$
$$k^{(2)} = (+1 \quad -1 \quad +1 \quad -1)$$
$$k^{(3)} = (+1 \quad +1 \quad -1 \quad -1)$$

Yielding the three learned memories: $k^{(1)}x^{(1)}$, $k^{(2)}x^{(2)}$, and $k^{(3)}x^{(3)}$

The combination of the individual outer products yields the transition matrix:

$$
\begin{bmatrix}
0 & -1 & -1 & -1 & +1 & -1 & -1 & +3 & +3 & -3 \\
-1 & 0 & -1 & -1 & +1 & -1 & +3 & -1 & -1 & +1 \\
-1 & -1 & 0 & -1 & -3 & -1 & -1 & -1 & -1 & +1 \\
-1 & -1 & -1 & 0 & +1 & +3 & -1 & -1 & -1 & -1 \\
+1 & +1 & -3 & +1 & 0 & +1 & +1 & +1 & +1 & -1 \\
-1 & -1 & -1 & +3 & +1 & 0 & -1 & -1 & -1 & +1 \\
-1 & +3 & -1 & -1 & +1 & -1 & 0 & -1 & -1 & +1 \\
+3 & -1 & -1 & -1 & +1 & -1 & -1 & 0 & +3 & +3 \\
+3 & -1 & -1 & -1 & +1 & -1 & +3 & +3 & 0 & -3 \\
-3 & +1 & +1 & +1 & +1 & +1 & -3 & -3 & -3 & 0
\end{bmatrix}
$$

To   examine   how   the   new   method   works,   consider   the   probe:

$$p = (1, -1, 1, -1, 1, 1, 1, 1, 1, 1)$$

Notice that this probe is indexed with the second key and therefore
should stabilize at the second memory. In fact the probe consists of the second
key followed by an arbitrarily chosen sequence of all 1's. The application of the
probe to the transition matrix yields a result of (1, 1, -1, -1, -1, -1, -1, 1, 1,
-1).   Comparing this with the original probe gives a set S = (2,3,5,6,7,10) of
positions in p as possible choices for change. Of these, positions 2 and 3 are
eliminated as elements of the key and so not allowed to change. Of the remaining
four positions, five may be chosen at random. The sequence of probe changes that

follows after such random choices is:

$$
\begin{array}{cccccccccc}
1 & -1 & 1 & -1 & 1 & 1 & 1 & 1 & 1 & 1 \\
1 & -1 & 1 & -1 & -1 & 1 & 1 & 1 & 1 & 1 \\
1 & -1 & 1 & -1 & -1 & 1 & 1 & 1 & 1 & -1 \\
1 & -1 & 1 & -1 & -1 & -1 & 1 & 1 & 1 & -1 \\
1 & -1 & 1 & -1 & -1 & -1 & -1 & 1 & 1 & -1
\end{array}
$$

The final state represents the memory $x^{(2)}$.

9. **Capacity Questions**: We note that for a set of m fundamental memories, each key is m+1 bits long. Therefore, if the number of neurons was originally n, the new number will be n+m+1. The total number of stored memories would be of the order of

$$
\frac{n + m + 1}{\log (n + m + 1)}
$$

This implies that the additional bits of the key are not entirely an overhead burden and they contribute somewhat to the capacity. To see this from a different perspective note that the (k+1) key bits are always 2 bits apart and this adds somewhat to the distance between the stored memories.

As a rough estimate one might still use the figure of 15 percent of the number of neurons to estimate the capacity of the network.

Another interesting question is the minimum size of a key that is necessary to recall a memory. The answer to this question appears to be that a key will always work so long as the memories could be stored by themselves. This may appear surprising. As example, consider 2 50-bit long memories that are indexed by 2 keys, each of length 3, both with the first bit fixed at 1. Such a 3-bit key would suffice to recall the 50 bit long memory using our method.

It should be noted however that there exists a tradeoff between speed of recall and the size of the key. In the example of 50-bit long memory, if the original random sequence is not a good choice, one might need to randomize several times along the way before arriving at the right solution.

To speed up the retrieval process, keys of length much longer than k+1 bits proposed earlier might be used. The key separation can be made 2ℓ bits by using patterns of ℓ 1's and -1's instead of single 1's and -1's proposed earlier. As before a single leading -1 suffices to identify complements. Thus to ensure separation of 4 bits (ℓ=2) for 3 memories the keys would be:

$$
\begin{array}{ccccccc}
(1 & 1 & 1 & -1 & -1 & -1 & -1) \\
(1 & -1 & -1 & 1 & 1 & -1 & -1) \\
(1 & -1 & -1 & -1 & -1 & 1 & 1)
\end{array}
$$

Of course using a higher dimensional model a separation of 2 suffices as shown earlier [13],[14].

10. <u>Conclusions</u>: When human beings recall information where a part is completely known, a search for the likely associations based on the known key is made. Thus if one saw a friend drive by and did not explicitly remember what model the vehicle was, one would run through several model types to see what image fitted the vague memory best. This is identical in its essence to the mechanism proposed in this paper.

Note, further, that the key bits are separate from the memory bits. Therefore, if this were indeed the mechanism at the basis of human recall then certain injuries would not cause loss of memory; rather the ordering of the memories might be effected. In reality, that does happen as a result of certain brain traumas.

It has been established that the bicameral neural network can be endowed with powerful ordering characteristics. This is in addition to the capacity of the model to exploit local as well as certain global characteristics of the memories for faster convergence and better recall. The use of clustering algorithms as well as other techniques in this model improves the performance further [15]. The new mechanism for memory and recall wherein a key suffices to retrieve a memory can be used to develop a data base management sytem (DBMS). This is the first demonstration of the use of neural networks in DBMS. For this use the key would consist of the fields that define the information classes.

REFERENCES

[1] J.A. Anderson and E. Rosenfeld, <u>Editors</u>, <u>Neurocomputing</u>. Cambridge: MIT Press, 1988.

[2] D.E. Rumelhart, J.L. McClelland, and PDP Research Group, <u>Parallel</u> <u>Distributed</u> <u>Processing</u>, Cambridge: MIT Press, 1986.

[3] S.P. Springer and G. Deutsch, <u>Left Brain, Right Brain</u>. New York: W.H. Freeman, 1985.

[4] S.C. Kak, <u>The Nature of Physical Reality</u>. New York: Peter Lang, 1986.

[5] S.C. Kak, <u>Patanjali and Cognitive Science.</u> Baton Rouge: Vitasta, 1987.

[6] M.C. Stinson and S.C. Kak, "Techniques to improve convergence of neural networks", <u>Proceedings of the 22nd Annual Conference on Information Sciences and Systems</u>, Princeton, p. 275, March 1988.

[7] M.C. Sintson and S.C. Kak, "Asynchronous controller to improve the convergence of neural nets", <u>Proceedings, of the 26th Annual ACM Conference</u>, Mobile, pp. 410-413, April 1988.

[8] M.C. Stinson and S.C. Kak, "On bicameral neural nets", <u>First International Neural Networks Society (INNS) Conference</u>, Boston, Sept. 1988.

[9] M.C. Stinson and S.C. Kak, "A new mechanism for memory and recall in neural networks", <u>Technical Report #88-042</u>, Dept. of Computer Science, LSU, August 17, 1988.

[10] J.J. Hopfield, "Neural Networks and physical systems with emergent collective computational abilities". Proc. Nat. Acad. Sci.. U.S.A.. vol. 79. pp.

[11] R.J. McEliece, E.C. Posner, E.R. Rodemich, and S.S. Venkatesh, "The capacity of the Hopfield associative memory", IEEE Trans. on Information Theory, vol. IT-33, 1987, pp. 461-482.

[12] S.C. Kak, "Threshold detection error bounds", Journal of the Inst. of Electronics and Telecomm. Engineers, vol. 30, pp. 29-35, 1984.

[13] D. Prados, "The capacity of a neural network", Electronics Letters, vol. 24, pp. 454-455, 1988.

[14] D. Prados and S.C. Kak, "Shift invariant associative memory", Proceedings of the International Workshop on VLSI for Artificial Intelligence, Oxford, U.K., July 1988.

[15] C.H. Youn and S.C. Kak, "New learning and control algorithms for neural networks", Technical Report, LSU, October 1988.

See also

S.C. Kak and M.C. Stinson, "Bicameral neural network where information can be indexed", Electronics Letters, vol. 25, no. 3, 1989, pp. 203-205.

NON-BINARY NEURAL NETWORKS

Donald Prados   Subhash Kak

Louisiana State University

Introduction: Current neural network models barely describe the complexity of biological neurons. These models allow neurons to take on only two values, either 1 and -1 or 1 and 0. In biological neurons, however, one comes across a continuum of values. Furthermore, it appears that not only is the number of spikes per time period important, but so is the distribution of the spikes in the spike train[1]. Optican and Richman[1] have, therefore, hypothesized that neurons may be encoding multidimensional spatial properties of visual stimuli with multivariate temporal modulation as the information code. It appears that neurons characterized by two output variables may lead to interesting results. Furthermore, biological neurons are subject to extensive time-delay and modulatory effects[2], and so assigning neurons only two possible output states is an oversimplificaion. From a computing science point of view, some of these effects may be of secondary importance since a simple neural network model may be capable of the same qualitative cognitive behavior as a more complex model. But whether such is the case is certainly not known at this time. Also note that part of the complex behavior of the human mind may be asscribed to specialized brain structures that includes feedback (such as hemispheric) between them.

Our objective in this paper is to touch upon just one kind of generalization from the current models. Most images are not binary and can lose a great deal of information when converted to a binary form. In addition, the choice of threshold one chooses when converting to binary greatly affects the amount of information lost. In this paper we propose a generalized neural network model that allows neurons to take on more than two values. Very few changes to existing binary neural network models are required to obtain an n-ary neural network model that can successfully store and retrieve patterns. We propose a learning rule for such a model and discuss the choice of the learning constant. We then compare the capacity of this model to that of the binary model.

A Quaternary Neural Network Model: Most neural network models use the following or similar equations to determine the next state of neuron i:

$$x_i = \sum_j T_{ij} V_j \tag{1}$$

$$V_i = \begin{cases} 1 & x_i > 0 \\ -1 & x_i \leq 0 \end{cases} \tag{2}$$

To convert such a binary neural network to the more general n-ary neural network, Equation 1 need not be changed; only Equation 2 needs to be changed. For instance, for a quaternary neural network, Equation 2 can be modified to

$$
V_i = \begin{cases}
3 & x_i > t \\
1 & t \geq x_i > 0 \\
-1 & 0 \geq x_i > -t \\
-3 & -t \geq x_i
\end{cases}
\tag{3}
$$

where t, 0, and -t are thresholds. The synaptic weights can be initialized in the same manner as in the binary model. To store m patterns, one can use the Hebbian equation:

$$
T_{ij} = \sum_{s=0}^{m} V^s_i V^s_j \qquad i \neq j
\tag{4}
$$

where V^s_i is the ith bit of the sth pattern.

If application of this equation does not store all patterns correctly, the synaptic weights can be modified using a learning algorithm similar to the delta rule. This generalized delta rule can be summarized as follows: Suppose that bit i of the pattern to be stored, V^s, changes when V^s is applied to the net. Since the ith row of the weight matrix determines the next state of neuron i (see Equation 1), each weight in the ith row can be changed in the direction that will cause V_i to approach V^s_i:

$$
\Delta T_{ij} = c \, (V^s_i - V_i) \, V^s_j
\tag{5}
$$

where V_i is the output of neuron i when pattern V^s is applied to the net, and c is a learning constant. If $V_i = V^s_i$, none of the weights in the ith row will be changed, and Equation 1 will produce the same V_i the next time V^s is applied to the network. If V_i is less than V^s_i, then each term $(T_{ij} V^s_j)$ of Equation 1 will increase; and, if V_i is greater than V^s_i, then each term of Equation 1 will decrease. Repeated application of Equations 5, 1 (with input V^s_i), and 3 will lead to a convergence of V_i to V^s_i provided that the convergence ratio, t/c, is large enough. If the convergence ratio is very large, the Equation may have to be applied many times to learn a pattern. On the other hand, if the ratio is too small, it may not be possible to learn the pattern at all.

On the Choice of the Convergence Ratio: This section will discuss the choice of the convergence ratio, t/c. First, let us suppose Equation 5 is applied to a binary neural network with outputs limited to +1 and -1. The change in x_i caused by application of Equation 5 can be calculated by combining Equations 5 and 1:

$$\Delta x_i = \sum_{j \neq i} c \left(V^s_i - V_i \right) V^s_j V^s_j \qquad [6]$$

$$\Delta x_i = c \left(V^s_i - V_i \right) \sum_{j \neq i} \left(V^s_j \right)^2 \qquad [7]$$

Since V^s_i and V_i can take on only the values of +1 and -1, $V^s_i - V_i$ will be either 0, +2, or -2. If $c = 1/2$, each of the N - 1 weights in row i will be changed by either 0, -1, or +1; and x_i will be changed by 0, -(N - 1), or N - 1. Usually this is enough to cause the desired change in sign of x_i and the correct value of V_i. Repeated application of this algorithm guarantees the storage of any binary pattern, although learning one pattern may cause the net to forget others.

Now suppose one uses the generalized neural network model. In this case the desired change in x_i not only has a minimum but also has a maximum. Normally, one would not want x_i to change by more than the distance between thresholds, t, since such a change could overshoot the desired value. Suppose, for, example, that the neurons can take on the values {3, 1, -1, -3}, as in Equation 3. If the desired value $V^s_i = 1$ and the actual value $V_i = -1$, too large a change in the weights will result in too large a change in x_i causing an output of $V_i = 3$ the next time V^s is applied to the network.

The greatest change in x_i will occur if the pattern to be stored, V^s, consists of only the values +3 and -3. In this case Equation 7 can be written as

$$\Delta x_i = 9 \, c \left(V^s_i - V_i \right) (N - 1) \qquad [8]$$

and the amount of change in x_i will depend on the difference between the desired output and the actual output. A reasonable choice of t/c in this case is 18 (N - 1). With such a choice, if V^s_i and V_i differ by 2, the change in x_i will be at most 18 c (N - 1) and will never be so much that more than one threshold will be crossed at a time. Also, if V^s_i and V_i differ by 4, then x_i will never change so much that more than two thresholds will be crossed at a time. For the {3, 1, -1, -3} quaternary model described above, repeated application of Equation 5 with t/c = 18 (N - 1) is guaranteed to store any pattern.

For any n-ary neural network, it is easy to calculate a ratio of t/c that is guaranteed to allow any pattern to be stored. Let Vmax be the maximum allowable magnitude for the output of a neuron (Vmax = 3 for the quaternary example given above). Let Vdiff be the maximum difference between any two "adjacent" output values (Vdiff = 2 for the example above). The following convergence ratio will allow any pattern to be stored:

$$t/c = Vmax^2 \, Vdiff \, (N - 1) \qquad [9]$$

This result stems from the fact that

$$\Delta x_i \leq c \, Vmax^2 \, Vdiff \, (N - 1) \qquad\qquad [10]$$

For the quaternary model discussed above, this formula gives the minimum value of t/c that will guarantee that any pattern can be stored. For other n-ary models, there are values of t/c below the one calculated from this formula that will also allow any pattern to be stored, as will be seen shortly.

Tests: Several tests were run to determine experimentally the optimal choice of the convergence ratio and to compare the capacity of the quaternary model to that of the binary. Sets of patterns to be learned were generated randomly; however, each time a pattern was generated, it was compared to each of the patterns already in the set to be sure that it differed in more than one place from each of them. If two patterns differ in only one place, they cannot both be stored simultaneously in most neural networks (networks with no direct self-feedback of neurons[3]). For example, patterns (3 1 -1 -1 -3) and (-1 1 -1 -1 -3) cannot both be stored simultaneously using our model. Tests were run on the following quaternary neural network models {3, 1, -1, -3}, {2, 1, -1,-2}, and {4, 1, -1, -4}. For each model, c was set to 1, and the value of t was varied over a large range. In all tests, the number of neurons, N, was set to 7. For each value of t, 100 attempts were made to store 1 random pattern, 100 attempts were made to store 2 random patterns, and so on up to 6 random patterns. As a reference, such tests were also run using a binary model.

Results: Table 1 shows the results for the binary model. Notice that all 100 attempts to store one random pattern were successful.

Number of Patterns	1	2	3	4	5	6
Trials Successful	100	97	86	76	57	36

Table 1. Binary model with N = 7.

The binary model that uses {1, -1} as the set of possible neuron states and Equations 1 and 2 to calculate the next state of a neuron can always store any one pattern by using the delta rule of Equation 5 with V_i calculated as

$$V_i = \begin{cases} 1 & \sum_j T_{ij} \, V^s_j > 0 \\ -1 & \sum_j T_{ij} \, V^s_j \leq 0 \end{cases} \qquad\qquad [11]$$

Table 2 shows results for the quaternary model with possible neuron states (2, 1, -1, -2). Using t = 48 (thresholds of 48, 0, and -48) as suggested by Equation 9 guarantees that any one random pattern can be stored successfully; however, t = 24 actually provides slightly greater capacity.

t/c	Number of Patterns					
	1	2	3	4	5	6
12	37	13	2	1	0	0
24	100	100	100	97	88	50
36	100	99	100	96	91	46
48	100	100	100	94	80	49
60	100	100	99	96	79	38
72	100	99	98	97	74	37

Table 2. Quaternary Model: (2, 1, -1, -2), N = 7.

Notice, with t = 24, we were successful about 50% of the time at storing random sets of six patterns. For example, applying the delta rule to the six patterns

$$
\begin{array}{rrrrrrrr}
[& -1 & 1 & -1 & -1 & 2 & 2 & -2 &] \\
[& -2 & 2 & 1 & 2 & 2 & -1 & -1 &] \\
[& -2 & 2 & -1 & -1 & 2 & 1 & -2 &] \\
[& 1 & -1 & -2 & 1 & -2 & -2 & 2 &] \\
[& 1 & 1 & -1 & -2 & -1 & 2 & 2 &] \\
[& 2 & -2 & 2 & -2 & -1 & 2 & -2 &]
\end{array}
$$

produces the weight matrix

$$
T = \begin{bmatrix}
0 & -8 & 1 & -6 & -8 & 1 & 0 \\
-15 & 0 & -7 & 1 & 13 & 5 & 7 \\
14 & -8 & 0 & 13 & 15 & 13 & -7 \\
-11 & 1 & 13 & 0 & -1 & -10 & 9 \\
-7 & 7 & 2 & -3 & 0 & 0 & -8 \\
-8 & 20 & 21 & -17 & -19 & 0 & 6 \\
6 & 8 & -11 & 10 & -17 & 2 & 0
\end{bmatrix}
\qquad [12]
$$

If t = 24, all six patterns will be attractors of this weight matrix. Tables 3 and 4 show results for the quaternary models (3, 1, -1, an -3) and (4, 1, -1, -4) respectively. For the former, t/c = 108 produced the best results. This is the choice suggested by Equation 9 and is the minimum value of t/c that is certain to store any one pattern. For the latter model, t/c = 288 is suggested by Equation 9; however, values of t/c less than this produced better results.

t/c	Number of Patterns					
	1	2	3	4	5	6
54	35	15	3	2	1	0
72	65	42	28	21	8	0
90	88	83	70	59	47	18
108	100	100	100	95	81	37
126	100	99	99	93	77	30
144	100	100	99	93	76	32
162	100	100	97	92	54	20

Table 3. Quaternary Model: {3, 1, -1, -3}, N = 7.

t/c	Number of Patterns					
	1	2	3	4	5	6
48	28	21	15	6	1	0
96	100	100	100	96	85	47
144	100	100	97	97	65	39
192	100	100	100	84	64	26
240	100	99	98	90	44	19
288	100	100	93	79	45	7
336	100	98	92	50	29	6
384	100	97	86	53	14	2

Table 4. Quaternary Model: {4, 1, -1, -4}, N = 7.

Notice that all three quaternary models can store more patterns for a given number of neurons than can the binary model. More importantly, each pattern contains twice as much information as a binary pattern of the same length, even though the same number of synaptic weights are used in both cases.

The work of Optican and Richman suggests that biological neurons may be characterized by two output variables. If both variables are binary, the four possible output states could be described as (-1, -1), (-1, 1), (1, -1), and (1, 1); however, this would be equivalent to simply doubling the number of neurons of the binary neural network model leading to a quadrupling of the number of weights. By assigning neurons four possible outputs, as we have done, each neuron conveys as much information as the two-variable neuron without the need to increase the number of weights. Also, note that a binary model of N neurons used to store N-bit patterns can be converted to a quaternary model of N/2 neurons used to store patterns of length N/2. This would reduce the number of weights by 1/4.

Capacity Comparisons: While we have devoted ourselves primarily to quaternary neurons, the question of a further generalization to analog neurons can be raised. For such a case the fixed points are the eigenvectors of T:

$$\lambda\, x = T\, x \qquad\qquad [13]$$

For distinct eigenvalues, we can store N patterns in an N-neuron network.

The capacity of a neural network in terms of patterns is bounded by the number of neurons, but the information capacity increases as the size of the neuron alphabet increases. For neurons that take K different values, the total information is the number of patterns (N) times the information in each pattern ($N \log_2 K$). In other words, the total information in bits is $N^2 \log_2 K$ bits. This may be compared to the capacity of the binary Hopfield network which is approximately N patterns if delta learning and random unlearning are used[4]. In terms of information, the capacity is N^2 bits.

Perhaps a more meaningful comparison between binary and non-binary neural networks can be made by placing restrictions on the synaptic weight values. It appears that the capacity of a binary neural network is reduced by a factor of $2/\pi$ if the T values are constrained to 0, +1, and -1[5]. One' would expect that a larger spectrum of real numbers for T would be required so that the capacity of a non-binary neural network was maintained to a corresponding measure.

Conclusions: On capacity considerations it is clear that it is advantageous to use non-binary neural networks.

References

1. Lance M. Optican and Barry J. Richmond, "Temporal Encoding of Two-Dimensional Patterns by Single Units in Primate Inferior Temporal Cortex. III. Information Theoretic Analysis," *Journal of Neurophysiology*, vol. 57, no. 1, pp. 163-178, January 1987.

2. Subhash C. Kak, "CHAOTIC STATES OF MIND--Neural Networks with Nonclassical Synaptic Function," Louisiana State University, Technical Report #50-88, Baton Rouge, LA, April 29, 1988.

3. Donald L. Prados, "The Capacity of a Neural Network," *Electronics Letters*, vol. 24, pp. 454-455, 1988.

4. T. W. Potter, "Storing and Retrieving Data in a Parallel Distributed Memory System," SUNY at Binghamton, Ph.D. Dissertation, 1987.

5. R. J. McEliece, E. C. Posner, E. R. Rodemich, and S. S. Venkatesh, "The Capacity of the Hopfield Associative Memory," *IEEE Transactions on Information Theory*, vol. IT-33, pp. 461-482, 1987.

NEW LEARNING & CONTROL ALGORITHMS FOR NEURAL NETWORKS

Chung H. Youn   Subhash C. Kak
Louisiana State University

1. <u>Introduction</u>: Neural networks offer inherent distributed processing power, error correcting capability and structural simplicity of the basic element, namely neuron. Neural networks have been found to be attractive for application such as associative memory, robotics, image processing and speech understanding. These current studies have focused on monolithic structures. On the other hand, it is well known that the human brain has a "bicameral" nature at the gross level and it also has several specialized structures. This paper investigates computing characteristics of neural networks that are not monolithic being enhanced by a controller that can run algorithms that take advantage of the known global characteristics of the stored information. Such networks have been called bicameral neural networks.

Our study is focused on the application area of content addressable memory and it is shown that our algorithms enhance performance. Since these networks approach the functioning of the human brain in a certain sense, we expect that our models would be useful in A.I. problems that require higher level reasoning.

McCulloch and Pitts [1] were the first to study neuron structures for computing in 1943. The next four decades saw applications of layered neural networks for pattern recognition. An influential recent work was the 1982 paper by Hopfield [2] who proposed a model that could be worked asynchronously. In this model, about $0.15N$ memories, where N is the number of neurons, can be simultaneously remembered before error in retrieval is severe. Some of the inherent drawbacks of the Hopfield model are listed below.

1. The retrieved memory may not be the nearest memory to the input in terms of Hamming distance.
2. The Hopfield model has complement memories to the originally stored memories. Consequently, it might converge to these wrong, stable points.
3. Spurious memories are created because of interference among stored memories.

In the Hopfield model, each stable point, either an original or a spurious memory, forms an <u>attraction basin</u> that influences the dynamics of the probe. If the probe takes a wrong direction influenced by the attraction basin of a spurious memory, it might converge to a spurious memory. The need to have an asynchronous controller which guides the sequence of update operations to the correct fixed point with the help of local and global statistics was pointed out by Stinson and Kak [3].

There are two ways of improving performance of neural networks. One is to improve the learning algorithm. Two popular algorithms are the Hebbian-rule [4] and the Delta-rule [5]; another new algorithm involving the unlearning of spurious memories will be presented in this paper. The other is to control the direction of convergence as in a bicameral structure. With an asynchronous controller in the feedback loop, whenever there is more than one neuron to update, the controller makes a choice considering global characteristics of the memories. New control algorithms will also be presented.

The asynchronous controller of a bicameral neural network can be implemented on either a conventional computer or by another neural network. Stinson and Kak [3] considered an elementary bicameral model that used asynchronous control but it was not exactly a bicameral network, because one of two subnetworks was not implemented as a neural network. In a later paper [6] they did show how information can be indexed in a neural network which makes it possible for neural networks to implement many standard algorithms like the ones that are run on the synchronous controller. We emphasize again that the two separate neural networks of a bicameral network have different structures and cannot be replaced by a monolithic structure.

2. <u>Representation of a Bicameral Network</u>: Early studies of neural network considered either a monolithic feedback structure or hierarchical or layered structures in which information (or signals) flows only in one direction. The visual processing systems of the brain is often seen to be layered [7]. On the other hand a feedback structure may be viewed as a concatenation of a series of layered structures. The bicameral network is essentially a feedback structure. In this paper, we study advanced bicameral networks which can be used for pattern recognition problems. Since it is possible to use several subnetworks to perform various control computations, the result may be viewed as a multicameral network.

3. <u>Learning Rules</u>: We begin by storing information in the neural network so as to find responses to probes (or inquiries) later. Since the weight matrix represents all the stored information, although not in the original form, the construction algorithm is crucial to the performance of the network. Several different learning rules have been described in the literature [8,9], but the most important ones are the Hebbian rule and the Delta-rule.

Generally, the Delta-rule adjusts the entries of weight matrix in small steps (or increments), whereas the Hopfield model with the Hebbian rule uses +1 or -1 as the amount of change. The Delta-rule performs better than the simple Hebbian rule because the weight matrix carries only the necessary amount of strength rather than multiples of +1's and -1's. Furthermore, with the simple Hebbian rule, one might erase the previous stored memory when there are too many memories due to interference between memories which are too close to each other.   Notice that matrix T of the Delta-rule may not be symmetric and can have non-zero diagonals. The Delta-rule significantly improves the performance of the Hopfield model for randomly chosen memories.   When the simple Hebbian rule deals with orthogonal memories, the Delta-rule behaves exactly as the Hebbian rule.

One disadvantage of the Delta-rule over the simple Hebbian rule used in the original Hopfield network is that we have to keep all stored memories in the controller to check whether they remain as fixed points or not.

4. <u>The Method of Continuous Unlearning</u>: The algorithms discussed so far, deal with learning new information.   In this section, we propose a new method of learning which structures the attraction basins differently. This method is based on the concept of "unlearning".   The idea of "unlearning" was noted earlier by Crick,Mitchison,Hopfield [10,11].   Hopfield's method picks probes at random and weakly unlearns the stable state reached regardless of whether it is an original memory or a spurious state.   In other words, this previous method is not tailored to spurious memories.   Furthermore, one has to experiment with the number of unlearning events before achieving optimal performance.   In a recent dissertation, Potter [12] claimed that one can, with unlearning and self-feedback loops (i.e. T_{ii}'s are not zeros), increase the capacity of the Hopfield model up to N, where N is the number of neurons in the network.   Our proposed model is based on unlearning the undesirable fixed points of the network and it does not place any new constraint on the structure.   Our method is therefore different from the methods of Hopfield and Potter.

Loading a Hopfield network with too many memories can end up in a loss of the previously stored memories and creation of spurious memories that result from interference among stored memories. There are four classes of fixed points in the Hopfield model:

1). original memories
2). complements to original memories
3). non-complement spurious memories
4). complements to fixed points in class 3

When neural networks are "lightly" loaded, they can retain all of original memories. The fact that the Hopfield model operates normally when it has 0.15N memories can be used as a measure to determine whether or not the model is lightly loaded. But when the model is "heavily" loaded with many memories, it only has some of original memories as fixed points.

The main reason which causes the poor performance of the Hopfield model is the correlation (or interference) between the original and the spurious memories. If the memories are too close, the chance of misguiding the direction of a probe is increased. Furthermore the existence of spurious memories in the network does not improve dynamics. They stretch and distort attraction basins of the memories which makes it more probable for a probe to take a wrong direction.

The presence of attraction basins around these spurious memories causes serious problems. When we have mutually orthogonal memories, even the simple Hebbian rule performs very well, since the outer-product of any two memories is zero, resulting in zero correlation among memories. The question is how to remove these spurious memories. As mentioned before, there are two kinds of spurious memories, complements and others which are linear combinations of stable points.

When we make an outer-product of a memory, its complement has the same outer-product. The stable points of Hopfield model are related to the eigenvectors of the weight matrix, T.

$$\text{sgn } [Tx] = \text{sgn } [x] \text{ where x is the eigenvector}$$

But at the same time, the following equation also holds.

$$\text{sgn } [T(-x)] = \text{sgn } [-x]$$

Consequently, we cannot remove complements from the Hopfield model. To remove it by subtracting its outer-product from the weight matrix also removes the corresponding original memory. Therefore, we will concentrate on the idea of removing the non-complement spurious memories (Class 3) that are encountered while information is being learned.

Basically, this method works similar to the Delta-rule except in the process of adjustment. Instead of adjusting entries of the weight matrix as in the Delta-rule, we subtract the outer-product of a spurious memory multiplied by the unlearning rate from the weight matrix.

Step 1) Initialize T with zero's
        Set V = [ ] and unlearning-rate, U.
Step 2) Construct the outer-product of a memory to store and
        and the memory to V. Add the outer-product to T.
Step 3) Set V' equal to V.
Step 4) If V' is empty and there is another memory to store
        then go to Step 2. Otherwise, pick a state, S, out of
        V' and run the Hopfield model to get an answer, A.

$$S = (s_1, s_2, ---, s_N)^t$$
$$A = (a_1, a_2, ---, a_N)^t$$

Step 5) If S = A, then go to Step 4. Otherwise, check to see if A is a spurious memory. If A is a complement, then go to Step 4. Otherwise, make the outer-product of A, multiply it with the unlearning-rate, U, and subtract it from T. Go to Step 3.

109

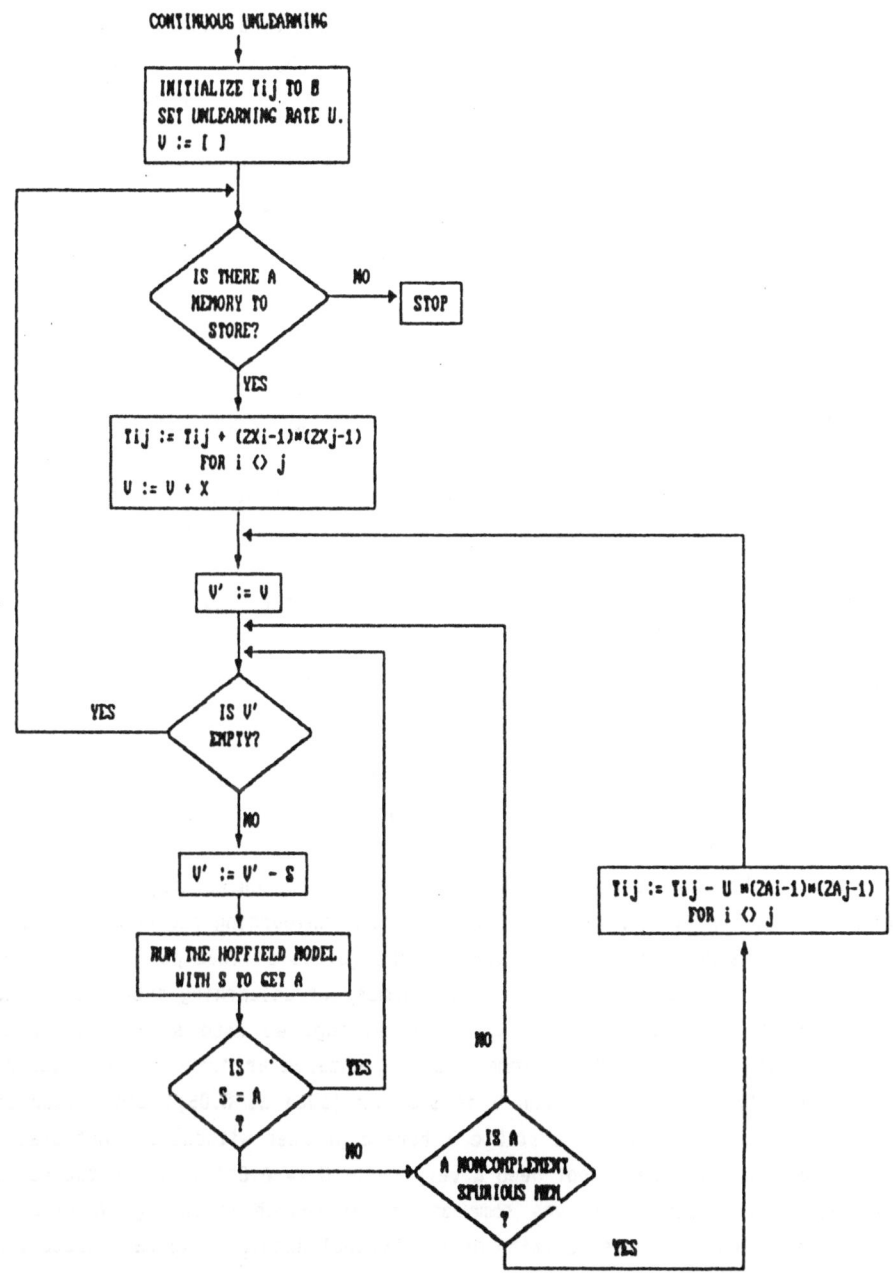

CONTINUOUS UNLEARNING

INITIALIZE Tij TO 8
SET UNLEARNING RATE U.
V := [ ]

IS THERE A
MEMORY TO
STORE?

NO → STOP

YES

Tij := Tij + (2Xi-1)*(2Xj-1)
FOR i <> j
V := V + X

V' := V

IS V'
EMPTY?

YES

NO

V' := V' - S

RUN THE HOPFIELD MODEL
WITH S TO GET A

IS
S = A
?

YES

NO

IS A
A NONCOMPLEMENT
SPURIOUS MEM
?

NO

YES

Tij := Tij - U *(2Ai-1)*(2Aj-1)
FOR i <> j

Flow Diagram for The Continuous Unlearning Algorithm

Note that we do not subtract the exact outer-product of a spurious memory from T. Instead, we multipy it with predefined parameter, U, the "unlearning rate". The main purpose of this parameter is to adjust the matrix T in small steps to reflect the incremental unlearning.

Another difference between the Delta-rule and the method of continuous unlearning will now be considered. In Step 4, we send a vector, S, to the Hopfield model to get an answer, A.

There are four types of answers we can get.

1) A is same as S.

2) A is another original memory.

3) A is a complement memory to one of stored memories.

4) A is a spurious memory which is a linear combination of stable points.

In the Delta-rule, we adjust entries of T when we have answers of type 2, 3, and 4. But in the method of continuous unlearning, we modify T only when we get answers of type 4.

We provide the comparisons of performance in the following using the unlearning rate of 0.1.

Examples	N	Improvement Over Delta
1	6	42%
2	9	52%
3	9	50%
4	12	21%

As shown in the table above, the method of continuous unlearning improves the performance of the Hopfield model. We have seen that unlearning reduces the number of spurious memories which are linear combinations (Class 3 fixed points). In our method, we do not have to experiment with the number of unlearning event as in the Hopfield's method [10]. During our extensive testing, we used several learning rates. For large U (such as 0.2), some examples entered an infinite loop during the process of adjustment of T. When U is smaller (such as 0.05), the number of unlearning events to converge to a stable T became greater without any noticeable increase in performance. We do not need a very small U (= 0.01) which is the value Hopfield used. Our method involves some amount of searching through a list of original memories, but with the addition of "orthogonal handles", we can reduce the amount of searching, because we can determine whether or not an answer is a non-complement spurious memory by looking at its handle.

There should be two different modes of operation for neural networks. One is the learning mode and the other is the retrieval mode which represents the normal operation. Currently the method of continuous unlearning is applied only in the learning mode. One possible extention to this method can be described as follows:

During the retrieval mode, whenever the model finds a spurious memory as one can do using "handles", it notifies the controller to resume the learning mode and "unlearn" that spurious memory.

5. <u>The Bicameral Classifier:</u> This model consists of two neural networks, a Hopfield model and an image classifier built on top of a network that computes Hamming distance from stored memories (Hamming network). There are two modes of operation, the learning mode and the retrieval mode.

In the learning mode, the model is initialized according to incoming memories. The Hopfield model will initialize its weight matrix using either the Hebbian rule, the Delta-rule or the new CU-rule (Continuous Unlearning). At the same time, memories are sent to the classifier, and the classifier determines the number of classes according to a chosed threshold. After classification, the Hamming network is initialized using those resulting classes.

In the retrieval mode, the Hopfield model receives an inquiry (or a probe) and begins operation unless there is more than one neuron to update. When there are several choices, it asks the classifier to make a decision.

The Hopfield model and the classifier work on different configurations of a memory. In other words, the number of neurons in the data vectors used by the two subnetworks is different. The Hopfield model will have a vector of size N where N is the number of neurons in the network.

$$X = (x_1, x_2, ---, x_N)$$

The classifier works with two different configurations of a vector. In the learning mode, it works with a vector of size N when it classifies incoming memories. Once it finishes classification, it reduces the size of a vector and initializes the Hamming network with smaller vectors. The reason is that in real-time applications such as the image recognition problem the size of an image can be huge. The major contribution of the classifier is to assist the Hopfield model in making decisions. Therefore, the response time of the classifier is an important factor. It can be reduced when it works with smaller configuration, even though the quality of assistance will degrade somewhat.

Generally, the bicameral classifier shows better performance than the original Hopfield model with random selection. For instance when the number of memories is 5 and the number of neurons is 9, the bicameral classifer shows an improvement that varies from 50% to 100% for Hebbian and Delta-rule learning measured in terms of the number of probes that lead to the correct memory. For details see [13].

6. <u>The Method of Iteration:</u> We have discussed the bicameral approach to improve performance of the Hopfield model. We know that the Hopfield model does not work very well, because it can converge to wrong and spurious memories. To improve convergence of the Hopfield model, we now propose another method based on the Hamming distance that we call the method of iteration. Later, we apply the concept of "Hidden-bit" and "Handles" described in [6] to this new method.

Basically, this method iterates until the Hamming distance between the probe and the response from the Hopfield model is less than or equal to the given threshold. In the Hopfield model, each stable point, either an original memory or a spurious state, forms an "attraction basin" that influences the dynamics of the probe. If the probe takes a wrong direction influenced by the attraction basin of a spurious state, it might converge to a spurious state. The bicameral model tries to guide the probe to take the right direction, although it is not always successful.

So far when a probe converges to a stable point, all that can be done is to output that stable point as an answer hoping that it is the right answer. But that may not happen. We propose that if the response is not close enough to the probe in terms of Hamming distance, several bits be changed randomly and the Hopfield model run again with the changed probe. Note that during the execution of Hopfield

Step 1) Initialize Hopfield model with original memories. Set the threshold, T, and the number of neurons to change randomly, R.

Step 2) Obtain the probe, P.
$$P = (p_1, p_2, ---, p_N)^t$$

Step 3) Run Hopfield model to get the answer, A.

Step 4) Compute the Hamming distance, D, between P and A. If D <= T, then output A as an answer. Otherwise, select R neurons randomly and change corresponding neurons of A to get P'.. Go to Step 3 with P'.
$$A = (a_1, a_2, ---, a_N)^t$$

One problem with this method is that it can enter an infinite loop which consists of Step 3 and 4. Therefore, we must set the limit on the number of iterations to a predetermined value, e.g. 100. Obviously, the performance of this method depends on the choice of threshold, T, and the number of neurons to change, R. If we select too small a T, this method will have many unnecessary iterations and possibly wrong answers. We find that T=N/3 and R = N/4 seem reasonable choice for most examples.

In the following, we provide comparisons of the Hopfield model and the method of iteration. Note that the simple Hebbian rule and the Delta-rule are used in the Hopfield model. C: number of probes that lead to the correct memory, W: number of probes that lead to one of the memories but not the closest one; S: number of probes that lead to spurious memories.

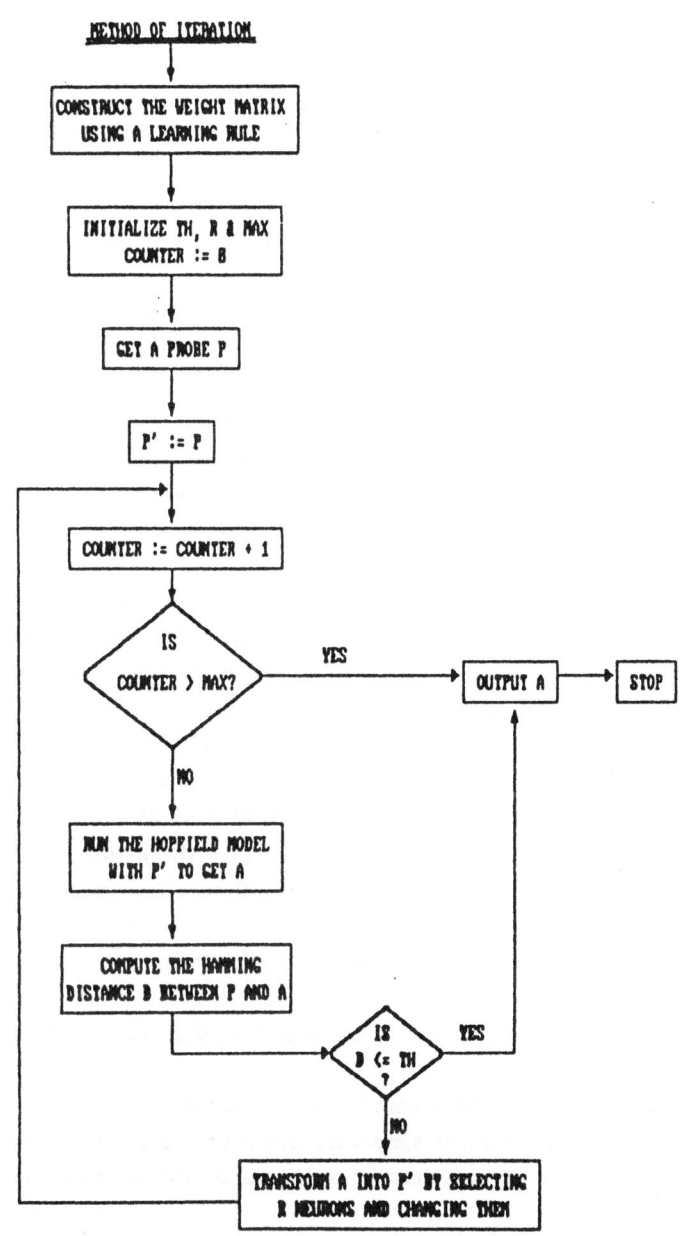

Flow Diagram for The Method of Iteration

Example: N = 12, Number of memories = 7, T = 4, R = 3

	Hebbian rule		Delta-rule	
	Hopfield	Iteration	Hopfield	Iteration
C	243	477	842	1442
W	310	179	571	356
S	3543	3440	2683	2298
Total Number of iterations	4096	55,792	4096	34,885

The method of Iteration with "Hidden-bit": Stinson and Kak introduced the concept of "hidden-bit" to detect complement states [6]. As shown in previous examples, the number of spurious answers are enormous. The Hopfield model does not have a way of detecting spurious states of which there are two kinds:

Complement states which are complements of some fixed points, and other spurious states which are linear combinations of some stable states.

With the addition of one extra hidden-bit, the complement states may be detected. For now, we deal only with the detection of the complement states. Instead of outputting a complement state as an answer, we can notify the user that this answer is considered worthless. Otherwise, the user might consider the answer to be correct.

The detailed procedure of this method is same as before except Step (1). All the original memories has one extra bit with the value of +1. These memories of size (N+1) are stored in the Hopfield model. And the probe with +1 as a hidden bit will be presented to the Hopfield model. Also, in Step (4), if the hidden-bit has +1, the rest of the answer is outputted as the response. Otherwise, the user is notified that the answer is deemed spurious.

[C : number of probes that lead to the correct memory with correct
        hidden-bit of +1.
 W : number of probes that lead to one of original memories, but not the closest
        one.
 SC : number of probes that lead to spurious memories with hidden-bit of -1.
 SP : number of probes that lead to spurious memories, but with hidden-bit of +1.
 WK : number of probes that lead to original memories, but with hidden-bit of -1.
 ***: Not applicable]

Example: N = 12 + 1, Number of memories = 7, T = 4, R = 3.

| | Hebbian rule | | Delta rule | | Continuous Unlearning |
	Hopfield	Hidden bit	Hopfield	Hidden bit	Hidden bit
C	243	924	842	1552	1741
W	310	535	571	295	393
SC	***	1075	***	1200	1254
SP	3543	1562	2683	1049	706
WK	***	0	***	0	3
Total number of iterations	4096	113,538	409	75,103	86,338

Notice that there is a significant reduction in the number of spurious statesfrom the Hopfield model to the. method of hidden-bit. This is because the complement spurious states are detected and considered as semi-correct answers.

The Method of Iteration With "Handles": As mentioned before, the "hidden-bit" method can detect only complement states. In [6] Stinson and Kak proposed the use of "orthogonal handles" to detect other spurious states. With the addition of orthogonal vectors to original memories, we can detect all spurious states which are linear combinations of stable points.

There is a large reduction in the number of spurious states from the Hopfield model and the hidden-bit method to the method of handles. But there is a trade-off between the methods of hidden-bit and handles. In the case of a wrong answer, the hidden-bit method performs better than the method of handles. When a network is heavily loaded the original memories may not be stored and therefore inclusion of handles may erase some memories.

7. Conclusion: This paper has provided experimental evidence that establishes that a neural network with a controller performs significantly superior to a conventional Hopfield neural network. Several new algorithms   are presented. These include a new learning algorithm, control using the method of iteration, and bicameral classifier techniques. For a detailed exposition of these results please consult the longer Technical report [13].

REFERENCES

[1] W.S. McCulloch and W. Pitts, "A Logical Calculus of the Ideas Imminent in Nervous Activity," Bulletin of Mathematical Biophysics, Vol. 5, pp. 115-133, 1943.

[2] J.J. Hopfield, "Neural Networks and Physical systems With Emergent Collective Computational Abilities," Proc. Nat. Acad. Sci. USA, vol. 79, pp. 2554-2558, 1982.

[3] M.C. Stinson and S.C. Kak, "Asynchronous Controller to Improve the Convergence of Neural Nets," *Proceedings of the 26th Annual ACM Conference*, *Mobile*, pp. 410-413, April 1988.

[4] D.O. Hebb, *The Organization of Behavior*, John Wiley & Sons, New York 1949.

[5] G. Widrow and M.E. Hoff, "Adaptive Switching Circuits," *1960 IRE WESCON Con. Record*, part 4, pp. 96-104, August 1960.

[6] M.C. Stinson and S.C. Kak, "On Bicameral Neurocomputering," LSU Technical Report #90-99, 1988.

[7] P.C. Jackson, *Introduction to Artificial Intelligence*, Dover Publications, New York, pp. 12-28, 1985.

[8] D.E. Rumelhart and J.L. McClelland, *Parallel Distributed Processing*, the MIT Press, Cambridge, pp. 53-54.

[9] M.L. and S.A. Papert, Perceptrons, *The MIT Press*, Cambridge, MA 1988.

[10] J.J. Hopfield et al, "Unlearning Has A Stablizing Effect In Collective Memories," *Naturem* vol. 304, pp. 158-159, July 1983.

[11] F. Crick and G. Mitchinson, "The Function of Dream Sleep," *Nature*, vol. 3044, pp. 111-114, July 1983.

[12] T. W. Potter, "Storing and Retrieving Data in a Parallel Distributed Memory System," *Ph.D. Dissertation*, SUNY at Binghamton, 1987.

[13] C.H. Youn and S.C. Kak, "New Learning and Control Algorithms for Neural Networks," *Technical Report #88-052*, Dept. of Computer Science, LSU, October 11, 1988.

A Fixed-Size Systolic System for the Transitive Closure and Shortest Path Problems†

H. Barada and A. El-Amawy

Louisiana State University

1. Introduction. The transitive closure and the shortest path problems are clearly two of the most important problems in graph theory. However, solving these two problems is computationally expensive. For a graph of n vertices, Warshall algorithm for performing the transitive closure or, equivalently, Floyd algorithm for the shortest path problem require $O(n^3)$ operations [1]-[2]. Therefore, it is desirable to build special architectures which can run these algorithms and many compute-bound algorithms at high speed.

One of the recent and very promising specialized architectures is based on the systolic array concept. A systolic array consists of a set of locally interconnected processing elements (PEs). They feature simple control and timing, regular data flow, modularity and most significantly compatibility with VLSI technology [3]-[10].

The transitive closure/shortest path problems have received the attention of several researchers who have proposed different structures of systolic arrays [6]-[10]. All the proposed arrays use approximately n^2 processors to find the transitive closure/shortest paths of an n-vertex directed graph G in $O(n)$ time. Obviously such arrays may not be implementable when n is large.

In this paper we propose a fixed-size systolic system to solve the transitive closure/shortest path problems of arbitrarily large sizes. We show that the product AT (A is the number of PEs and T is the execution time) is always better than that of the full size optimal array and that AT^2 can be minimized at a particular array size. This minimum AT^2 is also better than that of the full-size array. These results provide enough evidence that in (at least) some cases a fixed-size systolic system can be more efficient than its full-size (larger) counterpart.

2. Problem Definition. Let G be a directed graph. The graph G′ which has the same vertex set as G, but has an edge from u to w iff there is a path from u to w in G, is called the transitive closure of G [1].

Every simple graph G can be represented by an adjacency matrix. The adjacency matrix of an n-vertex directed graph is defined as an n×n binary matrix $A = [a_{ij}]$ in which $a_{ij} = 1$ iff there is a directed edge from i to vertex j or i=j. The transitive closure problem is to compute the transitive closure binary matrix $A' = [a'_{ij}]$

where $a'_{ij} = 1$ iff vertex j is reachable from vertex i through a sequence of directed edges [2].

The Warshall algorithm for solving the transitive closure accepts an n×n binary matrix A and performs the sequence of transformations described below [2]:

$$For\ i = 1\ to\ n$$
$$For\ j = 1\ to\ n$$
$$For\ k = 1\ to\ n$$

$$a_{jk} = a_{jk} + a_{ik} \cdot a_{ji} \ ; \tag{1}$$

where "+" and "." are Boolean OR and AND, respectively.

Equivalently, the Floyd algorithm for finding the shortest paths of every pair of vertices of an n-vertex directed graph can be described using the above equation where $A=[a_{ij}]$ is the distance matrix, "+" is the minimum operation, and "." is the addition operation [10]. The distance matrix of an n-vertex directed graph G is defined as an n×n matrix $A = [a_{ij}]$ in which

    1. a_{ij} = the cost of the edge from vertex i to vertex j if there is an edge.

    2. $a_{ij} = \infty$ if no edge exists from i to j.

    3. $a_{ij} = 0$ if i = j.

3. A Fixed-Size System Design. A fixed-size system is designed such that any arbitrarily large matrices can be processed by partitioning the matrix into smaller matrices that fit the size of the system. However, if the matrix is relatively small, it can be processed by a full-size systolic array. We first describe an array for small matrices. The array is then modified to show the steps taken toward the design of a fixed-size system that can process large matrices.

3.1 Arrays for Small Matrices. The structure of the full-size array and the input sequence of A (n=4) are shown in Figure 1. The array consists of two types of PEs, namely circular and square. The circular PEs are assigned the function of distributing the elements inputted on the south-bound link. The square PEs' function is to update the incoming elements. We assume that each PE (of either type) contains a register R for temporary storage.

The operation of the array is mainly divided into two major phases. In the first phase, the input data is partially processed to varying degrees depending on which row it is in, and then loaded into the R registers of the respective rows of the array. Row 1 is simply stored in the first row of PEs. The second row of inputs is processed by the 1st PE row and then loaded into the R registers of the 2nd row of the array, and so on... until the nth row of inputs is updated by the PE rows 1 to n-1 and stored in the R registers of the nth row of the array.

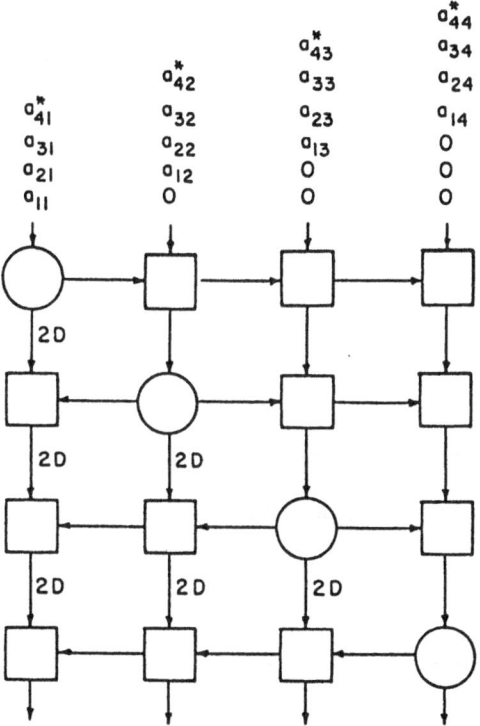

Figure 1. The full-size array with the input sequence.

The second phase starts when the last row of inputs crosses the 1st row of PEs. At this time, the PEs switch modes, and the stored elements begin to propagate southward. Consecutive PE rows will then switch modes in a staggered fashion, one per time unit. The input data which were partially processed and stored in the R registers (in phase 1) now move downwards and get completely transformed by the time they get out of the array.

From the above discussion it becomes evident that a control mechanism is needed. All PEs should be able to store the first element they receive. This can be done by including a flag bit in the PEs. This bit is represented by a flip-flop that is initially cleared (=0). If the flag is not set yet, the PE saves the input element and sets the flag. Once the flag bit is set (=1), the PE will operate in two different modes. These modes shall be referred to as Mode 1 and Mode 2. All PEs switch from the first to the second mode upon receiving the last input element.

To indicate the switching time, we choose a simple control mechanism based on a single control bit.

We associate a token bit with each input element a_{ij}. A set token (=1) will be represented by a *, and a clear token (=0) is represented by the absence of a *. Only the last element of every input column will initially have a set token as illustrated in Figure 1.

The arrival of a set token to the north input of any PE indicates that the associated element is the last element to be updated. One time unit after receiving a set token, the PE sends the stored element on the south-bound output link unchanged with an associated set token. Furthermore, the flag bit inside the PE is cleared so that the PE is ready to receive the new input matrix. A clear token at the south-bound input triggers Mode 1. At the mode, the circular PE distributes (systolically) the input element to the other neighbor PEs while the square PE updates the elements in accordance with eq. (1) given in Section 2. The two modes of operation for each of the two PE types are defined and illustrated in Figure 2.

The full-size array finds the transitive closure/shortest paths of n-vertex directed graph in 5n-4 time units. This time has been proven to be optimal in [10].

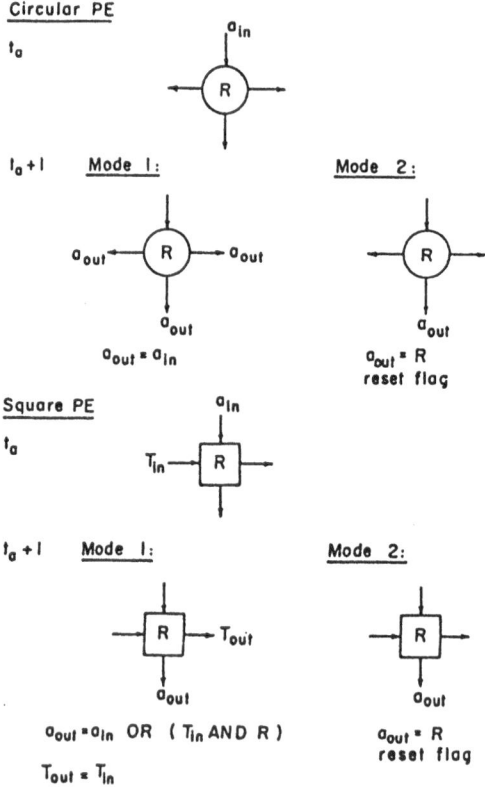

Figure 2. The modes of operation of the two PE types

3.2 Toward the Design of Fixed-Size System. The array described above can be modified by adjusting the number of delays between the PEs, to reduce execution time, while maintaining proper operation. Figure 3 shows the structure of the modified array. This nonplanar (cylindrical) array solves the transitive closure/shortest path problem in $\frac{9n}{2}-2$ time units instead of 5n-4 units. In [10], the 5n-4 completion time was proved to be optimal for 2D arrays implementing the Warshall/Floyd algorithm.

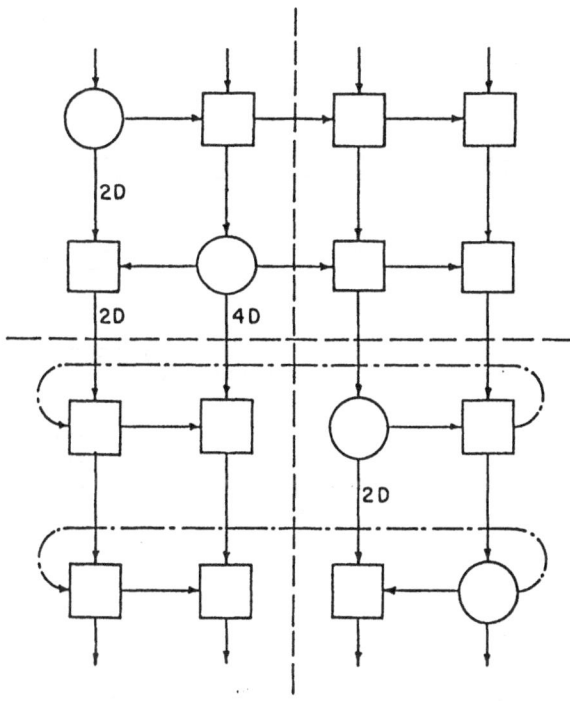

Figure 3. The modified array.

It can be easily seen that the modified array can be partitioned into 4 subarrays of size $\frac{n}{2}\times\frac{n}{2}$. One array is similar to the full-size array while the second array consists of only the square PEs described in Figure 2. This idea can be extended so that the full size array is divided into subarrays of size m×m, $2\leq m\leq\frac{n}{2}$. Therefore, one array of type A and another of type B and a buffer memory are enough to build a fixed-size system capable of solving the transitive closure/shortest path problems with arbitrarily large sizes. The diagram of the system is shown in Figure 4. The type A and type B arrays with their input sequences are presented in Figure 1 and Figure 5, respectively.

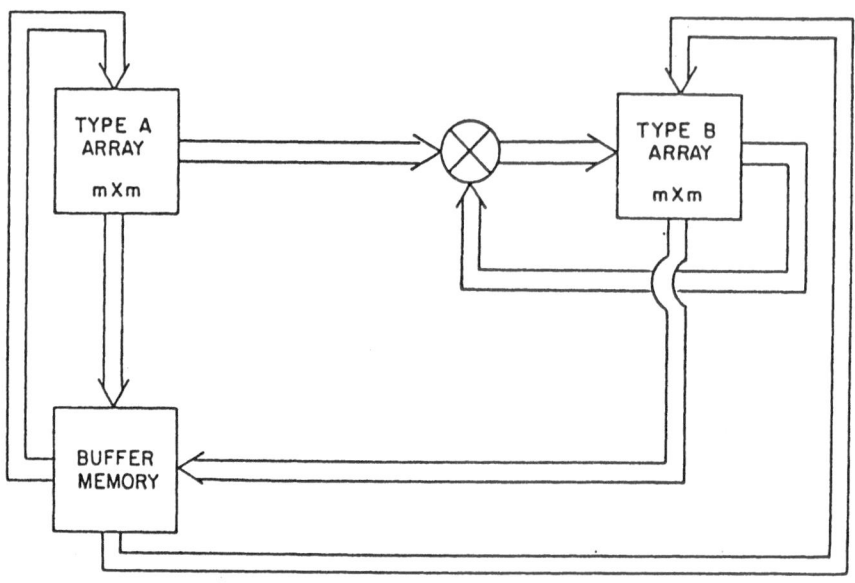

Figure 4. The fixed-size systolic system.
(⊗ is a multiplexer)

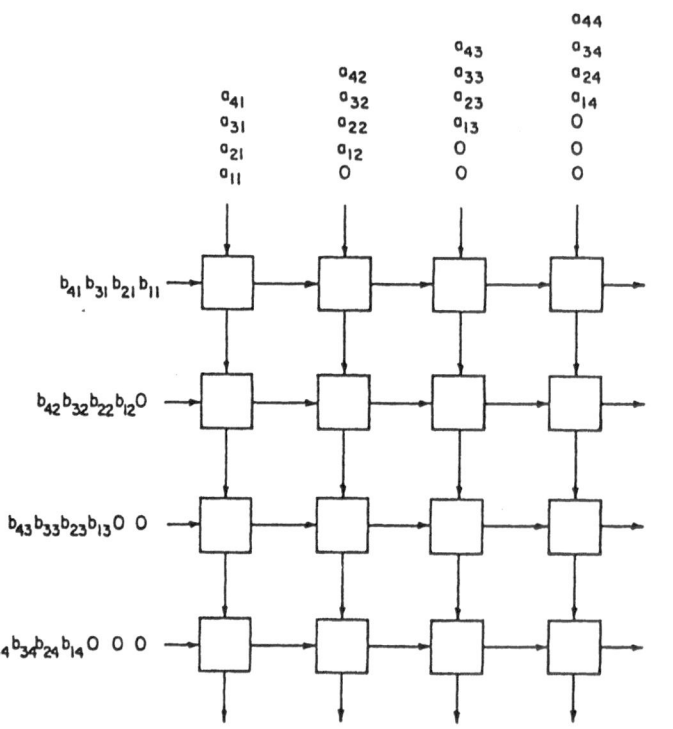

Figure 5. The type B array and the input sequence (n=4).

3.3 Fixed-Size System Operation. To run on the fixed-size system, an arbitrarily large n×n input matrix has to be partitioned into q^2 m×m submatrices, where $q= \left\lceil \dfrac{n}{m} \right\rceil$; m×m is the size of each of the two systolic arrays in the system. If the input matrix is denoted by A = [a_{ij}] (i,j = 1 to n), then the submatrices are represented by A_{lk} = [a_{ij}] where

> l and k = 1 to q,
> i = (l-1)m+1 to lm, and
> j = (k-1)m+1 to km.

At all time one of the diagonal submatrices A_{ii} (i = 1 to q) will have to be stationed in the type A array while one of the submatrices A_{ik} (k = 1 to q, k≠i) is positioned in the type B array. If, for example, A_{11} is in type A array, then type B array will have one of the submatrices A_{12}, A_{13},....A_{1q}.

For convenience, let us define iteration k as the time interval during which A_{kk} is in the array of type A. A subiteration (k,l) is defined as the time interval during which A_{kl} resides in the array B while A_{kk} is in the array A. At subiteration (1,2) (A_{11} is in array A, A_{12} in array B), submatrices A_{j1} (j = 2 to q) are updated by array A while the submatrices A_{j2} (j = 2 to q) are updated by array B. During subiteration (1,3) (A_{13} is in array B), the submatrices A_{j3} (j = 2 to q) are updated by the type B array. This process is repeated for all diagonal submatrices; a total of q iterations or q(q-1) subiterations.

During the system operation, when a submatrix is not used by the two systolic arrays, it resides in the buffer memory. This buffer memory could be either a part of the host main memory or could be a dedicated module added to the fixed-size system to save the partially updated elements.

The functions of the PEs in the size-independent system are the same as the functions in the full-size array described in Figure 2. All PEs operate in two modes. To switch to the second mode, the PEs will have to receive a set token with an input element. During the second mode, a PE sends the content of its R register southward and gets ready to save a new element. Thus a set token will accompany each element of the last row in a given iteration.

3.4 Performance Analysis. Without loss of generality, assume that n is a multiple of m, where n×n is the size of the input matrix and m×m is the size of each of the arrays in the fixed system. Let $q=\dfrac{n}{m}$. As stated before, the total number of subiterations needed to solve the problems is q(q-1). During each subiteration n elements are processed vertically. This makes the execution time nq(q-1) time units. But, we should add (3m-2) units to account for extracting the elements from the array at the last stage. Also, if q>2, m time units should be included because the type B array starts operating m units of time after the Type A array begins its operation. In the case where q=2, 2m time units should be added because the second iteration has to be delayed until the submatrix A_{22} is ready to enter the type A array. So, the total execution time is

$$T_p = \begin{cases} 2m + nq(q-1) + 3m-2 & \text{for } q=2 \\ m + nq(q-1) + 3m-2 & \text{for } q>2 \end{cases}$$

Excluding the buffer memory, the area of the system may be directly related to the number of processing elements. This makes the area of the size-independent system

$$A_p = 2m^2$$

If the problems were solved in a full size array, the computation time is

$$T_f = 5n-4$$

and the area of the systolic array is

$$A_f = n^2$$

Figure 6 shows the total execution time, the area-time product and the area-time^2 product versus m (for n = 100,500,1000), respectively. From the figure, the following remarks can be made:

1. The execution time is minimum for $m=\frac{n}{2}$. This minimum is achieved because of the non-planar structure of the system when m<n.

2. The fixed-size system is more efficient than the large full-size array in terms of area-time product; that is $A_p T_p < A_f T_f$ for all m.

3. $A_p T_p^2 \leq A_f T_f^2$ for $m \geq \frac{n}{4}$. Thus the area-time^2 product can also be reduced using the fixed-size systolic system.

In comparing the size-independent systolic system to the full-size array, we have not taken into account the size of the buffer memory. A buffer memory is almost always needed when an arbitrarily large matrix problem is to be solved on a smaller (fixed-size) array. Its purpose is to save the partially updated data. This memory could be a part of the host's main memory or could be a dedicated module that is an integral part of the fixed-size system. Either way a memory of size n^2-2m^2 words is needed for the operation of the system. The word size is one bit for the transitive closure problem, and depends on the desired accuracy for the shortest path problem.

4. Conclusion. The paper presented a fixed-size systolic system for the transitive closure and shortest path problems. The system can solve an arbitrarily large size matrix by partitioning it into smaller submatrices that fit the fixed-size systolic arrays. Equations for the execution time, area-time product and

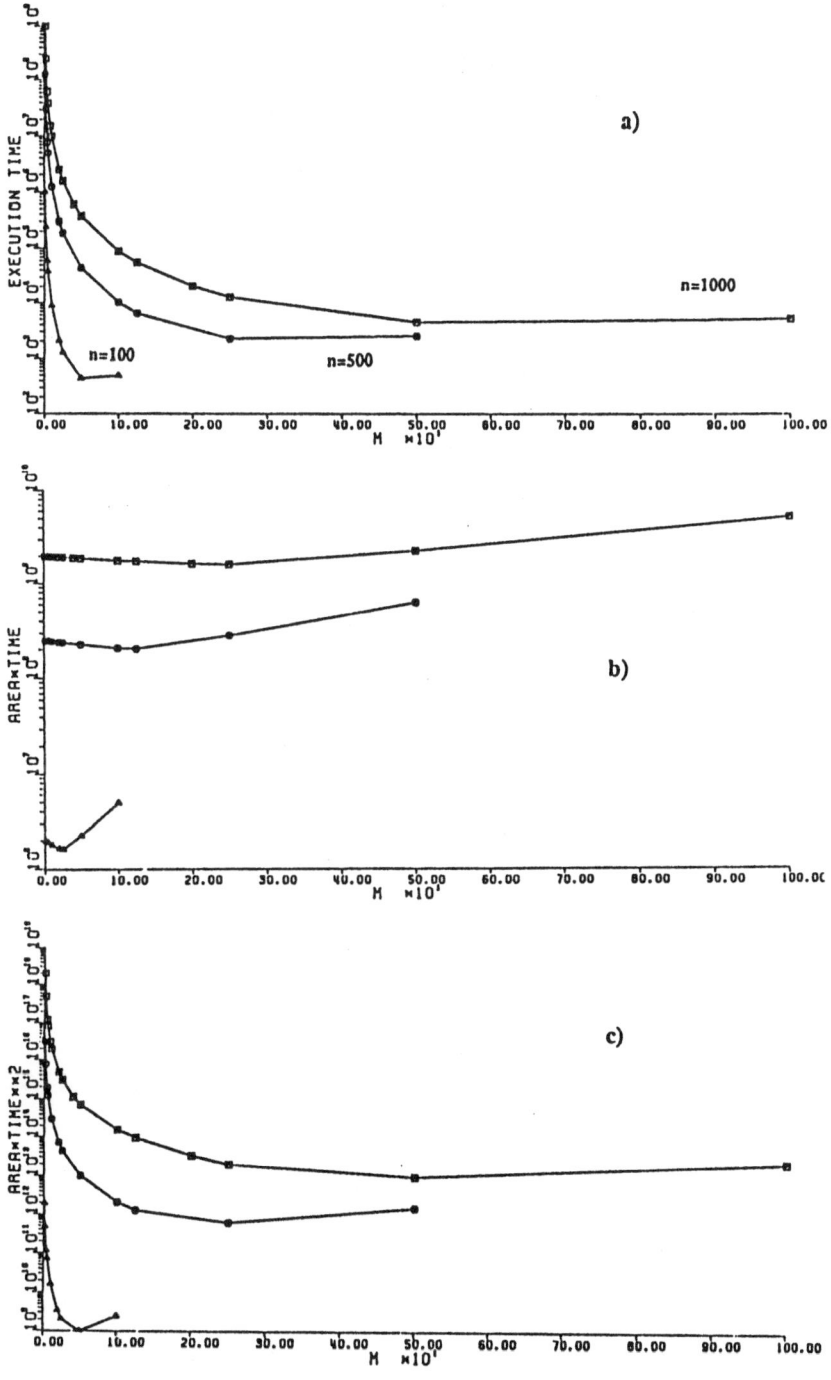

Figure 6. a) T vs M.
b) AT vs M.
c) AT² vs M.
(M is the size of the arrays in the system)

area–time^2 product for the fixed-size system were derived and plotted. A comparison with the full-size array showed that the fixed-size system is more efficient in terms of area-time product. It also showed that the area–time^2 complexity of the algorithms can also be reduced using the fixed-size systolic system.

Acknowledgment. This work was supported in part by the State of Louisiana under Grant LEQSF-RD-A-17. The authors are with the Department of Electrical and Computer Engineering, Louisiana State University, Baton Rouge, LA 70803.

References

[1] A. V. Aho, J. E. Hopcroft, and J. D. Ullman, *The Design and Analysis of Computer Algorithms*, Addison-Wesley, 1974.

[2] M. N. Swamy and K. Thulasiraman, *Graphs, Networks and Algorithms*, New York: Wiley, 1981.

[3] H. T. Kung, *"Why Systolic Architectures"*, Computer, vol. 15, Jan. 1982, pp. 37-46.

[4] H. Barada and A. El-Amawy, *"Systolic Architecture for Matrix Triangularisation with Partial Pivoting"*, IEE Proc., vol. 135, pt. E, no. 4, July 1988, pp. 208-213.

[5] A. El-Amawy, *"A Systolic Architecture for Optimal Filter Design Support"*, Circuits Systems and Signal Processing, vol. 7, no. 2, 1988, pp. 151-172.

[6] L. J. Guibas, H. T. Kung, and C. D. Thompson, *"Direct VLSI Implementation of Combinatorial Algorithms"*, Proc. Caltech Conf. VLSI, Jan 1979, pp. 509-525.

[7] F. C. Lin and Wu, *"Systolic Arrays for Transitive Closure Algorithms"*, Proc. Int. Symp. VLSI Sys., Nov. 1986.

[8] Y. Robert and D. Trystram, *"an Orthogonal Systolic Array for the Algebraic Path Problem"*, Res. report 553, TIM3/IMAG, Grenoble, France, 1985.

[9] G. Rote, *"A Systolic Array Algorithm for the Algebraic Path Problems (Shortest Paths, Matrix Inversion)"*, Computing, vol. 34, pp. 192-219.

[10] S. Y. Kung, S. Lo, and P. Lewis, *"Optimal Systolic Design for the Transitive Closure and the Shortest Path Problem"*, IEEE Trans. on Computers, vol. C36, no. 5, May 1987, pp. 603-614.

SYSTEMATIC DESIGN OF FAST DISCRETE CONTROLLERS

Jorge L. Aravena
Louisiana State University

1. _Introduction_: Array architectures which can effectively exploit the capabilities of VLSI techonolgy offer a possible solution to the requirements of real time large scale systems and signals applications ([1]-[4]). In particular, efficient array structures for matrix operations form the basis for many linear filter designs ([5]-[8]).   These architectures are not, however, without drawbacks. One limitation addressed in this paper is their inability to vary the class of algorithms implemented. We show that by using the execution time as the only parameter, a given architecture can actually implement different families of algorithms.

The perceived requirements of high performance strategic real time applications suggest that any computational capability may, eventually, be overloaded. Thus, it becomes imperative to consider computational efficiency as part of the design constraints for on-line operation. In this study, we incorporate this efficiency constraint in two complementary ways. We use the variable execution time to model the condition when the computing unit must deliver its commands within a limited time. In the worst cases, partial results would have to be used. As an additional constraint, the computing unit is assumed to be a speciazlied architecture. As a case study, we use systolic array implementations.

Engineering applications have more flexibility than standard numerical algorithms. Digital systems with similar performance may be equally acceptable. The additional degree of freedom is used in this paper to improve computational speed. Time and complexity constraints are included as constraints in an optimized design.

To be more specific about our problem class, consider the block diagram of Fig. 1. This block diagram has been used over the years to model a very general system optimization problem. In our case, we let L_1 denote the sensor complex while A is a controller. The block L_3 represents a fixed plant, L_2 represents an ideal response, and L_4 represents an output weighting. The signal β models disturbances or sensor errors.

A classic optimization problem sets the model in Hilbert spaces and takes all maps to be linear. One then chooses A to minimize

$$J(A) = E\{\|e\|^2 + \|r\|^2\}$$

over a class of operators.

Under very general conditions, the expression for the cost function can be shown to be

$$J(A) = tr\{(L_4^*L_4 + L_3^*L_3)AQ_vA^* + L_2Q_aL_2^* - L_2^*L_3AQ_{va} - L_3^*L_2Q_{av}A^*\}.$$

Here, the various Q indicates the respective covariance operators, the symbol (*) denotes the adjoint operation and tr{.} is the trace function. The solution to for this problem is well known.[9] The basic result is stated below.

Result 1. The linear operator, A is optimal if and only if

$$tr\{(Q_LAQ_v - L_3^*L_2Q_{av})\delta A^*\} \geq 0,$$

all δA such that $A + \delta A$ is admissible.

This study considers the inclusion of time and computational complexity in the characterization of admissible systems. For simplicity in the developments, the paper will discuss only the case when the Hilbert space is the finite dimensional Euclidian space, E^n. The various linear operator are represented by matrices of appropriate dimensions.

The next section introduces the concept of time constrained array operation. This constraint assumes that the computational structure is given but its execution time must be reduced. Section 3 will discuss the structure constrained design where the designer may select a priori a fast computational structure and must design the best system for that structure.

2. Time Constrained Controllers: The concept of time constrained operation will be developed using the well known orbital array architecture [3]. This is the fastest known planar array for matrix multiplication.

The array is depicted in Figure 2, for the case n=4. It is well known [3], [4], that, when the nxn matrices, A, B, can be preloaded, the array can compute the product, C = AB, in n cycles. The processing elements are assumed capable of storing the partial sums while the factors are transmitted along the indicated paths.

If a result is required after k<n cycles the matrix product is not yet completed. Instead, a different linear operation is executed. The description of the various linear transformations will assume the form $Y_k = A_k(X)$, where X contains the data points to be processed. For convenience, the matrix will be represented as an n^2 vector arranging the data columnwise. For n = 4 the possible transformations $A_k(.)$ are listed in Figure 3.

The linear transformation on the data points, X, and corresponding to the operation, Y = AX, can be described by the $n^2 \times n^2$ matrix, $\hat{A}$, for the k = n case. Clearly

$$\hat{A} = AoI,$$

where 'o' indicates the tensor or direct matrix product. The algorithms for each value of k are represented by the corresponding matrices, $\hat{A}_k$.

By simple examination, one can see that none of the partial algorithms corresponds to a matrix-matrix multiplication. However, one can see that each A_k can be derived from the previous one by adding non-zero entries. The added non-zero entries describe the multiplications executed on the orbital array during the k-th cycle.

From a linear algebraic point of view, as the coefficient, a_{ij}, take on all possible values, the matrix, AoI, describing the algorithm, defines a linear subspace in the space of $n^2 \times n^2$ matrices. The partial approximation, A_k, on the other hand, are contained in disjoint proper subspaces.

The situation described in the example is characteristic of systolic array operation, and can be modeled with a great degree of generality. The model assumes that the <u>implementable transformations</u> can be given the representation

$$A = \sum a_\ell E_\ell .$$

The coefficients, a_ℓ, and the basic matrices, E_ℓ, completely define the transformations. The basic matrices have entries in $\{-1,0,1\}$. The number of nonzero entries indicates the number of times the respective coefficient is used by the array. The location of the nonzero entries determines which input component uses that particular value. The matrices are completely determined by the interconnection among processing elements. We shall further assume that the basic matrices, E_ℓ, are mutually orthogonal. Hence

$$tr\{E_\ell E_s^*\} = 0, \quad all \; s \neq \ell .$$

This condition holds for the traditional systolic arrays. For our purposes it is a svmplification that could easily be removed.

The model enables a simple generation of implementable linear transformations. For this, let G be an arbitrary linear transformation, and let

$$\gamma_\alpha = \frac{tr\{GE_\alpha^*\}}{tr\{E_\alpha E_\alpha^*\}} .$$

We can easily prove:

Lemma 1: Given the transformation, G, the linear transformation $\Psi(G)$ defined by

$$\Psi(G) = \sum \gamma_\alpha E_\alpha$$

satisfies:

a) $\Psi(G)$ is implementable in the given array;

b) $tr\{G - \Psi(G)]E_\alpha^*\} = 0$; all α.

<u>Remark</u>: The lemma establishes that $\Psi(G)$ is the closest implementable transformation in the mean square sense. In fact, $\Psi(G)$ is the orthogonal projection of G on the subspace of implementable transformations. This result will be of use in the analysis of complexity constraints.

<u>Remark</u>: It is apparent that a similar analysis can be applied after each computational cycle. Thus, the collection of transformations can be modeled as the range of a suitable orthogonal projector.

For the example of the orbital array, if A is the matrix representation of the linear algorithm and A_k the approximation after k cycles, it follows that

$$A_k = \xi_k A. \tag{1}$$

Here, ξ_k represents the linear transformation necessary to determine A_k. It must be noted that ξ_k is uniquely determined by the configuration of the systolic array.

The formulation of the time constrained operation can now be stated in a general form. For this, let X denote the data space and x∈X be a vector of data points to be processed using the linear filter $A:X{\rightarrow}X$. Clearly A∈L(X), the linear space of linear operators on X. If the operation, A, is represented by its matrix form, the example shows that in general the collection of filters, implemented in the array, form a proper subspace, V, in L(X). In the same form, the collection of possible linear transformations implemented after k cycles will form a (different) subspace, V_k. The operation, ξ_k, is interpreted as the orthogonal projector, in L(X), with range V_k. In the literature, ξ_k, is often called a transformator.

<u>Remark</u>: For the case under study, all subspaces are disjoint. In general, however, two, or more, of them may have nonempty intersection. In these cases, it is possible to find particular realizations of an algorithm which can be exactly evaluated in less time than the general case. This result has important implications for the design of fast architectures for specific cases.

The design of a controller with a time constrained specification can be carried out in a straightforward manner. If the linear controller is implemented in a given structure, such as an array, the operation after k cycles will be of the form, $\xi_k A$, where ξ_k is the linear transformator determined by the computing structure and A is a linear operator in a suitable subspace V L(X).

The fact that the collection of time constrained operators form a linear space makes the determination of the optimal solution particularly simple. If the unconstrained operator, A, belongs to the subspace V L(X), we can define the orthogonal projector, ψ, with range, V, in such a way that for any arbitrary linear transformation, N, the transformation

$$A = \psi(N)$$

can be implemented in the computing array. Furthermore, the linear transformation

$$A_k = \xi_k \psi(N)$$

is implemented in k cycles. The range of the projector $\xi_k \psi$ defines the

By simple examination, one can see that none of the partial algorithms corresponds to a matrix-matrix multiplication. However, one can see that each A_k can be derived from the previous one by adding non-zero entries. The added non-zero entries describe the multiplications executed on the orbital array during the k-th cycle.

From a linear algebraic point of view, as the coefficient, a_{ij}, take on all possible values, the matrix, AoI, describing the algorithm, defines a linear subspace in the space of $n^2 \times n^2$ matrices. The partial approximation, A_k, on the other hand, are contained in disjoint proper subspaces.

The situation described in the example is characteristic of systolic array operation, and can be modeled with a great degree of generality. The model assumes that the _implementable transformations_ can be given the representation

$$A = \sum a_\ell E_\ell.$$

The coefficients, a_ℓ, and the basic matrices, E_ℓ, completely define the transformations. The basic matrices have entries in $\{-1,0,1\}$. The number of nonzero entries indicates the number of times the respective coefficient is used by the array. The location of the nonzero entries determines which input component uses that particular value. The matrices are completely determined by the interconnection among processing elements. We shall further assume that the basic matrices, E_ℓ, are mutually orthogonal. Hence

$$tr\{E_\ell E_s^*\} = 0, \quad all \; s \neq \ell.$$

This condition holds for the traditional systolic arrays. For our purposes it is a svmplification that could easily be removed.

The model enables a simple generation of implementable linear transformations. For this, let G be an arbitrary linear transformation, and let

$$\gamma_\alpha = \frac{tr\{GE_\alpha^*\}}{tr\{E_\alpha E_\alpha^*\}}.$$

We can easily prove:

Lemma 1: Given the transformation, G, the linear transformation $\Psi(G)$ defined by

$$\Psi(G) = \sum \gamma_\alpha E_\alpha$$

satisfies:

a) $\Psi(G)$ is implementable in the given array;

b) $tr[G - \Psi(G)]E_\alpha^*\} = 0$;   all α.

Remark: The lemma establishes that $\Psi(G)$ is the closest implementable transformation in the mean square sense. In fact, $\Psi(G)$ is the orthogonal projection of G on the subspace of implementable transformations. This result will be of use in the analysis of complexity constraints.

collection of admissible time constrained linear filters.   In this case, the necessary and sufficient condition for optimality, stated in Result 1, becomes

$$A_k \in W_k$$

$$Q_L A_k Q_v - L_3^* L_2 Q_{av} \perp W_k.$$

Here, W_k, denotes the range space of the transformator, $\xi_k \psi$.   If ζ_k is the orthogonal projector with range W_k, we can write

$$\zeta_k (Q_L \zeta_k (N) Q_v - L_3^* L_2 Q_{av}) = 0.$$

This equation is solvable using standard projection techniques [10].

Remark: the approach proposed here allows, easily, the inclusion of additional constraints in the controller design.

The time constrained case that has been discussed leads to families of algorithms which form linear subspaces.   This condition makes the solution of the optimization problem particularly simple.   Unfortunately, other types of representations lead to more complex situations.   Because of the similarity of the problems, this more complex case will be discussed in terms of structural constraints.

3. Structure Constrained Controllers: We consider first, for perspective, an elementary special case.   For this let y = Tx denote a linear operation.   Assume for example that the input vector x (respectively output y) can be packed as a two dimensional array of mxn elements (N = mxn) and that the filtering operation can be described as the matrix operation

$$Y = F X \tag{2}$$

with F an mxm matrix.   The output signal 'Y' can be computed at a speed proportional to m.   If m=n the increase in speed of the stacked model is in the order of $\sqrt{N}$.

The increase in speed is available because of the particular, and restricted, form of the hypothetical filter.   Indeed as an operator in X the filter T must be representable by the block diagonal NxN matrix

$$T = diag(F).$$

While block diagonality is a highly specialized form it may on occasion be available.   We note also that the collection of matrices of the form, T = diag(F), form a linear supspace in the space of operators on X.   A least square approximation to general linear maps can be accomplished with standard orthogonal projections [10].   Similar observations and techniques are available for approximations of the form Y = FX + XG.

Our attention turns now to FFT type realizations and/or approximations of the form Y = LXR.   The efficiency of the triple matrix product architecture is best illustrated by a simple example. Let x be an N-dimensional vector with $N = m^2$. Let X be the mxm matrix formed by mapping the vector on a two dimensional array using segments of x as columns in X. Consider the linear operations

$$y = Ax, \qquad Y = LXR.$$

The evaluation of Ax requires m^4 multiplications while that of LXR requires only $2m^3$ multiplications. Using a special systolic array implementation [4] the triple product can be computed in only 3m time without intermediate I/O operations. The advantages of the triple product representation are evident both for sequential computers and specialized architectures.

Tensor notation is useful to describe the transformations of interest here. If L and $R = [r_{kj}]$ are respectively mxm and nxn matrices then LoR denotes the mnxmn block partitioned matrix with $[LxR]_{ik} = r_{ik}L$. The matrix LoR defines a linear transformation in E^{mn}. Our next result relates tensor product to bilinear representations. The proof is direct and will be omitted.

Lemma 2: The linear transformation

$$y = [LoR]x$$

is equivalent to the transformation

$$Y = LXR^t$$

provided that y,x are column by column orderings of the matrices Y,X respectively and R^t is the transpose of R. Therefore, the class of systems that can be represented by a matrix triple product corresponds to the class of matrices which can be expressed as a tensor product.

The tensor product approximation offers a completely different situation to the one described in the time constrained case. Here the structure constrained controller is not determined by a linear transformator. The problem is still solvable however. For illustration purposes we give below a particular case where a complete solution is known.

3.1 The Best Array Approximation: Given a linear filter, T, we want to find the best, in some sense, approximation of the form LoR. Since the linear maps describing the filters can be provided with a suitable inner product, using the trace function, we shall define

$$J(L,R) = \|T - LoR\|^2$$
$$= tr\{(T-LoR)(T-LoR)^*\}. \qquad (3)$$

For given L_a the minimization with respect to R can be accomplished using standard variational techniques [10]. We have

$$J(L_a, R+\delta R) - J(L_a, R) = tr\{(T-L_a oR)(-L_a o\delta R)^* + (-L_a o\delta R)(T-LoR)^* +$$
$$(L_a o\delta R)(L_a o\delta R)^*\}$$

Hence the necessary and sufficient condition for optimality is given by

$$tr\{(T-L_a oR)(L_a o\delta R)^*\} = 0 \qquad all \ \delta R.$$

This orthogonality condition can be solved directly. Writing the matrices in block partitioned form we have

$$tr\{(T-L_aoR)(L_ao\delta R)^*\} = tr\{\Sigma_i\Sigma_k(T_{ik}-L_ar_{ki})(L_a^*\delta r_{ki}^*)\}$$

Since δr_{ik} can be completely arbitrary it follows that each trace in the double summation must be zero. Hence

$$r_{ik} \ tr\{L_aL_a^*\} = tr\{T_{ik}L_a^*\}.$$

For any given nonzero L_a one can write

$$r_{ik} = \frac{tr\{T_{ik} \ L_a^*\}}{tr\{L_aL_a^* \}} = \frac{<T_{ik},L_a>}{\|L_a\|^2} . \qquad (4)$$

Using again the orthogonality condition in Eq. (3)

$$J(L_a,R(L_a)) = tr\{TT^* - (L_aoR(L_a)(L_aoR(L_a)^*\}$$

$$= tr\{TT^* - L_aL_a^* \ o \ R(L_a)R^*(L_a)\}$$

where the elements of $R(L_a)$ are given in Eq(4). Carrying out the indicated tensor products one can show

$$J(L_a,R(L_a)) = tr\{TT^*\} - tr\{L_aL_a^*\} \ tr\{R(L_a)R^*(L_a)\}.$$

Since $tr\{RR^*\} = \|R\|^2 = \Sigma_{ki}|r_{ki}|^2$, using Eq. (4)

$$J(L_a,R(L_a)) = tr\{TT^* - \Sigma_{ki} \frac{|<T_{ik},L_a>|^2}{\|L_a\|^2}\} . \qquad (5)$$

Minimization of $J(L_a,R(L_a))$ is equivalent to the maximization of the double summation in Eq (5).

The summation is actually homogeneous in L_a. The maximization can be carried out on the unit ball $\|L_a\| = 1$. With an abuse of notation we use the same symbol (T_{ik},L_a) to denote the matrix or the vector obtained by stacking the columns of the corresponding matrix. With this convention we have

$$\Sigma_{ki} \frac{|<T_{ik},L_a>|^2}{\|L_a\|^2} = \frac{L_a^*(\Sigma_{ik} \ T_{ik} \ T_{ik}^*)L_a}{\|L_a\|^2} .$$

Hence the optimal $\hat{L}$ satisfies

$$(\Sigma_{ik}T_{ik}T_{ik}^*)\hat{L} = \mu\hat{L}$$

where μ is the maximum eigenvalue of the matrix $\Sigma_{ik}T_{ik}T_{ik}^*$. The minimum value for the cost function will be given by $J(\hat{L},\hat{R}) = tr\{TT^*\}-\mu$.

The solution of this example depends on the finite dimensional nature of the operators involved. From a practical point of view this is not a severe constraint. This case is a very special case of the general optimization problem. The solution of the general case is available for the finite dimensional is also available. We give below the outline of the procedure.

The cost function is of the form

$$J(A) = tr\{[Q_L AQ_v - L_3^* L_2 Q_{av}]A^* + L_2 Q_\alpha L_2^* - L_2 L_3 AQ_{av}\};$$

where $A = L \circ R$. The variational technique establishes (Result 1) that

$$tr\{Q_L AQ_v - L_3^* L_2 Q_{av}]\delta A^*\} \geq 0.$$

The collection of admissible variations <u>is not</u> a subspace and no immediate orthogonality condition can be derived. However, if we consider the special cases

$$\delta A_1 = L \circ \delta R,$$

$$\delta A_r = \delta L \circ R,$$

we can obtain necessary conditions for optimality. The resulting set of equations can be solved by numerical techniques.

<u>Conclusion</u>: The paper has developed a systematic approach to improve computational efficiency of linear controllers. The improvement is obtained by directly constraining the execution time or by constraining the structure of the system. Both alternatives can be approached with similar techniques. In either case, the solution is derived using optimality criteria. The generalized statement of the problem makes the results easily applicable to filtering and prediction problems.

Acknowledgment

This research is supported in part by the State of Louisiana under Research Contract LEQSF-A-17.

References

[1] Kung, H.T., "Why systolic architectures?", <u>Computer</u>, Vol. 15, January 1982, pp. 37-46.

[2] Wold, E.H., A.M. Despain, "Pipeline and parallel-pipeline FFT processors for VLSI implementation", <u>IEEE Trans. Comp.</u>, Vol. C-33, No. 5, May 1984.

[3] Aravena, J.L., W.A. Porter, "Nonplanar Switchable Array", <u>J. Circ. Syst. Sig. Proc.</u>, Vol. 7, No. 1, 1988.

[4] Aravena, J.L., "Triple Matrix Product Architectures for Fast Signal Processing", <u>IEEE Trans.</u>, Vol. CAS-35, No. 1, 1988.

[5] Kung, H. T. , L. M. Ruane, D. W. L. Yen, "A Two Level Pipelined Systolic Array for Convolutions", <u>VLSI Systems and Computations</u>, pp. 255-264, Computer Science Dept. Carnegie-Mellon, 1981.

[6] Liu, K. Y., "A Pipelined Tree Machine Architecture for Computing A Mulidimensional Convolution", _IEEE Trans. CAS, vol._ C-33, No. 5, May 1984.

[7] Zhang, Ch., Yun, D., "Multidimensional Systolic Networks for Discrete Fourier Transforms", _Proc. ICCD_, pp. 215-222, 1984.

[8] Owens, R. M., Ja' Ja', J. "A VLSI Chip for the Winograd Prime Factor Algorithm

[9] DeSantis, R. M., Saeks, R. and Tung, L., "Basic Optimal Estimation and Control Problems in Hilbert Space", Math. Syst. Theory, Vol. 12, 1978.

[10] Aravena, J. L., and Porter, W. A., "Multidimensional Aperture Control", _IEEE Trans. Aut. Contr._, Vol AC-29, No. 12, pp. 1114-1118, December 1984.

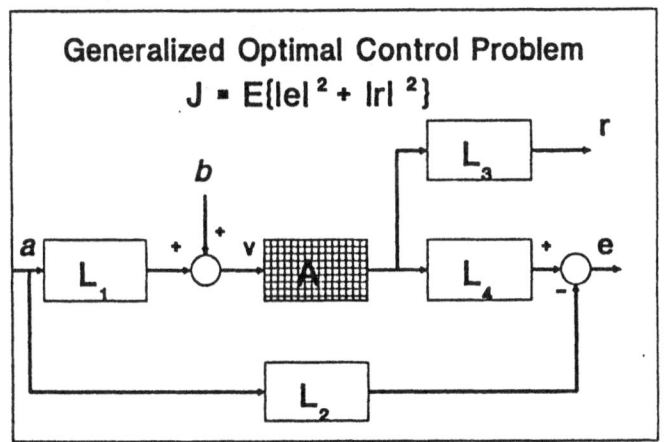

Generalized Optimal Control Problem

$$J = E\{|e|^2 + |r|^2\}$$

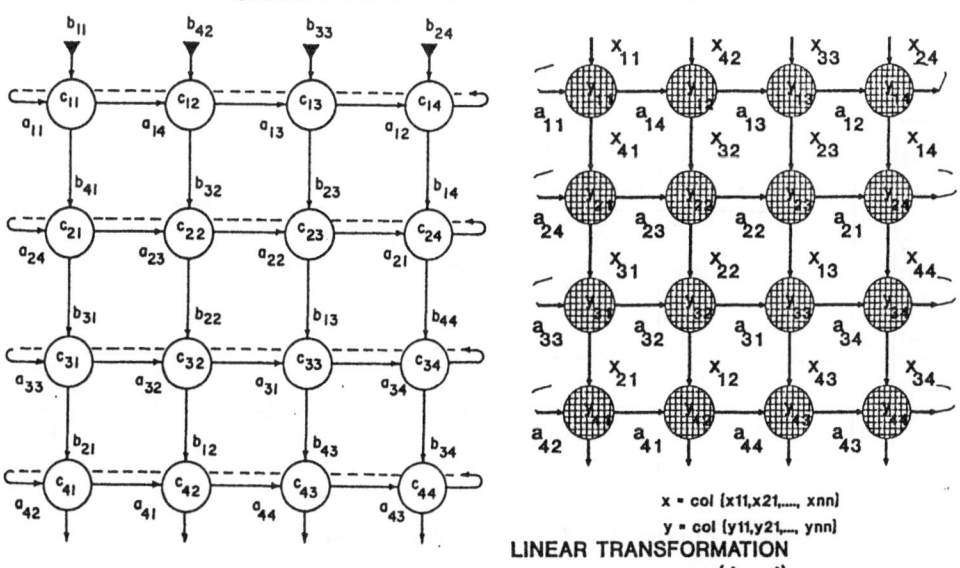

ORBITAL ARRAY FOR MATRIX MULTIPLICATION

$x = \text{col } [x11, x21, ..., xnn]$

$y = \text{col } [y11, y21, ..., ynn]$

LINEAR TRANSFORMATION

$$y = (A \circ I)x$$

Transformation for k = 1

$$
\begin{bmatrix} Y_{11} \\ Y_{21} \\ Y_{31} \\ Y_{41} \\ Y_{12} \\ Y_{22} \\ Y_{32} \\ Y_{42} \\ Y_{13} \\ Y_{23} \\ Y_{33} \\ Y_{43} \\ Y_{14} \\ Y_{24} \\ Y_{34} \\ Y_{44} \end{bmatrix}
=
\begin{bmatrix}
a_{11} & 0 & 0 & 0 & & & & & & & & & & & & \\
0 & a_{22} & 0 & 0 & & & 0 & & & & 0 & & & & 0 & \\
0 & 0 & a_{33} & 0 & & & & & & & & & & & & \\
0 & 0 & 0 & a_{44} & & & & & & & & & & & & \\
& & & & 0 & a_{12} & 0 & 0 & & & & & & & & \\
& & 0 & & 0 & 0 & a_{23} & 0 & & & 0 & & & & 0 & \\
& & & & 0 & 0 & 0 & a_{34} & & & & & & & & \\
& & & & a_{41} & 0 & 0 & 0 & & & & & & & & \\
& & & & & & & & 0 & 0 & a_{13} & 0 & & & & \\
& & 0 & & & & 0 & & 0 & 0 & 0 & a_{24} & & & 0 & \\
& & & & & & & & a_{31} & 0 & 0 & 0 & & & & \\
& & & & & & & & 0 & a_{42} & 0 & 0 & & & & \\
& & & & & & & & & & & & 0 & 0 & 0 & a_{14} \\
& & 0 & & & & 0 & & & & 0 & & a_{21} & 0 & 0 & 0 \\
& & & & & & & & & & & & 0 & a_{32} & 0 & 0 \\
& & & & & & & & & & & & 0 & 0 & a_{43} & 0
\end{bmatrix}
\begin{bmatrix} X_{11} \\ X_{21} \\ X_{31} \\ X_{41} \\ X_{12} \\ X_{22} \\ X_{32} \\ X_{42} \\ X_{13} \\ X_{23} \\ X_{33} \\ X_{43} \\ X_{14} \\ X_{24} \\ X_{34} \\ X_{44} \end{bmatrix}
$$

Transformation for k = 2

$$
\begin{bmatrix} Y_{11} \\ Y_{21} \\ Y_{31} \\ Y_{41} \\ Y_{12} \\ Y_{22} \\ Y_{32} \\ Y_{42} \\ Y_{13} \\ Y_{23} \\ Y_{33} \\ Y_{43} \\ Y_{14} \\ Y_{24} \\ Y_{34} \\ Y_{44} \end{bmatrix}
=
\begin{bmatrix}
a_{11} & 0 & 0 & a_{14} & & & & & & & & & & & & \\
a_{21} & a_{22} & 0 & 0 & & & 0 & & & & 0 & & & & 0 & \\
0 & a_{32} & a_{33} & 0 & & & & & & & & & & & & \\
0 & 0 & a_{43} & a_{44} & & & & & & & & & & & & \\
& & & & a_{11} & a_{12} & 0 & 0 & & & & & & & & \\
& & 0 & & 0 & a_{22} & a_{23} & 0 & & & 0 & & & & 0 & \\
& & & & 0 & 0 & a_{33} & a_{34} & & & & & & & & \\
& & & & a_{41} & 0 & 0 & a_{44} & & & & & & & & \\
& & & & & & & & 0 & a_{12} & a_{13} & 0 & & & & \\
& & 0 & & & & 0 & & 0 & 0 & a_{23} & a_{24} & & & 0 & \\
& & & & & & & & a_{31} & 0 & 0 & a_{34} & & & & \\
& & & & & & & & a_{41} & a_{42} & 0 & 0 & & & & \\
& & & & & & & & & & & & 0 & 0 & a_{13} & a_{14} \\
& & 0 & & & & 0 & & & & 0 & & a_{21} & 0 & 0 & a_{24} \\
& & & & & & & & & & & & 0_{31} & a_{32} & 0 & 0 \\
& & & & & & & & & & & & 0 & a_{42} & a_{43} & 0
\end{bmatrix}
\begin{bmatrix} X_{11} \\ X_{21} \\ X_{31} \\ X_{41} \\ X_{12} \\ X_{22} \\ X_{32} \\ X_{42} \\ X_{13} \\ X_{23} \\ X_{33} \\ X_{43} \\ X_{14} \\ X_{24} \\ X_{34} \\ X_{44} \end{bmatrix}
$$

Figure 3. Partial Transformations for Y = AX.

Transformation for k = 3

$$
\begin{bmatrix}
Y_{11} \\ Y_{21} \\ Y_{31} \\ Y_{41} \\
Y_{12} \\ Y_{22} \\ Y_{32} \\ Y_{42} \\
Y_{13} \\ Y_{23} \\ Y_{33} \\ Y_{43} \\
Y_{14} \\ Y_{24} \\ Y_{34} \\ Y_{44}
\end{bmatrix}
=
\begin{bmatrix}
\begin{matrix} a_{11} & 0 & a_{13} & a_{14} \\ a_{21} & a_{22} & 0 & a_{24} \\ a_{31} & a_{32} & a_{33} & 0 \\ 0 & a_{42} & a_{43} & a_{44} \end{matrix} & 0 & 0 & 0 \\
0 & \begin{matrix} a_{11} & a_{12} & 0 & a_{14} \\ a_{21} & a_{22} & a_{23} & 0 \\ 0 & a_{32} & a_{33} & a_{34} \\ a_{41} & 0 & a_{43} & a_{44} \end{matrix} & 0 & 0 \\
0 & 0 & \begin{matrix} a_{11} & a_{12} & a_{13} & 0 \\ 0 & a_{22} & a_{23} & a_{24} \\ a_{31} & 0 & a_{33} & a_{34} \\ a_{41} & a_{42} & 0 & a_{44} \end{matrix} & 0 \\
0 & 0 & 0 & \begin{matrix} 0 & a_{12} & a_{13} & a_{14} \\ a_{21} & 0 & a_{23} & a_{24} \\ a_{31} & a_{32} & 0 & a_{34} \\ a_{41} & a_{42} & a_{43} & 0 \end{matrix}
\end{bmatrix}
\begin{bmatrix}
X_{11} \\ X_{21} \\ X_{31} \\ X_{41} \\
X_{12} \\ X_{22} \\ X_{32} \\ X_{42} \\
X_{13} \\ X_{23} \\ X_{33} \\ X_{43} \\
X_{14} \\ X_{24} \\ X_{34} \\ X_{44}
\end{bmatrix}
$$

Exact Transformation for k = n (n = 4)

$$
\begin{bmatrix}
Y_{11} \\ Y_{21} \\ Y_{31} \\ Y_{41} \\
Y_{12} \\ Y_{22} \\ Y_{32} \\ Y_{42} \\
Y_{13} \\ Y_{23} \\ Y_{33} \\ Y_{43} \\
Y_{14} \\ Y_{24} \\ Y_{34} \\ Y_{44}
\end{bmatrix}
=
\begin{bmatrix}
\begin{matrix} a_{11} & a_{12} & a_{13} & a_{14} \\ a_{21} & a_{22} & a_{23} & a_{24} \\ a_{31} & a_{32} & a_{33} & a_{34} \\ a_{41} & a_{42} & a_{43} & a_{44} \end{matrix} & 0 & 0 & 0 \\
0 & \begin{matrix} a_{11} & a_{12} & a_{13} & a_{14} \\ a_{21} & a_{22} & a_{23} & a_{24} \\ a_{31} & a_{32} & a_{33} & a_{34} \\ a_{41} & a_{42} & a_{43} & a_{44} \end{matrix} & 0 & 0 \\
0 & 0 & \begin{matrix} a_{11} & a_{12} & a_{13} & a_{14} \\ a_{21} & a_{22} & a_{23} & a_{24} \\ a_{31} & a_{32} & a_{33} & a_{34} \\ a_{41} & a_{42} & a_{43} & a_{44} \end{matrix} & 0 \\
0 & 0 & 0 & \begin{matrix} a_{11} & a_{12} & a_{13} & a_{14} \\ a_{21} & a_{22} & a_{23} & a_{24} \\ a_{31} & a_{32} & a_{33} & a_{34} \\ a_{41} & a_{42} & a_{43} & a_{44} \end{matrix}
\end{bmatrix}
\begin{bmatrix}
X_{11} \\ X_{21} \\ X_{31} \\ X_{41} \\
X_{12} \\ X_{22} \\ X_{32} \\ X_{42} \\
X_{13} \\ X_{23} \\ X_{33} \\ X_{43} \\
X_{14} \\ X_{24} \\ X_{34} \\ X_{44}
\end{bmatrix}
$$

Figure 3. (Continued)

NONLINEAR STABILIZING CONTROL
OF HIGH ANGLE OF ATTACK FLIGHT DYNAMICS

Eyad H. Abed

Department of Electrical Engineering
and the Systems Research Center
University of Maryland
College Park, MD 20742

1. Introduction. This paper discusses a new approach to the feedback control of aircraft at high angles of attack. This approach is based on recent results on control of nonlinear systems at bifurcation points [3, 4, 5] . The lateral dynamics of a slender aircraft studied by Ross [6] provides a convenient model for illustrating this application. It is shown how local bifurcation control can be used in the stabilization of the so-called "wing rock" phenomenon. The appropriateness of this approach derives from the fact that unstable motions at high incidence, such as stall, spin entry, wing rock and roll coupling, have been attributed to bifurcations in the vehicle dynamic equations. The instability can indeed be predicted via bifurcation analysis, and depends critically on the nonlinear aerodynamic force modeling. Bifurcation control results can facilitate the synthesis of nonlinear stabilizing feedback controls for aircraft at the onset of vortex separation; an angle of attack limiter is not employed in this approach. Possible means of achieving the required control laws include the control surface deflections and thrust vector control.

The bifurcation control technique considered here, developed in [3, 4] , provides an analytically-based, algorithmic approach to the stabilization of vehicles at the onset of flow separation. In classical terminology, the goal of the nonlinear control design is to render the stability boundary at high incidence "safe" (as opposed to "dangerous"). Practically the result would be to enlarge the flight envelope, by ensuring that any near-stall sustained deviation from trim is stable and of small amplitude, and therefore tolerable. Divergence cannot result from such a state without the introduction of a large disturbance. Thus, pre- and post-stall dynamics are stabilized.

The remainder of this paper is organized as follows. Section 2 presents a brief discussion of relevant background material on high angle of attack dynamic modeling and associated nonlinear effects. In Section 3, the bifurcation control framework of [3, 4] is reviewed. In Section 4, application of the bifurcation control technique to a model for the high incidence dynamics of the Handley Page 115 research aircraft model of Ross [6] is presented.

2. High Angle of Attack Flight Dynamics. Analytical study of rigid body flight at high angle of attack (α) is made difficult by essentially three main factors:

(i) *Strong dynamic coupling of longitudinal and lateral-directional motions* almost always necessitates the use of a six degree-of-freedom model, with state variables representing angular orientation and rates, as well as translational position and velocity;

(ii) *Nonlinearity of the dynamic motion equations.* Since the trim condition borders on instability in the high α regime, linearized equations of motion are of little use in predicting dynamic behavior; and

(iii) *Nonlinear dependence of aerodynamic forces on states.* Vortex shedding and flow separation complicate aerodynamic modeling in this flight regime, especially with regard to wing-body and tail-body interference effects. In the supersonic range, time-history effects also become important.

Although nonlinear aerodynamic modeling is difficult and highly configuration-dependent, the dynamics of rigid body flight are well understood. Viewing an aircraft as a rigid body of fixed mass, its motion is completely determined (once the aerodynamic forces are known) by the following six equations, which are expressed in the basic Euler form with respect to body axes:

$$\dot{u} = \frac{f_x}{m} + \frac{X}{m} + vr - wq \tag{2.1}$$

$$\dot{v} = \frac{f_y}{m} + \frac{Y}{m} + wp - ur \tag{2.2}$$

$$\dot{w} = \frac{f_z}{m} + \frac{Z}{m} + qu - vp \tag{2.3}$$

$$\dot{p} = \frac{L}{I_x} - (\frac{I_z - I_y}{I_x})qr + \frac{I_{xx}}{I_x}(\dot{r} + pq) + \frac{I_{xy}}{I_x}(\dot{q} - rp) + \frac{I_{yz}}{I_x}(q^2 - r^2) \tag{2.4}$$

$$\dot{q} = \frac{M}{I_y} + (\frac{I_x - I_z}{I_y})pr - \frac{I_{xx}}{I_y}(p^2 - r^2) + \frac{I_{xy}}{I_y}(\dot{p} + qr) + \frac{I_{yz}}{I_y}(\dot{r} - pq) \tag{2.5}$$

$$\dot{r} = \frac{N}{I_z} - (\frac{I_y - I_x}{I_z})qp + \frac{I_{xx}}{I_z}(\dot{p} - qr) + \frac{I_{xy}}{I_z}(p^2 - q^2) + \frac{I_{yz}}{I_z}(\dot{q} + rp). \tag{2.6}$$

Here u, v, w are the velocity components of the center of gravity along the x, y and z axes, respectively; p, q, r are the components of the angular velocity about these axes; m is the mass of the aircraft; f_x, f_y and f_z are the Eulerian components of the instantaneous gravity force; X, Y and Z are the total aerodynamic forces; and L, M, N are the total moments with respect to the standard (body) axes. The notation I_x, I_{xy}, etc. refers to the moments and products of inertia relative to the center of mass.

If the mass distribution of the aircraft is symmetrical with respect to, say, the xz plane, then $I_{xy} = I_{yz} = 0$ and the last two terms in each of the equations (2.4)-(2.6) vanish. This is a very good approximation for most aircraft in steady flight, but may not be appropriate if, for example, a significant amount of fuel motion occurs in a tank ("fuel sloshing").

The aerodynamic forces and moments acting on the aircraft are expressed in terms of the familiar dimensionless aerodynamic coefficients:

$$X = (1/2)\, C_s\, \rho V^2 S,\ \ Y = (1/2)\, C_y\, \rho V^2 S,\ \ Z = (1/2)\, C_s\, \rho V^2 S, \tag{2.7a}$$

$$L = (1/2)\, C_l\, \rho V^2 Sb,\ \ M = (1/2)\, C_m\, \rho V^2 Sc,\ \ N = (1/2)\, C_n\, \rho V^2 Sb; \tag{2.7b}$$

with C_s, C_l, etc. being the dimensionless coefficients; ρ the atmospheric density; S the reference area; c the mean aerodynamic chord; b the wing span; and V the magnitude of the velocity of the center of gravity. The aerodynamic coefficients C_s, C_y, etc. above are not constants but depend on the orientation and angular velocity of the vehicle. Accurate analytical evaluation of these coefficients is difficult and data from wind tunnel tests is usually relied upon (when available). Note that the effect of control surface deflections is incorporated into the model through the aerodynamic coefficients. The control surface deflections correspond to the values of the aileron, rudder and elevator angles δ_a, δ_r and δ_e, respectively. The thrust may also be viewed as a control variable, such as in thrust vector control.

Equations (2.1)-(2.3) may be recast in terms of the angle of attack α, the sideslip angle β and the velocity magnitude V, where

$$\alpha = \tan^{-1}\frac{w}{u},\ \beta = \sin^{-1}\frac{v}{V},\ \text{and}\ V = (u^2 + v^2 + w^2)^{1/2}\,. \tag{2.8}$$

This is facilitated by expressing the aerodynamic forces X, Y, Z as

$$X = L\sin\alpha - D\cos\beta\cos\alpha$$

$$Y = D\sin\beta \tag{2.9}$$

$$Z = -L\cos\alpha - D\cos\beta\sin\alpha$$

where L and D are the net lift and drag components of the aerodynamic force:

$$L = (1/2)\, C_L\, \rho V^2 S = \text{lift}$$

$$D = (1/2)\, C_D\, \rho V^2 S = \text{drag}$$

The transformed equations (for a symmetrical body) may be found in Adams [7] .

It appears that the first analytical study of bifurcation or jump phenomena in high α flight dynamics was due to Schy and Hannah [8] although predictions of the utility of the approach had previously been made (Phillips [9] , Pinsker [10] , Rhoads and Schuler [11]). This analysis benefited from the report of Adams [7] which used constrained minimization techniques to solve the nonlinear algebraic equations for an equilibrium spin. In [8] it was shown how the so-called "pseudosteady-state solutions" could be obtained by solving a single polynomial in roll rate with coefficients that were functions of the control inputs. This was followed by the paper Young, Schy and Johnson [12] in which the inclusion of aerodynamics nonlinear in the angle of attack led to the solution of two polynomials in roll rate whose coefficients are functions of angle of attack and the control inputs. Analytical results were compared with calculated time histories to demonstrate the validity of the method for predicting jump-like instabilities. This work led Carroll and Mehra [13] to develop a methodology called BACTM ("Bifurcation Analysis and Catastrophe Theory Methodology") intended mainly for application to the aircraft high α stability problem. This methodology entails the use of results from bifurcation theory for ordinary differential equations, elementary catastrophe theory, the Global Implicit Function Theorem of Palais, and continuation methods to analyze bifurcations and to determine whether a given control destroys a bifurcation. In contrast to the work of Carroll and Mehra, the recent work of Hui and Tobak [14] focuses on unsteady flow effects at the expense of narrowing the scope of the motion analysis.

Bifurcation to periodic solutions was also detected by Carroll and Mehra [13] for the F-4 and the F-80A. Indeed, the appearance of this type of bifurcation in aircraft dynamics was demonstrated for the first time by Carroll and Mehra [13] . The various nonlinear phenomena associated with high α dynamics, such as roll coupling, low rudder effectiveness, reduced dihedral effect, stall, wing rock, nose slice, departure, post-stall gyration, spin, incipient spin, recovery roll and pre-stall buffeting, have been described in Mehra, Kessel and Carroll ([15] , pp. 46-50).

3. Bifurcation Control. Bifurcation theory studies the qualitative changes in the solution sets of equations as a parameter μ is (quasistatically) varied. The theory as it applies to systems of ordinary differential equations

$$\dot{x} = f_\mu(x) \tag{3.1}$$

where μ is a real parameter and x is a real n-vector is of interest here. There are two basic types of bifurcation from equilibrium for Eq. (3.1). The first of these is caused by changes in the number of solutions to the algebraic equation

$$f_\mu(x) = 0 \tag{3.2}$$

as the bifurcation parameter μ is varied. Since (3.2) describes the set of equilibrium (or stationary) points of the differential equation (3.1), this first type of bifurcation is referred to as a *static* (or *stationary*) *bifurcation*. The second type of bifurcation is involves in a nontrivial way the dynamics of the differential equation (3.1), and is hence referred to as a *dynamic bifurcation*. This usually takes the form of a *Hopf bifurcation*, involving appearance of periodic solutions from an equilibrium point as the parameter μ is varied. It is necessary to study both types of bifurcation to obtain a complete description of the dependence on μ of solutions of Eq. (3.1) near an equilibrium point.

The basic theorems on static and Hopf bifurcations for Eq. (3.1) are easily stated but the detailed statements require notation and technicalities which preclude their inclusion here. It suffices to summarize a few important facts: (i) The assumptions are easy to check and hold for generic nonlinear models (3.1); (ii) The theorems provide means for the extraction of stability information on the new solutions; (iii) Public domain computer software tools exist allowing interactive application of these bifurcation results; and (iv) The importance of bifurcations in high α flight dynamics is well-established, as indicated by the brief historical remarks above.

In the papers [3, 4, 5] a systematic and versatile approach for the synthesis of stabilizing controls for nonlinear systems undergoing bifurcations has been introduced. The technique allows one to *parametrize* the set of local nonlinear stabilizing feedback control laws for a control system

$$\dot{x} = f_\mu(x,u). \tag{3.3}$$

In (3.3), u is a control function and μ is, as before, a bifurcation parameter. The main assumption of the approach is simply that the 'uncontrolled' system (obtained by setting $u \equiv 0$ in (3.3))

$$\dot{x} = f_\mu(x,0) \tag{3.4}$$

undergoes a bifurcation as μ is varied. The most important benefit of the approach is an algorithmic design aid which generates, in a computationally efficient way, all the purely nonlinear feedback control laws $u = u(x)$ ensuring stability of the new ("bifurcated") solutions for the controlled system. In the case of a Hopf bifurcation, for instance, one can design feedbacks ensuring that for any value of the parameter μ in some domain around the critical value, the state converges either to the original equilibrium solution or to a nearby (stable) periodic solution of small (nearly zero) amplitude. Since the amplitude of the bifurcated periodic orbits is small for small μ, this type of (stable) oscillatory behavior may be tolerable from a practical point of view, and is certainly preferable to divergence or uncontrolled wing rock.

Our approach to the local feedback stabilization problem has the novel feature that it facilitates the derivation of generally valid analytical criteria for stabilizability, as well as specific stabilizing feedback controls. This is possible through use of bifurcation formulae which involve only Taylor series expansion of the vector field and eigenvector computations. To illustrate the generality of the approach and the simplicity of the associated calculations, a representative result is given next.

Consider a one-parameter family of nonlinear control systems (3.3) where $x \in R^n$, u is a scalar control, μ is a real-valued parameter, and the vector field f_μ is sufficiently smooth. Suppose that for $u \equiv 0$, Eq. (3.3) has an equilibrium point $x_0(\mu)$ which depends smoothly on μ. The linearization of Eq. (3.3) at $x = 0$, $u = \mu = 0$ is given by

$$\dot{x} = Ax + bu \tag{3.5}$$

where $A := \dfrac{\partial f}{\partial x}(0,0)$ and $b := \dfrac{\partial f}{\partial u}(0,0)$. The nature of the stabilizing nonlinear feedack controls given by the theory depends on the controllability properties of this linear system. Rewrite Eq. (3.3) in the series form

$$\dot{x} = L_0 x + u\gamma + uL_1 x + Q_0(x,x)$$

$$+ C_0(x,x,x) + \cdots \tag{3.6}$$

where the terms not written explicitly are of higher order in x, u and μ than those which are. Thus L_0 and L_1 are square matrices, γ is a constant vector, $Q_0(x,x)$ is a quadratic form generated by a symmetric bilinear form $Q_0(x,y)$ giving the second order (in x) terms at $u = 0$, $\mu = 0$, and $C_0(x,x,x)$ is a cubic form generated by a symmetric trilinear form $C_0(x,y,z)$ giving the third order (in x) terms at $u = 0$, $\mu = 0$. (Note that L_0 is simply A of Eq. (3.5), and γ corresponds to b.) Denote by r the right (column) and by l the left (row) eigenvector of L_0 with eigenvalue $i\omega_c$. Normalize by setting the first component of r to 1 and then choose l so that $lr = 1$.

Only the linear, quadratic and cubic terms in an applied feedback $u(x)$ have potential for influencing the stability. Stability may be checked by computing so-called *bifurcation coefficients*. Public domain software packages exist which perform the bifurcation analysis, as well as follow the bifurcation branches in one or two parameters. We mention, in particular, the BACTM software of Carroll and Mehra [13] , the package BIFOR2 of B. Hassard [16] , SUNY Buffalo, and the interactive package AUTO [17] authored by E. Doedel at Concordia University, Montreal. Bifurcation analysis software can be used, in conjunction with simulation tools, to obtain a global picture of the efficacy of a particular control law design.

Theorem. *Let the uncontrolled system ((3.3) with $u = 0$) undergo an unstable Hopf bifurcation to periodic solutions. Then there is a smooth feedback $u(x)$ with $u(0) = 0$ which solves the local Hopf bifurcation stabilization problem for Eq. (3.3) provided that*

$$0 \neq \mathrm{Re} \left\{ -2lQ_0(r, \tfrac{1}{2}L_0^{-1}\gamma) + lQ_0(\bar{r}, \tfrac{1}{2}(2i\omega_c I - L_0)^{-1}\gamma) \right.$$

$$\left. + \tfrac{1}{4}l[\, 2L_1 r + L_1\bar{r}\,] \right\}. \tag{3.7}$$

A whole family of conditions similar to Eq. (3.7) may be derived, each corresponding to a particular family of nonlinear stabilizing control laws $u(x)$. The results are available for both static and Hopf bifurcations [18] . Eq. (3.7) indicates the relative computational ease with which the results may be applied; only a Taylor series expansion of the dynamic equations and standard spectral calculations are needed.

4. Wing Rock Stabilization for the HP115. Nonlinear feedback control laws ensuring a "safe" stability boundary at high angle of attack may be designed using the bifurcation control results discussed in Section 3. Use of this approach has several advantages over most standard approaches to high angle of attack flight control: (i) Rather than avoiding stall through angle of attack limiters, the aim is to use feedback to permit stable flight in the stall and post-stall regime; (ii) All nonlinear stabilizing feedback controls can be parametrized, allowing consideration of other criteria besides local asymptotic stability, such as size of the domain of stability. (The theory provides a way to estimate the achieved domain of stability.); and (iii) Stall stabilization may well be preferable to deferring stall to higher angles of attack, since, as noted by Babister [2] , the latter often leads to a much more violent stall at a higher angle of attack.

To indicate the type of aircraft control problem which can benefit directly from the bifurcation control approach, we mention thrust vector control as a means for stabilizing aircraft longitudinal and/or lateral dynamics at high incidence. We note that in a recent paper [19] , Ashley has concluded that thrust vector control will be crucial in subsonic maneuvering at extreme angles of attack.

Consider the following model [6, 20] for the lateral dynamics of the HP115 aircraft at high incidence:

$$\dot{\beta} = \sin\alpha\, p\ -\ \cos\alpha\, r\ +\ \frac{g\cos\alpha}{V}\phi \tag{4.1a}$$

$$+\ \frac{\rho SV}{m}\{\, y_1\beta\ +\ y_3\beta^3\ +\ y_p\frac{ps}{V}\ +\ y_r\frac{rs}{V}\,\}$$

$$\dot{p} = \frac{\rho SV^2 sI_s}{I_s I_s - I_s^2}\{\, l_1\beta\ +\ l_3\beta^3\ +\ l_p\frac{ps}{V}\ +\ l_r\frac{rs}{V} \tag{4.1b}$$

$$+\ \frac{I_{ss}}{I_s}(\, n_1\beta\ +\ n_3\beta^3\ +\ n_p\frac{ps}{V}\ +\ n_r\frac{rs}{V}\,)\}$$

$$\dot{r} = \frac{\rho SV^2 sI_s}{I_s I_s - I_s^2}\{\, n_1\beta\ +\ n_3\beta^3\ +\ n_p\frac{ps}{V}\ +\ n_r\frac{rs}{V} \tag{4.1c}$$

$$+\ \frac{I_{ss}}{I_s}(\, l_1\beta\ +\ l_3\beta^3\ +\ l_p\frac{ps}{V}\ +\ l_r\frac{rs}{V}\,)\}$$

$$\dot{\phi} = p\ +\ \tan\alpha\, r \tag{4.1d}$$

Here, α is the (assumed constant) angle of attack, β and ϕ are the angles of sideslip and roll, respectively, and p and r are the rates of roll and yaw, respectively. The reader is referred to [6, 20] for definitions of the parameters appearing in (4.1), though we note that aerodynamic coefficients cubic in the sideslip angle β appear, with various parameters depending as well on the angle of attack α. Precise forms of these dependencies are given by Ross [6], which are derived by approximating flight test data. Specifically, for a particular configuration and flight condition, we have

$$y_p = 0.014 + 0.505\alpha - 0.47\alpha^2 \tag{4.2a}$$

$$l_p = -0.132 + 0.08\alpha \tag{4.2b}$$

$$n_p = 0.0125\alpha - 0.938\alpha^2 \tag{4.2c}$$

$$y_r = 0 \tag{4.2d}$$

$$l_r = 0.006 + 0.54\alpha \tag{4.2e}$$

$$n_r = -0.351 - 0.089\alpha \tag{4.2f}$$

Other (nonaerodynamic) parameters are set as follows [6] : $\rho = 0.906$ kg/m^3, $V = 71.25$ m/sec, $m = 2154$ kg, $s = 3.05$ m, $S = 40.18$ m^2, $I_{ss} = 2182$ kg m^2, $I_{ss} = 25430$ kg m^2, $I_{ss} = 1615$ kg m^2.

The parameters which remain to be specified in Eq. (4.1) are the linear and cubic coefficients in the aerodynamic moments in sideslip, roll and yaw. These are y_1, y_3, l_1, l_3, and n_1, n_3, respectively. The Hopf bifurcation analysis package BIFOR2 [16] was used to numerically investigate the influence of these parameters on the stability of a Hopf bifurcation in the dynamics. The results showed a strong dependence of the stability coefficient β_2 on the nonlinear yaw parameter n_3. The values of these parameters obtained from [6, 20] are: $y_1 = -0.191$, $y_3 = -1.958$, $l_1 = -0.184$, $l_3 = -0.0$, $n_1 = 0.07$, $n_3 = 2.65$. In particular, we note that from Figure 6 of [6] , the values of n_1 and n_3 are positive for the nominal configuration. Thus, the negative values for these parameters quoted in [20] do not appear to agree with the actual data. It is therefore not surprising that the calculations we have performed do not coincide exactly to those of [20] . Note also that we have followed [6] in assuming an air density value of $\rho = 0.906$, while a value of 0.5 is quoted in [20] .

For the nominal parameter values given in the foregoing, the program BIFOR2 computes a Hopf bifurcation from the origin for Eq. (4.1) with a critical angle of attack $\alpha_c = 0.295$ rad $= 16.9$ deg, in basic agreement with the references. The bifurcation is supercritical, with the bifurcation stability coefficient $\beta_2 = -4.1$. The calculations for the nominal system also reveal a difficulty with the assumed aerodynamic model: the eigenvalues cross the imaginary axis from the right half of the complex plane into the left half plane as α is increased through its critical value. For the nominal values above, BIFOR2 determines that the rate of change of the real part of these eigenvalues with respect to α is -0.937. The noncritical eigenvalues of (4.1) at the bifurcation point are found to be -0.64 and -0.05, which compare well with [20] when the differences in the parameter values used are taken into consideration.

In order to illustrate application of the bifurcation control results discussed in Section 3, we consider the effect of the parameter n_3 on stability of the Hopf bifurcation. Using a formula for β_2 given in [3] , we have determined that β_2 for this problem depends *linearly* on the coefficient n_3. This fact is easily obtained since only cubic nonlinearities in one component of the state (β) appear in (4.1). In particular, we find that *increasing* the value of n_3 from 2.65 results in a more negative stability coefficient β_2, and hence in better local stability properties. In contrast, decreasing n_3 increases β_2, possibly leading to a subcritical (unstable) bifurcation. For example, taking $n_3 = -10$ results in $\beta_2 = 2.56$, and hence an unstable bifurcation.

Acknowledgment. This work has been supported in part by the National Science Foundation's Engineering Research Centers Program: NSFD CDR-8803012, as well as by the NSF under Grant ECS-86-57561 and by the Air Force Office of Scientific Research under Grant AFOSR-87-0073.

References

[1]   T. Hacker, *Flight Stability and Control*, American Elsevier, New York, 1970.

[2]   A.W. Babister, *Aircraft Dynamic Stability and Response*, Pergamon Press, Oxford, 1980.

[3]   E.H. Abed and J.-H. Fu, "Local feedback stabilization and bifurcation control, I. Hopf bifurcation," *Systems and Control Letters* 7, pp. 11-17, 1986.

[4]   E.II. Abed and J.-H. Fu, "Local feedback stabilization and bifurcation control, II. Stationary bifurcation," *Systems and Control Letters* **8**, pp. 467-473, 1987.

[5]   E.II. Abed, "Local bifurcation control," pp. 224-241 in *Dynamical Systems Approaches to Nonlinear Problems in Systems and Circuits*, F.M.A. Salam and M. Levi, Eds., SIAM Publications, Philadelphia, 1988.

[6]   A.J. Ross, "Investigation of nonlinear motion experienced on a slender-wing research aircraft," *J. Aircraft* **9**, pp. 625-631, 1972.

[7]   W.M. Adams, Jr., "Analytic prediction of airplane equilibrium spin characteristics," NASA TN D-6926, 1972.

[8]   A.A. Schy and M.E. Hannah, "Prediction of jump phenomena in roll coupled maneuvers of airplanes," *J. Aircraft* **14**, pp. 375-382 (First reported in the Proceedings of the AIAA Third Atmospheric Flight Mechanics Conference, Arlington, Texas, June 1976.), 1977.

[9]   W.H. Phillips, "Effect of steady rolling on longitudinal stability," NASA TN 1627, 1948.

[10]   W.J.G. Pinsker, "Aileron control of small aspect ratio aircraft in particular, Delta aircraft," R and M 3188, NASA Ames Research Center, 1953.

[11]   D.W. Rhoads and J.M. Schuler, "A theoretical and experimental study of airplane dynamics in large-disturbance maneuvers," *J. Astronautical Sciences* **24**, pp. 507-526, 532, 1957.

[12]   J.W. Young, A.A. Schy, and K.G. Johnson, "Prediction of jump phenomena in aircraft maneuvers, including nonlinear aerodynamic effects," *J. Guidance and Control* **1**, pp. 26-31 (First reported as Paper 77-1126 in the Proceedings of the AIAA 4th Atmospheric Flight Mechanics Conference, Hollywood, Fla., 1977.), 1978.

[13]   J.V. Carroll and R.K. Mehra, "Bifurcation analysis of nonlinear aircraft dynamics," *J. Guidance* **5**, pp. 529-536, 1982.

[14]   W.II. Hui and M. Tobak, "Bifurcation analysis of aircraft pitching motions about large mean angles of attack," *J. Guidance* **7**, pp. 113-122, 1984.

[15]   R.K. Mehra, W.C. Kessell, and J.V. Carroll, "Global Stability and Control Analysis of Aircraft at High Angles-of-Attack," ONRCR-215-248-1, U.S. Office of Naval Research, Arlington, VA, June 1977.

[16]   B.D. Hassard, N.D. Kazarinoff, and Y.-II. Wan, *Theory and Applications of Hopf Bifurcation*, Cambridge Univ. Press, Cambridge, 1981.

[17]   E. Doedel, *AUTO User's Manual*, Computer Science Dept., Concordia Univ., Montreal, May 1984.

[18]   J.-H. Fu, *Bifurcation Control and Feedback Stabilization*, Ph.D. Dissertation, Dept. of Elec. Eng., Univ. of Maryland, College Park, 1988.

[19]   H. Ashley, "On the feasibility of low-speed fighter maneuvers involving extreme angles of attack," Tech. Rept., Dept. of Aeronautics and Astronautics, Stanford Univ., to be published.

[20]   J.E. Cochran, Jr. and C.-S. Ho, "Stability of aircraft motion in critical cases," *J. Guidance* 6, pp. 272-279, 1983.

CONTROLLABILITY OF LARGE-SCALE NONLINEAR SYSTEMS WITH DISTRIBUTED DELAYS IN CONTROL AND STATES

A. S. C. Sinha

Purdue University

INTRODUCTION

Many authors have investigated the problem of the stability of large-scale interconnected systems and a number of significant results have been obtained [1 - 4]. However, null controllability of large-scale systems with distributed delays and coefficients of bounded variations have not been considered.

We consider a large-scale system of functional differential equations of the form

$$\dot{x}(t) = L(t, x_t) + \int_{t_o-h}^{t_o} d_s[\eta(t,s)] \, u(t + s) + f(t, x_t, u_t) \tag{1}$$

$x(t) = \phi(t)$, t in $[t_o-h, t_o]$, where the integral is in Lebesque-Stieltjes sense with respect to the variable 's'. $h > 0$ represents the delay. L is the linear operator defined by

$$L(.) = A(t) \, x(t - h) + \int_{-\gamma}^{0} B(\theta) \, x(t + \theta) \, d\theta \tag{2}$$

The system (1) is decomposed as

$$L(.) = \text{diag}\{L_1(.), \ldots, L_m(.)\}; \quad (1 < m \le n)$$

consisting of blocks of linear operator of order $r_k \ge 1$ as

$$L_k(.) = A_k(t) \, x(t - h) + \int_{-\gamma}^{0} B_k(\theta) \, x_k(t + \theta) \, d\theta \tag{3}$$

The remaining blocks of the linear operator $L(.)$ may consist of zero for $k = 1, \ldots, m$; $r_1 + r_2 +, \ldots, + r_m = n$.

$$u_k(.) = \int_{t_o-h}^{t_o} d_s[\eta^k(t,s)] \, u_k(t + s) \tag{4}$$

The vector function

$$f(t, x_t, u_t) = \{f_1(t, x_t, u_t), \ldots, f_m(t, x_t, u_t)\}$$

is defined and continuous for $t > 0$ and for all vectors $x_t = (x_t^1, \ldots, x_t^m)$ in the n-dimensional vector space E^n over the field of real numbers, $x \in E^{r_k}$; $k = 1, \ldots, m$ respectively and it satisfies conditions for uniqueness and continuability for solution of (1) together with the restraints on growth of coupling

$$f_i : I \times B \left([t_0 -h, t], E^{r_i}\right) \times U_i \to E^{r_i} \tag{5}$$

The following notational conventions will be used throughout the sequel. The symbol $L_2([t_0-h, t_1], E^m) = U$ denotes the class of square integrable funtions u: $[t_0-h, t_1] \to E^m$ with values in the unit cube. Any $u \in U$ is called admissible.

In equation (1) each A is a continuous nxn matrix function. $\eta(t,s)$ is also an nxn matrix whose elements are square integrable on $(-\infty, 0]$. Let $\gamma \geq h \geq 0$ be given real numbers (γ may be $+\infty$). E^n is an n-dimensional linear vector space with norm $|\cdot|$. The function q : $[-\gamma, 0] \to (0, \infty)$ is Lebesgue integrable on $[-\gamma, 0]$, positive and non-decreasing on $[-\gamma, 0]$. Let $B = B([-\gamma, 0], E^n)$ be the Banach space of functions which are continuous on $[-\gamma, 0]$ and such that

$$\|\phi\| = \sup_{s \in [-\gamma, 0]} |\phi(s)| + \int_{-\gamma}^{0} q(\tau) |\phi(\tau)| \, d\tau < \infty \tag{6}$$

For any $t \in I$, and any x : $[t-\gamma, t] \to E^n$, let x_t : $[-\gamma, 0] \to E^n$ be defined by $x_t(s) = x(t+s)$, $s \in [-\gamma, 0]$. Hale [5] obtained exponential estimates on the solutions of the linear equation of the form

$$\dot{x}_k(t) = L_k(t, x_t^k(\cdot, s)) = A_k(t) x(t-h) + \int_{-\gamma}^{0} B_k(\theta) x_k(t+\theta) \, d\theta$$

$$x_k(t) = \phi_k(t), \ t \text{ in } [t_0 -h, t_0] \tag{7}$$

Let $X_k(t,s)$ satisfy the equation (7) such that

$$\frac{\partial X_k(t,s)}{\partial t} = L(t, x_t^k(\cdot, s)), \quad t \geq s$$

$$x_k(t,s) = \begin{cases} 0, & s-h \le t \le s \\ I, & t = s \end{cases}$$

where $x_t^k \, (.,s)(\theta) = x^k(t+\theta,s)$, $-h \le \theta \le 0$.

Definition 1: The system (1) is said to be null controllable on I if for every complete initial state $z(t_o) = (x(t_o), \phi, u_{t_o})$, and every

vector $0 \, \epsilon \, E^n$ there exists a square integrable function $u:[t_o, t_1] \to E^m$

such that

$x(t_1, z(t_o), u(.))=0$.

In the paper we give results equivalent to that of Ref. [6 - 11] for the null controllability of large-scale system with distributed delays.

PRILIMINARY RESULTS

Next we introduce the notion of a proper control for the sub-system (8). The solution of eqn. (8)

$$\dot{x}_k(t) = L_k(t,x_t) + \int_{t_o -h}^{t_o} d_s[\eta^k(t,s)] \, u(t+s) \tag{8}$$

$x_k(t) = \phi_k(t)$, $t \, \epsilon \, [t_o -h, t_o]$

is given by

$$x_k(t) = x_k(t,t_o,\phi_k) + \int_{t_o}^{t} x_k(t,s) \{ \int_{t_o-h}^{t_o} d_r[\eta^k(s,r)] u_k(r+s) \, ds \} \tag{9}$$

$x_k(t) = \phi_k(t)$, $t \, \epsilon \, [t_o-h, t_o]$

The last term in equation (9) contains values of $u(s)$ for $t < t_o$. The values for $u(t)$ for $t < t_o$ enter into the definition of complete state. To separate them, the last term can be transformed by changing the order of integration where the unsymmetric Fubini theorem is used. From the Fubini theorem, we have

$$\int_{t_o}^{t} x_k(s,t) \{ \int_{t_o-h}^{t_o} d_r[\eta^k(s,r)] \, u_k(r+s) \, ds \}$$

$$= \int_{t_o-h}^{t_o} d_r \; \{ \int_{t_o}^{t} X_k(s,t) \; [\eta^k(s,r)] \; u_k(r+s) \; ds \}$$

$$= \int_{t_o-h}^{t_o} d_r \; \{ \int_{t_o+r}^{t+r} X_k(s-r, \; t) \; \eta^k(s-r,r) \; u_k(s) \; ds \}$$

$$= \int_{t_o-h}^{t_o} d_r \; \{ \int_{t_o+r}^{t_o} X_k(s-r, \; t) \; \eta^k \; (s-r, \; r) \; u_k(s) \; ds \}$$

$$+ \int_{t_o-h}^{t_o} d_r \; \{ \int_{t_o}^{t+r} X_k(s-r, \; t) \; \eta^k(s-r, \; r) \; u_k(s) \; ds \} \tag{10}$$

Since $r \leq 0$ so that $t+r \leq t$ and $X_k(s-r,t) = 0$ for $s > t+r$, the integrals in brackets can be extended to $s = t$. The second bracket in eqn. (10) can be written as

$$\int_{t_o-h}^{t_o} d_r \; \{ \int_{t_o}^{t} X_k(s-r,t) \; \eta^k(s-r,r) \; u_k(s) \; ds \}$$

$$= \int_{t_o}^{t} \; \{ \int_{t_o-h}^{t_o} X_k(s-r,t) \; [d_r \; \eta^k(s-r,r)] \; u_k(s) \; ds \} \tag{11}$$

We now isolate the expression

$$\int_{t_o-h}^{t_o} X_k(s-r,t) \; [d_r \; \eta^k(s-r,r)] = Y_k(s,t) \tag{12}$$

and define it as the index of the control system (8). We also observe that the first bracket in eqn. (10) can be written as

$$\int_{t_o-h}^{t_o} d_r \; \{ \int_{t_o+r}^{t_o} X_k(s-r,t) \; \eta^k(s-r,r) \; u_k(s) \; ds \}$$

$$= \int_{t_o+r}^{t_o} \; \{ \int_{t_o-h}^{t_o} X_k(s-r,t) \; [d_r \; \eta^k(s-r,r)] \} \; u_{t_o}^k \; ds \tag{13}$$

Using eqn. (11) and (13) in eqn. (9) gives

$$x_k(t, \; z(t_o) \; , \; u(.)) = V_k \; (t, \; z(t_o) \; , \; u_{t_o}) + \int_{t_o}^{t} Y_k(s,t) \; u_k(s) \; ds, \tag{14}$$

where

$$V_k \; (t, \; z(t_o), u_{t_o}) = x_k(t,t_o, \phi_k) + \int_{t_o+r}^{t_o} \; \{ \int_{t_o-h}^{t_o} X_k(s-r,t)$$

$$\cdot [d_r \; \eta^k(s-r,r)] \} \; u_{t_o}^k \; ds \tag{15}$$

Lemma 1. The sub-system (8) is proper on $[t_o, t_1]$ if and only if the origin is an interior point of

$$R^k(t_1) = \int_{t_o}^{t_1} Y_k(s,t) \, u_k(s) \, ds \tag{16}$$

where

$$Y_k(s,t) \stackrel{\text{def.}}{=} \int_{t_o-h}^{t_o} X_k(s-r,t) \, [d_r \, \eta^k \, (s-r,r)] \tag{17}$$

The proof of this lemma follows by the similar argument as in the proof given in [11].

Definition 2 The sub-system (8) is null controllable on $[t_o, t_1]$ if and only if the grammian of (8), namely

$$W_k(t_o, t_1) = \int_{t_o}^{t_1} Y_k(s,t_1) \, Y_k^T(s,t_1) \, ds \tag{18}$$

where

$$Y_k(s,t) = \int_{t_o-h}^{t_o} X_k(s-r,t) \, [d_r \, \eta^k(s-r,r)] \tag{19}$$

is nonsingular for each $k = 1, \ldots, m$.

MAIN RESULTS

We now state the main results of this paper. Let **B** be the Banach space of all functions

$$(x,u) \; : \; [t_o-h, t_1] \times [t_o-h, t_1] \to E^n \times E^m$$

where $x \varepsilon B([t_0 - h, t_1] \, E^n)$; $u \varepsilon L_2([t_o, -h, t_1], E^m)$ with norm defined by

$$|| (x,u) ||_2 = ||x||_2 + ||u||_2 \tag{20}$$

where

$$||x||_2 = \{ \int_{t_o-h}^{t_1} | x(s) |^2 ds \}^{\frac{1}{2}}$$

$$||u||_2 = \{ \int_{t_o-h}^{t_1} | u(s) |^2 ds \}^{\frac{1}{2}}$$

The coupling function f_i ($i = 1, \ldots, m$) satisfies the following conditions:

$$f_i \; : \; I \times B \, ([t_o-h, t], E^{r_i}) \times L_2 \to E^{r_i}, \tag{21}$$

$$\sup \{f_i \ (t, x(.), u(.)) : t \ \varepsilon \ I\} \le F_i(r) < \infty \tag{22}$$

<u>Theorem 1.</u> Suppose that the assumptions of Lemma 1 hold, and in addition if

(i) $x_k(s,t)$, $k = 1, \ldots, m$ is the unique solution of eqn. (7) so that $|X_k(s,t)| \le M$, $M > 0$ for each $k = 1, \ldots, m$

(ii) the continuous function f_i (i = 1, ..., m) satisfies

$$\sup \{f_i \ (t, x(.), \ u(.)) : t \ \varepsilon \ I\} \le F_i(r) < \infty$$

for all (x, u) ε B such that $\lim_{r \to \infty} F_i(r)/r \to 0$ for each

i = 1, ..., m.

Then the large-scale system (1) is null controllable.

<u>Proof:</u> Let $u(t) = \mathrm{col} \ (u_1(t), \ldots, u_m(t))$, $x(t) = \mathrm{col}(x_1(t), \ldots, x_m(t))$, $Y(s,t) = \mathrm{diag} \ \{Y_1(s,t), \ldots, Y_m(s,t)\}$, and $X(s, t) = \mathrm{diag} \ \{X_1(s, t), \ldots, X_m(s, t)\}$ consisting of blocks with square matrices of order $r_k \ge 1$ $(k = 1, \ldots, m, \ r_1 + \ldots + r_m = n)$. Suppose the pair of functions x, u form a solution pair to the set of integral equation

$$u(t) = -Y^*(t, t_1) \ W^{-1}(t_0, t_1) \ [V \ (t_1, z(t_0), u_{t_0}) + \int_{t_0}^{t_1} X(s, t_1) \ f(.)ds$$

$$\tag{23}$$

where $V(t_1, \ z(t_0), \ u_{t_0})$ is defined from eqn. (15) as

$$V(t_1, \ z(t_0), \ u_{t_0}) = \mathrm{col} \ (V_1(.), \ldots, V_m(.))$$

$$f(.) = \mathrm{col}(f_1(.), \ldots, f_m(.)),$$

$$W^{-1}(t_0, t_1) = \mathrm{diag} \ \{W_1^{-1}(t_0, t_1), \ldots, W_m^{-1}(t_0, t_1)\};$$

and

$$x(t) = V(t, \ z(t_0), \ u_{t_0}) + \int_{t_0}^{t} Y(s,t) \ u(s) \ ds + \int_{t_0}^{t} X(s,t) \ f(.) \ ds \tag{24}$$

$$x(t) = \phi(t), \ t \ \varepsilon \ [t_0 - h, \ t_0]$$

Then u is square integrable on $[t_0 - h, \ t_1]$ and $x(t)$ is a solution of (1) corresponding to u with the initial state $z(t_0) = (x(t_0), \ \phi, \ u_{t_0})$.

Also,

$$x(t_1) = V(t_1, z(t_0), u_{t_0}) + \int_{t_0}^{t_1} Y(s, t_1) \, u(s) \, ds + \int_{t_0}^{t_1} X(s, t_1) \, f(.) \, ds$$

Substituting $u(t)$ from eqn. (23) we have

$$x(t_1) = V(t_1, z(t_0), u_{t_0}) - \int_{t_0}^{t_1} Y(s, t_1) \, Y^*(s, t_1)$$

$$. W^{-1}(t_0, t_1) \; [V(t_1, z(t_0), u_{t_0}) + \int_{t_0}^{t_1} X(s, t_1) \, f(.)]$$

$$+ \int_{t_0}^{t_1} X(s, t_1) f(.) \, ds = 0$$

We now show that $u: [t_0, t_1] \rightarrow U$ is in the arbitrary compact constraint

subset of E^m, that is, $|u(t)| \le r$ for some constant $r > 0$. In eqn. (23)

$$u(t) = - Y^*(t_1, t) \, W^{-1}(t_0, t_1) \; [V(t_1, z(t_0), u_{t_0}) + \int_{t_0}^{t_1} X(s, t_1) \, f(.) \, ds]$$

choose constants as

$\alpha_1 = \sup | Y(t, t_1)| \; , t \; \varepsilon \; I$

$\alpha_2 = |W^{-1}(t_0, t_1)|$

$\alpha_3 = \sup | V(t, z(t_0), u_{t_0})| \; ; \; t \; \varepsilon \; I$

$\alpha_4 = \sup |X(s,t)| \; ; \; (s,t) \varepsilon \; IxI$

$\delta_1 = 8 \, \alpha_1 \alpha_2 \alpha_4 \, (t_1 - t_0)$

$\delta_2 = 8 \alpha_4 \, (t_1 - t_0)$

$v = \max \{1, (t_1 - t_0) \alpha_1, (t_1 - t_0 + h)^{\frac{1}{2}}\}$

$\beta_1 = 8 \alpha_1 \, \alpha_2 \, \alpha_3 v$

$\beta_2 = 8 \alpha_3$

$\delta = \max (\delta_1, \delta_2)$

$\beta = \max (\beta_1, \beta_2)$

Then,

$|u(t)| \le \alpha_1 \alpha_2 \, [\alpha_3 + \alpha_4 \, (t_1 - t_0) \sup |f|]$

$\le \dfrac{\beta_1}{8v} + \dfrac{\delta_1}{8v} \sup |f|$

$$\leq \frac{1}{8v} \; [\beta_1 + \delta_1 \; \sup \; |f|]$$

$$\leq \frac{1}{8v} \; [\beta + \delta \; \sup \; |f|] \tag{25}$$

Next we show that $\beta + \delta \; \sup \; |f| < r$, $r > 0$. To that end from the

assumption (ii) of Theorem 1, if $(x,u) \varepsilon B$ and $\lim\limits_{r \to \infty} \frac{F(r)}{r} = 0$, then there

exists $r > 0$ such that

$$\sup_{t_o - h \leq t \leq t_o} x_{t_o} \; (t) \leq r/4 \; ; \qquad \sup_{t_o - h \leq t \leq t_o} u_{t_o} \; (t) \leq \frac{r}{8v}$$

and for

$r \geq 2\beta$, we have $\frac{F(r)}{r} < \frac{1}{2\delta}$

Therefore from above inequality, we have

$$\beta + \delta \; \sup \; |f| \leq \beta + \delta \; F(r) \leq \frac{r}{2} + \frac{r}{2} \leq r$$

Hence the inequality (25) can be written as

$$|u(t)| \; \leq \frac{r}{8v} \leq \frac{r}{4} \tag{26}$$

Also from eqn. (24)

$$|x(t)| \; \leq \alpha_3 + \alpha_1 \int_{t_o}^{t_1} |u(s)| \; ds + \alpha_4 (t_1 - t_o) \; \sup|f|$$

$$\leq \frac{\beta}{8} + \alpha_1 \int_{t_o}^{t_1} |u(s)| \; ds + \frac{\delta}{8} \; \sup|f|$$

$$\leq \frac{\beta}{8} + \alpha_1 \frac{r}{8v} \; (t_1 - t_o) + \frac{\delta}{8} \; \sup|f|$$

$$\leq \frac{r}{16} + \frac{rv}{8v} + \frac{r}{16} \leq \frac{r}{4} \tag{27}$$

Now, define the operater $T: B \to B$ by $T(x,u) = (y,w)$. From eqn. (26)
we have

$$|w(t)| \; \leq \frac{r}{4}, \; t \; \varepsilon \; [t_o - h, t_1]$$

hence

$$||w(t)||_2 = \frac{r}{4} \; (t_1 + h - t_o)^{\frac{1}{2}}$$

and from eqn. (27), we have

$$|y(t)| \leq \frac{r}{4}, \; t \; \varepsilon \; I$$

$$\|y(t)\|_2 = \frac{r}{2}(t_1 + h - t_0)^{\frac{1}{2}}$$

so that $\|(x,u)\|_2 \leq \frac{r}{2} (t_1 + h - t_0)^{\frac{1}{2}} < \infty$.   Then if we let

$$Q(r) = \{ (x,u) \ \varepsilon \ Q: \ \|x\|_2 \leq r, \ \|u\|_2 \leq r \}$$

we have proven that T: $Q(r) \rightarrow Q(r)$.   Since $Q(r)$ is closed, bounded and convex by Riez's theorem [13], it is compact under T, then by Schander's theorem T has a fixed point $(x,u) \ \varepsilon \ Q(r)$.   This fixed point (x,u) of T is a solution pair of equations (23) and (24).   Hence the large-scale system (1) is null controllable.

   Example: Let us give an example to illustrate the theorem.

$$\ddot{x} \ (t) + b\dot{x}(t) + a\dot{x}(t - h) + kx(t)$$
$$= \mu_1 \ y(t - h) + \sin[x(t - h)y(t - h)] + u(t)$$

$$\dot{y}(t) = -\frac{1}{e} y(t - h) + \frac{x(t - h)}{1 + x^2(t - h)} + e^{-t} u(t) \tag{28}$$

   Let us decompose the system (28) into two subsystems as follows:

$$\ddot{x} \ (t) + b\dot{x}(t) + q\dot{x}(t - h) + kx(t) = 0 \tag{29}$$

$$\dot{y}(t) = -\frac{1}{e} y(t - h) \tag{30}$$

   If we convert Eq. (29) into first-order system by introducing

$$x_1(t) = x(t), \ x_2(t) = \dot{x}(t)k^{\frac{1}{2}}, \ \text{we get the subsystem I and II as}$$

$$\dot{X}_I(t) = AX_I(t) + BX_I(t - h) \tag{31}$$

where

$$A = \begin{bmatrix} -0 & k^{\frac{1}{2}} \\ -k^{\frac{1}{2}} & -b \end{bmatrix}, \ B = \begin{bmatrix} -0 & 0 \\ 0 & -q \end{bmatrix}, \ X_I(t) = \begin{bmatrix} x_1(t) \\ x_2(t) \end{bmatrix}, \ X_{II}(t) = y(t)$$

$$\dot{X}_{II}(t) = -\frac{1}{e} X_{II}(t - h) \tag{32}$$

   In order to obtain the matrix $X_I(t,s)$, we write the characteristic equation of (31)

$$\det(A + Be^{-\lambda h} - \lambda I) = 0 \tag{33}$$

This leads to the equation

$$\lambda^2 + b\lambda + q\lambda e^{-\lambda h} + k = 0 \tag{34}$$

Let λ_1 and λ_2 (assumed distinct) be two roots of Eq. (34) each

having $|\lambda| < \frac{1}{h}$ and Re $\lambda < 0$. The solution matrix $X_I(t,s)$ for Eq. (31) is

given by

$X_I(t,s)$

$$= \begin{cases} \dfrac{e^{\lambda_1(t-s)}}{(\lambda_1 - \lambda_2)} \begin{bmatrix} -\lambda_2 & 1 & k^{\frac{1}{2}} \\ -\lambda_1\lambda_2/k^2 & \lambda_1 \end{bmatrix} + \dfrac{e^{\lambda t - s)}}{(\lambda_1 - \lambda_2)} \begin{bmatrix} \lambda_1 & -k^{\frac{1}{2}} \\ \lambda_1\lambda_2/k^2 & -\lambda_2 \end{bmatrix}; \quad s<t, \\[4mm] 1 \quad \text{when } s=t \\[2mm] 0 \quad \text{when } t<s<t+h \end{cases} \tag{35}$$

Estimating the solution of subsystem I, where λ is the largest negative
root.

$$|x_1(t), \ x_2(t)| \le |x_1(t_o), \ x_2(t_o)| \ e^{\lambda(t-t_o)} \tag{36}$$

If we set $x_1(t) = x_1(t_o)$ and $x_2(t) = x_2(t_o)$ for $t \le t_o$, we obtain

$$|x_{1t}, \ x_{2t}|_o = \max \{ \ x_1(s), \ x_2(s) \ , \ s\epsilon[-h,0]\}$$

$$= e^{-\lambda h} \ |x_{1t_o}, \ x_{2t_o}| \ e^{\lambda(t-t_o)}; \quad Re\lambda < 0 \tag{37}$$

Now consider the subsystem II, the fundamental solution matrix
$X_{II}(t,s)$ has the form

$$X_{II}(t,s) = \begin{cases} e^{-(t-s)} & \text{when } s < t \\ 1 & \text{when } s = t \\ 0 & \text{when } t < s \le t + h \end{cases} \tag{38}$$

If we set $y(t) = y(t_o)$ for $t \le t_o$, we obtain

$$|y_t|_o = \max\{ \ y(s) \ ; \ s\epsilon[-h,0]\} = e^h |y_{t_o}| \ e^{-(t-t_o)}; \quad t \ge t_o$$

Now, taking into account the decompositon of (28) into subsystems I and
II, and writing $f_I(.)$ and $f_{II}(.)$ as

$$f_I(.) = \begin{bmatrix} 0 \\ \sin[x_1(t-h)\ y(t-h)] \end{bmatrix}; f_{II}(.) = \frac{x(t-h)}{1 + x^2(t-h)} \tag{39}$$

we see that $f_i(r) < \infty$ with $r < \infty$ and $\lim_{r \to \infty} \frac{f_i(r)}{r} = 0$; $i = I, II$.

The decomposed system where all the remaining terms included with the subsystems I and II yield

$$\dot{X}_I(t) = \lambda\ X_I(t) + \mu_1\ e^h\ X_{II}(t) + f_I(.) + u(t) \tag{40}$$

$$\dot{X}_{II}(t) = -\ X_{II}(t) + f_{II}(.) + e^{-t}\ u(t)$$

One verifies rather easily that $X_I(t,s)$, $X_{II}(t,s)$ and $f_I(.)$, $f_{II}(.)$ satisfy the conditions of the Theorem 1. The matrices $W_k(.)$, $k = I, II$ are non-singular. Hence the system is null controllable.

This paper deals with null controllability of a class of large-scale nonlinear systems with distributed delays in control and unbounded delays in states. Thus, our main result generalizes the results of Chukwu [11], to cover the case of the large-scale systems with unbounded distributed delays in states and control.

References

1. D. D. Siljak, "Stability of Large-scale System Under Structural Perturbulations", IEEE Trans. Systems Man. Cybernet, 2 (1972), pp. 657 - 661.

2. A. N. Michel and R. K. Miller, "Qualitative Analysis of Large-scale Dynamical Systems" (Academic Press, New York 1977).

3. D. D. Siljak, "Large-scale Dynamic Systems" (Elsevier/North-Holland, Amsterdam, 1978).

4. A. S. C. Sinha, "Asymptotic Behaviour for Some Large-scale Systems with Infinite Delay"; Int. J. Systems Sci., 1986, pp. 373 - 380.

5. J. K. Hale, "Functional Differential Equations with Infinite Delays", J. of Math. Anal. and Appl., Vol. 48, pp. 276 - 283, 1974.

6. M. Khambadkone, "Euclidean Null-Controllability of Linear Systems with Distributed Delays in Control", IEEE Trans. on Automatic Control, Vol. AC-27, pp. 210 - 211, 1982.

7. J. Klamka, "Controllability of Nonlinear Systems with Distributed Delays in Control", Int. J. of Control, Vol. 31, pp. 811 - 819, 1980.

8. J. P. Dauer and R. D. Gahl, "Controllability of Nonlinear Delay Systems", J. of Optimization Theory and Applications, Vol. 21, pp. 59 - 70, 1977.

9. H. T. Banks and G. A. Kent, "Control of Functional Differential Equations of Retarded and Neutral Type of Target Sets in Function Space", SIAM J. on Control, Vol. 10, pp. 567 - 593, 1972.

10. A. S. C. Sinha, "Controllability of Non-linear Delay Systems", Int. J. Control, 1986, Vol. 43, pp. 1305 - 1315.

11. E. N. Chukwu, "On the Null-Controllability of Non-linear Delay Systems with Restrained Controls", J. of Math. Anal. and Appl. 1980, Vol. 76, pp. 283 - 296.

12. H. Hermes and J. P. LaSalle, "Functional Analysis and Time optimal Control", Academic Press, New York, N.Y., 1969.

13. L. V. Kantorovich and G. P. Akilov, "Functional Analysis", Pergamen Press, Oxford, 1982.

ROBUST APPLICATION OF BEARD-JONES DETECTION FILTER

Paul S. Min
Bell Communications Research
Piscataway, New Jersey

1. Introduction

This paper is aimed at formulating an analytical framework for "robust" applications of Beard-Jones Detection Filter (BJDF). The theory of BJDF was devised by Beard [1] and Jones [2] for use in real time diagnosis of system failures.

BJDF, a state-estimator like device, incorporates a linear dynamic model of a system and compares the system's performance to the model's prediction. Unlike ordinary state estimators, however, the main task of a BJDF is to identify a set of failures in the system. Consequently, a BJDF is only a suboptimal state estimator.

Although there have been many theoretical developments in the BJDF concept since the introduction, the purpose of the present paper is justified by the fact that the lack of robustness has inhibited BJDF from physical implementations. It should be noted that a typical dynamic system can be a complex, mechanical-electrical-chemical plant with severe non-linearities, and as a result, the system dynamics are not often adequately understood.

This paper will establish a set of optimal procedures for designing and implementing BJDF for the systems with uncertain dynamics. The result should yield a maximal numerical stability against "unmodeled" parameter variations in the systems.

First, a general solution of BJDF will be reviewed in section 2. The approach taken here will be similar to that of Jones [2], but a strong emphasis will be placed on the implementation aspects.

In section 3, the scope of BJDF will be extended to the case of uncertain systems. This is done by employing the concept of "invariant subspaces," and maximal invariance will be attempted by assigning a proper eigen-structure to the dynamics of BJDF. The results presented in this section will include the optimal reference model for uncertain systems, the optimal BJDF feedback gain (detection operator), and the optimal decision/isolation operator for the set of hypothesized failures. The results will be in closed analytical forms, and any particular numerical solution can be achieved by simply solving a combination of linear and eigenvalue equations.

The concept of system diagnosis has been generally referred to as the failure detection and isolation (FDI) theory. The FDI theory is aimed at simplifying the systematic diagnosis of failures for the improvement of overall system reliability. In the literature, various forms of the FDI theory have been reported, and [3] - [5] present good general surveys.

2. Beard-Jones Detection Filter Problem

In this paper, dynamic systems will be defined in state space notation. A linear time-invariant system in discrete time domain can be described as,

$$\underline{x}(k+1) = A\underline{x}(k) + B\underline{u}(k) \tag{1}$$

$$\underline{y}(k) = C\underline{x}(k)$$

Note that the selection of the discrete time domain is only for the notational convenience, and the results of this paper can be easily extended to the continuous time cases.

Furthermore, the system (1) is assumed to be "observable", and

$$\underline{x}(k) \in X = R^n \tag{2}$$

$$\underline{u}(k) \in U = R^p$$

$$\underline{y}(k) \in Y = R^m$$

where R^n is the n dimensional vector space, etc. Each component of $\underline{x}(k)$, $\underline{y}(k)$, and $\underline{u}(k)$ will be subscripted, with $x_i(k)$, for example, being the i'th component of $\underline{x}(k)$.

Let $d(\cdot)$, $Im(\cdot)$, and $Ker(\cdot)$ denote the "dimension" (of the space), the "image" (of the operator), and the "kernel" (of the operator) in the argument.

It can be seen that the dynamic system (1) under a certain class of failures can be modeled as,

$$\underline{x}(k+1) = A\underline{x}(k) + B\underline{u}(k) + fn(k) \tag{3}$$

where f and $n(k)$ are referred to as the "event vector" and the "scalar magnitude function" defining the failure, respectively.

It should be noted that (3) is applicable to a broad class of failures in the system. The examples of system failures represented by (3) are illustrated in [2] (actuator failures and some system dynamic failures) and [6] (sensor failures in controlled systems).

For example, if (1) is written as,

$$\underline{x}(k+1) = A\underline{x}(k) + B\underline{u}(k) \tag{4}$$

$$= A\underline{x}(k) + \sum_{j=1}^{p} B_j u_j(k)$$

where B_j is the j'th component of B, then in the case of the i'th actuator failure, (4) becomes

$$\underline{x}(k+1) = A\underline{x}(k) + \sum_{j=1}^{p} B_j u_j(k) + B_i n(k) \tag{5}$$

Hence $f = B_i$. If the i'th actuator in (5) is completely out of commission, $n(k) = -u_i(k)$, whereas simple calibration errors will result in $n(k) = \alpha u_i(k)$ (α constant), etc.

However, some cases of sensor failures cannot be modeled by (3), and in such situations, the alternative techniques presented in [6] are recommended instead of BJDF.

Define BJDF as shown below.

$$\underline{v}(k+1) = A\underline{v}(k) + B\underline{u}(k) + D(\underline{y}(k) - \underline{w}(k)) \tag{6}$$

$$\underline{w}(k) = C\underline{v}(k)$$

$$\underline{e}(k) = \underline{x}(k) - \underline{v}(k)$$

$$\underline{q}(k) = \underline{y}(k) - \underline{w}(k)$$

where $\underline{q}(k)$ and $D : Y \rightarrow X$ in particular are referred to as the "failure signature" and the "detection operator" of BJDF, respectively. Note that although BJDF has the form identical to the ordinary state estimator, it differs from the ordinary state estimator according to the following definition.

Definition Let f be the event vector of a failure in the observable system (1). Then D is a "detection operator" for f if

(i) $d(CQ_f) = 1$ where

$$Q_f = \text{span}\{f, (A-DC)f, \cdots, (A-DC)^{n-1}f\}$$

(ii) D assigns the eigenvalues of $(A-DC)$ to any arbitrary self-conjugate set.

It can be shown by (3) and (6) that

$$\underline{e}(k+1) = (A-DC)\,\underline{e}(k) + fn(k) \tag{7}$$

$$\underline{q}(k) = C\underline{e}(k)$$

Hence, the definition above suggests that the failure signature of BJDF must be restricted to an one-dimensional subspace of the output. It is shown in [2] that such task can be accomplished by constraining the controllable subspace of $(A-DC, f)$ to the invariant subspace of minimal dimensionality. Subsequently, the process of failure identification becomes looking for the failure signature in CQ_f direction.

It has been shown in [2] and [6] that a detection operator D satisfying the definition always exists. The following steps describe the procedure. (See [6] for more detail.)

1) Let μ be the first positive integer such that $CA^{\mu-1}f \neq 0$. Find $\hat{D}$ such that
$$A^{\mu}f - \alpha f_0 = \hat{D}CA^{\mu-1}f$$
where $f_0 \in f \oplus Af \oplus A^2f \oplus \cdots \oplus A^{\mu-1}f$ with direct sum operator $\oplus$. This choice of $\hat{D}$ satisfies the condition (i) of definition. Such $\hat{D}$ can be always found for observable systems. Note that $Q_f = f \oplus Af \oplus A^2f \oplus \cdots \oplus A^{\mu-1}f$

2) Find a linear operator T such that,

$$TCQ_f = 0 \quad \text{where } T : Y \to Y$$
$$d(Ker(T)) = 1 \quad d(Im(T)) = m-1$$

3) Let W_f be the unobservable subspace of $(TC, A-\hat{D}C)$. Note that W_f contains Q_f. Then, if $d(W_f) = \nu$, there exists a unique generator vector g satisfying, (see [2])

$$\text{i)} \quad CA^k g = 0 \quad k=0,1, \cdots, \nu-2$$
$$\text{ii)} \quad CA^{\nu-1}g = CQ_f$$

Hence, the set $\{g, Ag, \cdots, A^{\nu-1}g\}$ is a basis for W_f.

4) It can be shown that the degree of the minimal polynomial associated with W_f is ν. Let $\Psi(\lambda)$ be the minimal polynomial for W_f with desired eigenvalues $\{\lambda_i, i=1,2,...,\nu\}$.

$$\Psi(\lambda) = \prod_{i=1}^{\nu} (\lambda - \lambda_i)$$
$$= \lambda^{\nu} + \alpha_1\lambda^{\nu-1} + \alpha_2\lambda^{\nu-2} + \cdots + \alpha_{\nu-1}\lambda + \alpha_{\nu}$$

Then, D_1 solving (8) below satisfies the eigenvalue assignment on W_f (hence Q_f).

$$D_1CQ_f = \Psi(A)g \tag{8}$$

5) Next find D_2 such that $(A-D_1C-D_2TC)$ has the remaining $n-\nu$ eigenvalues at proper locations. This completes the eigenvalue requirement (the condition (ii) of definition) and hence, $D = D_1-D_2T$.

If a single BJDF is desired to accommodate a set of failure events $\{f_i\}_{i=1}^{h}$ simultaneously, (9) below must replace (8). Q_i here is the controllable subspace of $(A-\hat{D}_iC, f_i)$, where $\hat{D}_i$ $i=1,2,...,h$ was found for each f_i using step 1).

$$D_1 [CQ_1 \ CQ_2 \ \cdots \ CQ_h]$$

$$= [\ \Psi_1(A)g_1 \ \Psi_2(A)g_2 \ \cdots \ \Psi_h(A)g_h] \tag{9}$$

Note that $\Psi_i(\lambda)$ and g_i are the minimal polynomial and the generator vector corresponding to the unobservable subspace W_i of $(T_iC, A - \hat{D}_iC)$, respectively.

The resulting operator D_1 will satisfy $d(CQ_i) = 1$ for $i = 1,2,...,h$ and will assign the desired eigenvalues for $\{ W_i \}_{i=1}^{h}$. It can be shown that such D_1 exists if

$$CQ_i \cap (\sum_{j \neq i} CQ_j) = \phi \quad i = 1,2,...,h$$

For some situations of multiple failure events, there exist eigenvalues that are not arbitrarily assignable. Fortunately, these unassignable eigenvalues do not correspond to the dynamics of the failure signatures. It is also possible to eliminate such situations by re-grouping the failure event set.

3. Design of BJDF under Uncertainties

Section 2 has shown the procedure of the BJDF design (i.e., determining D) when the given system presents precisely known dynamics. Once uncertainties are introduced in the system, however, the definition of the detection operator in section 2 cannot always be satisfied, forcing subsequent modifications.

This section will derive the optimal solutions for BJDF parameters when the dynamic model of a system is contaminated by uncertainties. It is assumed in this paper that all the uncertainties lie in A, assuming that the dynamics models for the peripheral devices (i.e., B and C) can be made exact.

Let Θ correspond to the finite set containing uncertainty hypotheses θ_j $j = 1,2,...,s$. Assume also that the probability $p(\theta_j)$ is known for each θ_j such that $\sum_{j=1}^{s} p(\theta_j) = 1$.

Then, for each θ_j, define a reference model for the given system, which is denoted as below.

$$\underline{x}(k+1) = A(\theta_j)\underline{x}(k) + B\underline{u}(k) \tag{10}$$

$$\underline{y}(k) = C\underline{x}(k) \qquad j = 1,2,...,s$$

3.1 Reference Model (A) Selection

The first step of the BJDF design now becomes determining the reference model A. The solution of A must optimally represent the entire set of $A(\theta_j)$, $j = 1,2,..,s$ such that the resulting BJDF will be minimally affected by model mis-matches. Let $\Delta A(\theta_j) = A(\theta_j) - A$, $j = 1,2,..,s$.

Then given an event vector f_i, (10) becomes,

$$\underline{x}(k+1) = A\underline{x}(k) + B\underline{u}(k) + \Delta A(\theta_j)\underline{x}(k) + f_i n(k) \tag{11}$$

$$\underline{y}(k) = C\underline{x}(k)$$

For each θ_j, define $\underline{e}(\theta_j; k)$, $\underline{q}(\theta_j; k)$ as,

$$\underline{e}(\theta_j; k) = \underline{x}(k) - \underline{v}(k) \big|_{\theta_j} \tag{12}$$

$$\underline{q}(\theta_j; k) = \underline{y}(k) - \underline{w}(k) \big|_{\theta_j}$$

Hence, from (6) and (11),

$$\underline{e}(\theta_j; k+1) = (A - DC)\, \underline{e}(\theta_j; k) + \Delta A\,(\theta_j)\underline{x}(k) + f_{in}(k) \tag{13}$$

$$\underline{q}(\theta_j; k) = C\underline{e}(\theta_j; k)$$

Note in (13) that $\underline{e}(\theta_j; \cdot)$ has two forcing functions: $\Delta A\,(\theta_j)\underline{x}(\cdot)$ and $f_{in}(\cdot)$. To minimize the effect of $\Delta A(\theta_j)\underline{x}(\cdot)$ (and hence, maximizing the relative effect of $f_{in}(\cdot)$) in $\underline{e}(\theta_j; \cdot)$, the reference model A must be selected such that,

$$A = \operatorname*{Argmin\ maximum}_{\text{all } \theta_j,\, k} \quad |\Delta A(\theta_j)\underline{x}(k)|^2 \tag{14}$$

where $|\cdot|$ denotes the Euclidean norm.

Note that A satisfying (14) will guarantee a minimum upper limit in $\operatorname*{Maximum}_{\text{all } \theta_j,\, k} |\underline{e}(\theta_j; k)|^2$ due to the bounded-input-bounded-output property of the stable system (i.e., the eigenvalues of $A - DC$ in the left half plane).

Assume that the states $\underline{x}(k)$ exhibits the bounded-region property. Then, there exist R and r such that,

$$\underline{x}^T(k)\, R\, \underline{x}(k) \le r$$

where R and r are a positive definite matrix and a positive constant, respectively. The bounded-region property is a characteristic intrinsic to stable systems, and such $\underline{x}(k)$ can be constrained within an ellipsoid centered at the origin for all k.

$$|\Delta A\,(\theta_j)\,\underline{x}(k)|^2 = \{\Delta A\,(\theta_j)\,\underline{x}(k)\}^T\{\Delta A\,(\theta_j)\,\underline{x}(k)\}$$

$$= \underline{x}^T(k)\, R_1(\theta_j)\, \underline{x}(k)$$

where $R_1(\theta_j) = \Delta A\,(\theta_j)^T\, \Delta A\,(\theta_j)$

Let $\lambda_R{}^{(t)}$ and $\lambda_{R_1(\theta_j)}{}^{(t)}$ $t = 1, 2, .., n$ denote the the eigenvalues of R and $R_1(\theta_j)$ such that,

$$0 < \lambda_R{}^{(1)} \le \lambda_R{}^{(2)} \le \cdots \le \lambda_R{}^{(n)}$$

$$0 \le \lambda_{R_1(\theta_j)}{}^{(1)} \le \lambda_{R_1(\theta_j)}{}^{(2)} \le \cdots \le \lambda_{R_1(\theta_j)}{}^{(n)}$$

Then, $\underline{x}^T(k)\,\underline{x}(k) \le \dfrac{r}{\lambda_R{}^{(1)}} = M$ where M is a known positive constant. Subsequently,

$$|\Delta A\,(\theta_j)\,\underline{x}(k)|^2 \le \frac{\lambda_{R_1(\theta_j)}{}^{(n)}\, r}{\lambda_R{}^{(1)}} = \lambda_{R_1(\theta_j)}{}^{(n)}\, M \tag{15}$$

Hence, (14) becomes

$$A = \operatorname*{Argmin\ maximum}_{\text{all } \theta_j} \lambda_{R_1(\theta_j)}{}^{(n)} \tag{16}$$

Unfortunately, A satisfying (16) cannot be easily found. It should be noted that even if such A is found, the probability of the worst case scenario (i.e., for all k, $\underline{x}(k)$ lies entirely in the direction corresponding to $\operatorname*{maximum}_{\text{all } \theta_j} \lambda_{R_1(\theta_j)}{}^{(n)}$) may be very small. For this reason, the following approximation is considered.

Since $R_1(\theta_j)$ for each θ_j is a non-negative definite matrix,

$$\frac{1}{n}\sum_{i=1}^{n} \lambda_{R_1(\theta_j)}{}^{(i)} \le \lambda_{R_1(\theta_j)}{}^{(n)} \le \sum_{i=1}^{n} \lambda_{R_1(\theta_j)}{}^{(i)}$$

$\sum_{i=1}^{n} \lambda_{R_1(\theta_j)}{}^{(i)}$ above provides the upper and the lower bounds for $\lambda_{R_1(\theta_j)}{}^{(n)}$, and hence it is a reasonable substitute for $\lambda_{R_1(\theta_j)}{}^{(n)}$ in (16). Furthermore, if the unequal probability of each θ_j is accounted for, then,

$$A = \text{Argmin} \sum_{j=1}^{J} \sum_{i=1}^{n} p(\theta_j) \lambda_{R_i(\theta_j)}^{(i)} \tag{17}$$

Let $\Delta A(\theta_j) = [\ \Delta a_{lp}(\theta_j)\ ]_{lp}$, where $[\ \Delta a_{lp}(\theta_j)\ ]_{lp}$ denotes the matrix with $\Delta a_{lp}(\theta_j)$ as the lth row pth column element. Then,

$$| \Delta A(\theta_j) |_F^2 = \sum_{l=1p}^{n} \sum_{p=1}^{n} \Delta a_{lp}(\theta_j)^2 \tag{18}$$

$$= \text{Trace} [\ \Delta A(\theta_j)^T \Delta A(\theta_j)\ ]$$

$$= \sum_{i=1}^{n} \lambda_{R_i(\theta_j)}^{(i)}$$

where $|\cdot|_F$ is the Frobenius norm.

Then using (18), (17) becomes,

$$A = \text{Argmin} \sum_{j=1}^{J} \sum_{l=1}^{n} \sum_{p=1}^{n} p(\theta_j) \Delta a_{lp}(\theta_j)^2 \tag{19}$$

$$= \underset{G:X \to X}{\text{Argmin}} \sum_{j=1}^{J} p(\theta_j) | A(\theta_j) - G |_F^2$$

$$= \sum_{j=1}^{J} p(\theta_j) A(\theta_j)$$

where the last step above is due to the fact that the minimum of the sum of the weighted-squared distances to all points is achieved at the statistical average.

3.2 Detection Operator (D) Selection

In section 2, the existence of D was guaranteed by the output separability of the failure event set (i.e., $CQ_i \cap (\sum_{j \neq i} CQ_j) = \phi$ $i = 1, 2, ..., h$). It should be noted, however, that the condition of the output separability becomes a difficult criterion to achieve once uncertainties are introduced in the system dynamics.

In this section, the procedure of selecting D will be generalized to extend its applicability to the systems with uncertain dynamics. To satisfy the definition for D (section 2) as closely as possible, the least-squared-error (LSE) criterion will be employed here.

As a result of section 3.1, for each θ_j and f_i,

$$\underline{v}(k+1) = A(\theta_j)\underline{v}(k) + B\underline{u}(k) + D(\underline{y}(k) - \underline{w}(k)) - \Delta A(\theta_j)\underline{v}(k)$$

$$\underline{\varepsilon}(\theta_j; k+1) = (A(\theta_j) - DC)\underline{\varepsilon}(\theta_j; k) + \Delta A(\theta_j)\underline{v}(k) + f_i n(k)$$

$$\approx (A(\theta_j) - DC)\underline{\varepsilon}(\theta_j; k) + f_i n(k)$$

Let each element f_i of the failure event set be normalized such that $|f_i| = 1$. Under the assumption of small system uncertainties, it is assumed here that there exists a positive integer $\hat{v}_i$ for each f_i such that for all θ_j,

(i) $\quad | C\hat{f}_i(\theta_j) | >> \delta$ (20)

(ii) $\quad | CA(\theta_j)^k f_i | < \delta \qquad$ for all $k < \hat{v}_i - 1$

where $\hat{f}_i(\theta_j) = A(\theta_j)^{\hat{v}_i - 1} f_i$. The positive constant δ should be determined according to a specific application.

Define each subspace $Q_i(\theta_j)$ as,

$$Q_i(\theta_j) \overset{\Delta}{=} f_i \oplus A(\theta_j) f_i \oplus A(\theta_j)^2 f_i \oplus \cdots \oplus A(\theta_j)^{\hat{v}_i - 1} f_i$$

where the direct sums are true by (20).

Let the desired eigenvalues of $Q_i(\theta_j)$ be $\{ \lambda_i(\theta_j)^{(p)} \quad p = 1, 2, ..., \hat{v}_i \}$. In addition to satisfying the eigenvalue requirements, the resulting D must also satisfy (21) below for the maximality of the uniqueness among the failure signatures. I.e., given f_i,

$$| \underline{\epsilon}(\theta_j; k) \bmod Q_i(\theta_j) | < \bar{\delta} \quad \text{for all } k > 0$$ (21)

for an arbitrarily small positive constant $\bar{\delta}$, where "mod" denotes the part of $\underline{\epsilon}(\theta_j; k)$ belonging to the space orthogonally complement to $Q_i(\theta_j)$. Note that (21) can be achieved if the dynamics of $(A(\theta_j) - DC)$ are stable, and in the literature, such $Q_i(\theta_j)$ is referred to as the "almost invariant" space. [7]

Define,

$$\Psi_i(\theta_j; \lambda) = \prod_{p=1}^{\hat{v}_i} (\lambda - \lambda_i(\theta_j)^{(p)})$$

$$= \lambda^{\hat{v}_i} + \alpha_i(\theta_j)^{(1)} \lambda^{\hat{v}_i - 1} + \alpha_i(\theta_j)^{(2)} \lambda^{\hat{v}_i - 2} + \cdots + \alpha_i(\theta_j)^{(\hat{v}_i - 1)} \lambda + \alpha_i(\theta_j)^{(\hat{v}_i)}$$

If D satisfies the eigenvalue requirement precisely, then $\Psi_i(\theta_j; A(\theta_j) - DC) f_i = 0$. Consequently,

$$- (A(\theta_j) - DC)^{\hat{v}_i} f_i = \alpha_i(\theta_j)^{(1)} (A(\theta_j) - DC)^{\hat{v}_i - 1} f_i + \alpha_i(\theta_j)^{(2)} (A(\theta_j) - DC)^{\hat{v}_i - 2} f_i + \cdots$$

$$\cdots + \alpha_i(\theta_j)^{(\hat{v}_i - 1)} (A(\theta_j) - DC) f_i + \alpha_i(\theta_j)^{(\hat{v}_i)} f_i$$

Using (20), the above equation can be approximated as

$$- (A(\theta_j) - DC) A(\theta_j)^{\hat{v}_i - 1} f_i \approx \alpha_i(\theta_j)^{(1)} A(\theta_j)^{\hat{v}_i - 1} f_i + \alpha_i(\theta_j)^{(2)} A(\theta_j)^{\hat{v}_i - 2} f_i + \cdots$$

$$\cdots + \alpha_i(\theta_j)^{(\hat{v}_i - 1)} A(\theta_j) f_i + \alpha_i(\theta_j)^{(\hat{v}_i)} f_i$$

Hence, D is a solution of,

$$DCA(\theta_j)^{\hat{v}_i - 1} f_i \approx \Psi_i(\theta_j; A(\theta_j)) f_i \overset{\Delta}{=} \phi_i(\theta_j)$$ (22)

Note that although the solution of (22) will satisfy (21), for each θ_j,

$$d(CQ_i(\theta_j)) \geq 1 \qquad i = 1, 2, ..., h$$

$$CQ_i(\theta_j) \cap \{ \sum_{p \neq i} CQ_p(\theta_j) \} \neq \phi$$

Hence, the solution of D solving (22) simultaneously for all f_i and θ_j may not exist.

The optimal solution for D in this paper will be the LSE solution of (22), and the eigenvalue assignments corresponding to such D will be limited to those of $Q_i(\theta_j)$ for $i = 1, 2, ..., h$ $j = 1, 2, ..., s$.

With $\phi_i(\theta_j)$ $i = 1, 2, ..., h$, define,

$$\Pi(\theta_j) = [\ C\hat{f}_1(\theta_j)\quad C\hat{f}_2(\theta_j)\quad \cdots\quad C\hat{f}_h(\theta_j)\ ]$$

$$\Phi(\theta_j) = [\ \phi_1(\theta_j)\quad \phi_2(\theta_j)\quad \cdots\quad \phi_h(\theta_j)\ ]$$

Then, for D to satisfy (22) for all f_i and θ_j,

$$D\ \Pi(\theta_j) = \Phi(\theta_j)\qquad j=1,2,...,s \tag{23}$$

As previously mentioned, the exact solution of (23) for all θ_j may not exist. Thus, LSE solution of (23) will be sought. Such a criterion will result in a closed form solution. Define,

$$F(D) \overset{\Delta}{=} \sum_{j=1}^{s}\sum_{l=1}^{n}\sum_{p=1}^{h} p(\theta_j)\ \beta_{lp}(\theta_j)^2 \tag{24}$$

where $[\ \beta_{lp}(\theta_j)\ ]_{lp} = [\ D\ \Pi(\theta_j) - \Phi(\theta_j)\ ]$

Assume that in minimizing (24), (23) has been normalized such that the Euclidean norm of each column in the right hand side of (23) is equal to 1. Then $F(D)$ is the mean-square-of-error function, and D minimizing $F(D)$ is the LSE solution. Define,

$$\Pi = [p(\theta_1)^{\frac{1}{2}}\ \Pi(\theta_1)\quad p(\theta_2)^{\frac{1}{2}}\ \Pi(\theta_2)\quad \cdots\quad p(\theta_s)^{\frac{1}{2}}\ \Pi(\theta_s)\ ]$$

$$\Phi = [p(\theta_1)^{\frac{1}{2}}\ \Phi(\theta_1)\quad p(\theta_2)^{\frac{1}{2}}\ \Phi(\theta_2)\quad \cdots\quad p(\theta_s)^{\frac{1}{2}}\ \Phi(\theta_s)\ ]$$

Then the LSE solution of (23) is the LSE solution of,

$$D\ \Pi = \Phi \tag{25}$$

As shown below, let d_i and ϕ_i be the i'th rows of D and Φ respectively.

$$D = \begin{bmatrix} d_1 \\ d_2 \\ \cdot \\ \cdot \\ d_n \end{bmatrix} \qquad \Phi = \begin{bmatrix} \phi_1 \\ \phi_2 \\ \cdot \\ \cdot \\ \phi_n \end{bmatrix}$$

Hence (25) becomes equivalent to,

$$\Pi^T\ d_i{}^T = \phi_i{}^T\qquad i=1,2,...,n \tag{26}$$

If the $rank(\Pi) = m$, then Π^T is one-to-one and thus,

$$d_i{}^T = (\ \Pi\ \Pi^T)^{-1}\ \Pi\ \phi_i{}^T\qquad i=1,2,...,n$$

$$\begin{bmatrix} d_1 \\ d_2 \\ \cdot \\ \cdot \\ d_n \end{bmatrix} = \begin{bmatrix} \phi_1\Pi^T(\ \Pi\ \Pi^T)^{-1} \\ \phi_2\Pi^T(\ \Pi\ \Pi^T)^{-1} \\ \cdot \\ \cdot \\ \phi_n\Pi^T(\ \Pi\ \Pi^T)^{-1} \end{bmatrix}$$

$$D = \Phi\ \Pi^T(\ \Pi\ \Pi^T)^{-1} \tag{27}$$

If the rank$(\Pi) < m$, $(\ \Pi\ \Pi^T)^{-1}$ does not exist. Such Π^T is neither one-to-one nor onto, and consequently the LSE solution of D is not unique. This ambiguity can be resolved by assigning D as the least effort solution among all the LSE solutions of (25). Assume that rank$[\ \Pi\ ] = \hat{m} < m$. Then Π^T can be decomposed into,

$$\Pi^T = \Pi_1\ \Pi_2$$

where $\Pi_1 \in M^{\hat{m} \times \hat{m}}$ is one-to-one, and $\Pi_2 \in M^{\hat{m} \times m}$ is onto. It is known that such decomposition is unique. [8] Then,

$$d_i = \phi_i \Pi_1 (\Pi_2 \Pi \; \Pi_1)^{-1} \Pi_2$$

$$D = \Phi \; \Pi_1 (\Pi_2 \Pi \; \Pi_1)^{-1} \Pi_2 \tag{28}$$

where $\{ \Pi_1 (\Pi_2 \Pi \; \Pi_1)^{-1} \Pi_2 \}^T$ is referred as the "pseudo-inverse" of Π^T. The solution of D satisfying (28) will result in the minimum amount of feedback gain, which is desirable in noisy environments.

3.3 Hypothesis Isolation

In what preceded, our attention has been devoted to generating a set of mutually distinct failure residuals with detectable magnitudes. Recall that this task has been accomplished via proper selection of D (and A in cases of uncertain systems). Once such failure residuals have been generated, the next step in the BJDF design becomes constructing a decision procedure to isolate the corresponding failure event when a pre-specified pattern is observed in the failure residual.

The approach taken in this section is to first evaluate the simple decision procedure for the case of a precisely known system, and then the procedure will be generalized to the case of a system with dynamic uncertainties.

Consider the dynamic system (1) and the set of failure events $\{ f_i \}_{i=1}^{h}$. Then, the procedure given in section 2 will constrain the failure residual $\underline{q}(k)|_{f_i}$ in the direction of CQ_i. Therefore, the detection of the i'th failure hypothesis can be declared once the residual constrained in the CQ_i direction is observed.

In a more precise mathematical sense, this procedure can be expressed as an automorphic transformation $P : Y \rightarrow Y$, which is no more than a coordinate transformation of $\underline{q}(k)$. Such P should be in the form of,

$$P = P_\Xi + P_{\Xi^\propto}$$

where P_Ξ $(P_{\Xi^\propto})$ is a projection operator on Ξ $(\Xi^\propto)$ along $\Xi^\propto$ (Ξ) such that,

$$R (P_\Xi) = \Xi$$

$$\eta (P_\Xi) = \Xi^\propto$$

$$R (P_{\Xi^\propto}) = \Xi^\propto$$

$$\eta (P_{\Xi^\propto}) = \Xi$$

where $\Xi^\propto$ is the orthogonal complement of the flag space $\Xi \subseteq Y$, which is the space containing the hypothesized failure residuals. Hence, $d(\Xi) = h$. If vectors $\{ \xi_i \}_{i=1}^{h}$ and $\{ \xi_i^\propto \}_{i=1}^{m-h}$ are the unit basis vectors for Ξ and $\Xi^\propto$ respectively, any $\underline{q}(k) \in Y$ can be uniquely written as,

$$\underline{q}(k) = \sum_{i=1}^{h} \rho_i(k) \xi_i + \sum_{i=1}^{m-h} \rho_{h+i}(k) \xi_i^\propto$$

Define $[\rho_1(k) \; \rho_2(k) \; \cdots \; \rho_m(k)]$ to be the set of "flag coefficients." Since these flag coefficients preserves all the information of $\underline{q}(k)$, the present design problem becomes equivalent to constructing the optimal basis set for Ξ such that the resulting basis coordinate yields the "maximal distances" between failure hypotheses. Note that since $\Xi^\propto$ is orthogonal to Ξ, the basis selection for $\Xi^\propto$ can be independently carried out.

In this paper, a criterion is chosen for the optimal basis set such that

$$\xi_i = \operatorname*{Argmin}_{|y|=1} \; |\; P_{y^*} q(k) \;|_{f_i}\;|^2 \qquad i=1,2,...,h \tag{29}$$

where P_{y^*} is an orthogonal projection operator with $KerP_{y^*} = y$. It is easy to see that

$$P_{y^*} = I_m - yy^T \tag{30}$$

Hence, the basis set satisfying (29) will result in the maximally distinguishable set of flag coefficients.

Let each ξ_i $i=1,2,...,h$ be the unit vector in CQ_i direction. Furthermore, let each ξ_i^∞ $i=1,...,m-h$ be arbitrarily chosen for Ξ^∞. Then if $\{ f_i \}_{i=1}^h$ is an output separable set, given f_i, the flag coefficient set becomes $[0 \; 0 \; .. \; 0 \; p_i(k) \; 0 \; .. \; 0]$. Hence, such basis set will satisfy (29).

Given uncertainties in the system dynamics, however, the flag coefficient set can neither be made all zeroes under the absence of failures, nor can be restricted to precisely one nonzero entry when a hypothesized failure is present. Instead, at best, the flag coefficient set can be made as near to zero as possible under normal conditions, while it closely correlates to an identifiable pattern when the corresponding hypothesis occurs.

To accomplish the cases of uncertain dynamics, (29) is modified to,

$$\xi_i = \operatorname*{Argmin}_{|y|=1} \; \operatorname*{maximum}_{\text{all } \theta_j} \; |\; P_{y^*} q_i(\theta_j) \;|^2 \qquad i=1,2,...,h \tag{31}$$

where $q_i(\theta_j)$ is the unit vector in $CA(\theta_j)^{\dot{v}_i-1} f_i$ direction.

Consider the figure below.

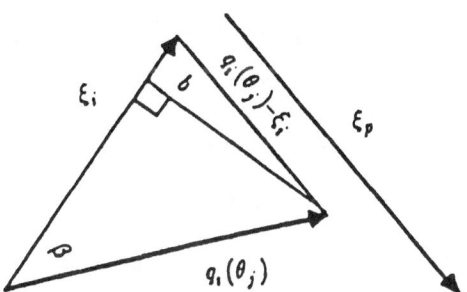

Figure 1. Worst basis case

If ξ_p $p \neq i$ becomes aligned in the direction of the vector $q_i(\theta_j) - \xi_i$ for some j, the effect of f_i is maximal along the ξ_p direction with magnitude $c > 0$. Since c is a monotone increasing function of $|\beta|$, minimizing another monotone increasing function $b > 0$, will simultaneously minimize c. Assumed here is small system uncertainty so that $-\frac{\pi}{2} \leq \beta \leq \frac{\pi}{2}$.

However, to get a closed form solution for (31), the problem must be modified to

$$\xi_i = \operatorname*{Argmin}_{|y|=1} \; \sum_{j=1}^{i} \; |\; P_{y^*} q_i(\theta_j) \;|^2 \tag{32}$$

Again, this approach is justified by the fact that

$$\frac{1}{s}\sum_{j=1}^{s} \mid P_{y^{\perp}}q_i(\theta_j)\mid^2 \; \le \; \max_{\text{all }\theta_j} \mid P_{y^{\perp}}q_i(\theta_j)\mid^2 \; \le \; \sum_{j=1}^{s} \mid P_{y^{\perp}}q_i(\theta_j)\mid^2$$

Thus (32) provides the upper and lower bounds for (31). From (32),

$$\xi_i = \text{Argmin}_{|y|=1} \sum_{j=1}^{s} q_i(\theta_j)^T P_{y^{\perp}}^T P_{y^{\perp}} q_i(\theta_j)$$

$$= \text{Argmin}_{|y|=1} \sum_{j=1}^{s} q_i(\theta_j)^T P_{y^{\perp}} P_{y^{\perp}} q_i(\theta_j)$$

$$= \text{Argmin}_{|y|=1} \sum_{j=1}^{s} q_i(\theta_j)^T P_{y^{\perp}} q_i(\theta_j)$$

$$= \text{Argmin}_{|y|=1} \sum_{j=1}^{s} q_i(\theta_j)^T (I - yy^T) q_i(\theta_j)$$

$$= \text{Argmin}_{|y|=1} \sum_{j=1}^{s} 1 - q_i(\theta_j)^T yy^T q_i(\theta_j)$$

$$= s - \text{Argmax}_{|y|=1} \sum_{j=1}^{s} y^T q_i(\theta_j) q_i(\theta_j)^T y$$

Note that symmetry $(P_{y^{\perp}}^T = P_{y^{\perp}})$ and idempotence $(P_{y^{\perp}} P_{y^{\perp}} = P_{y^{\perp}})$ of $P_{y^{\perp}}$ were used during the above derivation.

Then it is easy to see that (32) becomes,

$$\xi_i = \text{Argmax}_{|y|=1} y^T \{ q_i(\theta_1) q_i(\theta_1)^T + q_i(\theta_2) q_i(\theta_2)^T + \cdots + q_i(\theta_s) q_i(\theta_s)^T \} y$$

Define $Q_i = \sum_{j=1}^{s} q_i(\theta_j) q_i(\theta_j)^T$. Assume $0 \le \lambda_{Q_i}^{(1)} \le \lambda_{Q_i}^{(2)} \le \cdots \le \lambda_{Q_i}^{(m)}$ to be the eigenvalues of Q_i. Then, ξ_i becomes the unit eigenvector corresponding to the largest eigenvalue $\lambda_{Q_i}^{(m)}$. Therefore, ξ_i $i=1,2,...,h$ is a unit solution of,

$$(Q_i - \lambda_{Q_i}^{(m)} I_m) \xi_i = 0 \tag{33}$$

Such selection of $\{ \xi_i \}_{i=1}^{h}$ will (sub)minimize the probability p_{ij} $i \ne j$, where

$$p_{ij} = \text{probability (declare the i'th failure given } f_j) \quad i,j = 1,2,...,h$$

If the effect of the probability function $p(\theta_j)$ needs to be included, (32) must be further modified to,

$$\xi_i = \text{Argmin}_{|y|=1} \sum_{j=1}^{s} p(\theta_j) \mid P_{y^{\perp}} q_i(\theta_j)\mid^2 \tag{34}$$

Note that the solution for (34) will again be the unit solution of (33) with Q_i as,

$$Q_i \overset{\Delta}{=} \sum_{j=1}^{s} p(\theta_j) q_i(\theta_j) q_i(\theta_j)^T$$

4. Conclusion

The problem of BJDF is in fact equivalent to finding a post compensator that diagonalizes the transfer function matrix between the failure events and the corresponding flag coefficient set.

Consequently, the FDI task via BJDF becomes a sensitive process to the parameter variations in the dynamic systems. This fact is well recognized in the control problem of output decoupling, which is the dual problem to BJDF.

In the past, the lack of robustness has been the principal cause that prevented the widespread applications of BJDF. Hence in this paper, a strong emphasis was placed on developing design procedures for the cases of uncertain systems. As a result, solutions for the optimal reference model, the optimal detection operator, and the optimal basis set for the hypothesis isolation were derived in section 3, where each of the solution was presented in a closed form.

5. References

[1] R. V. Beard, "Failure Accommodation in Linear System Through Self-Reorganization," Rept. MIT-71-1, Man Vehicle Laboratory, Feb. 1971.

[2] H. L. Jones, "Failure Detection in Linear Systems," Ph.D. Thesis, MIT, Sept. 1973.

[3] A. S. Willsky, "A Survey of Design Methods for Failure Detection in Dynamic Systems," Automatica, vol. 12., 1976.

[4] R. Isermann, "Process Fault Detection Based on Modeling and Estimation Method, a Survey," Automatica, vol. 20, No. 4, 1984.

[5] A. S. Willsky, "Failure Detection in Dynamic System," AGARD-LS-109, 1980.

[6] P. S. Min, "Detection of Incipient Failures in Dynamic System," Ph.D. Thesis, Dept. of Electrical Engineering and Computer Science, University of Michigan, 1987.

[7] J. C. Willems, "Almost Invariant Subspaces: An Approach to High Gain Feedback Design - Part I: Almost Controlled Invariant Subspaces," IEEE Trans. Auto. Control, February 1981.

[8] C. N. Dorny, "A Vector Space Approach to Models and Optimization," Wiley-Interscience, 1975.

LIE GROUPS UNDERLYING FAULT AVOIDANCE
IN DYNAMICAL CONTROL SYSTEMS

Dahlard L. Lukes

1. Introduction. Sometimes situations arise requiring the avoidance of certain states x^0 of the implementation O^i of a prescribed input-output operator O; perhaps there are faults in the implementation. (See Figure 1.)

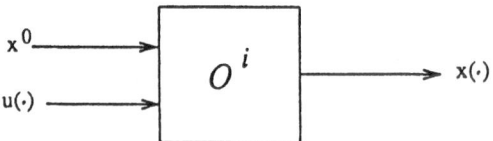

FIGURE 1. The implementation O^i of a
prescribed input-output operator O.

The critical applications envisioned disallow changing the implementation, faulty as it might be. Production of the intended response is crucial and O^i is the sole operator dedicated to the task.

An example of such an O^i is a control system containing components that malfunction in some operating states; another is a computer simulation containing programming errors. The troublesome states might be unknown. If an event, consisting of a collection of such states, seemed imminent then steps might be taken to avoid its occurrence. In a faster than real time computer simulation in which erroneous output did occur and were recognized as such, there might still be time to attempt an alternative procedure. These problems have been recognized in the area of computer software fault tolerance. (See Ammann and Knight,[1].)

The approach developed here embeds the operator O^i into an ambient control structure O^a as indicated in Figure 2. (Temporarily ignore the subscript A,B whose explanation follows the next paragraph.)

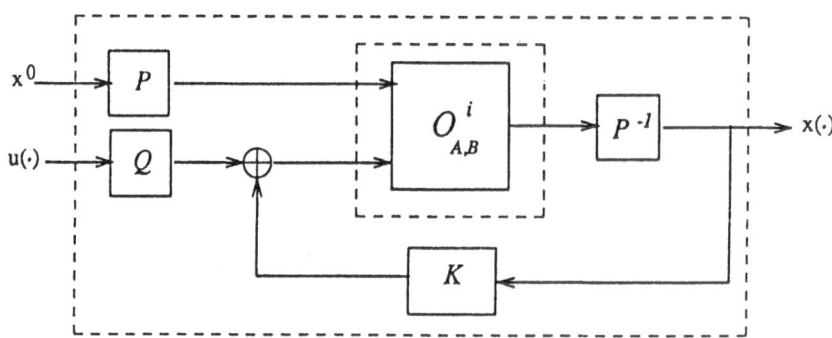

FIGURE 2. The embedding of the implementation O^i
into an ambient operator O^a.

For inputs into O^a, x^0 presumably known and most likely $u(\cdot)$ unknown, a potential now exists for producing the intended response while avoiding the fault triggering events, by appropriate manipulation of the operators P,Q and K in O^a. Precisely how to do the manipulation, an important issue that eventually must be faced, is not considered here. The goal is to inject a maximal set of parameters into the ambient system via P,Q and K for doing the avoidance and to describe their allowable range conforming to the constraint that O^a reduce to O when $O^i = O$. This formulation of the problem is intended to bear on various situations in which O^i responds to inputs in the same manner as O except for exceptional values of x^0. If the fault states can be avoided then O can be substituted for O^i in doing the analysis.

This article is confined to operators $O = O_{A,B}$ of the type defined by either differential control equations

$$\dot{x} = Ax + Bu \tag{1}$$

$$x(t_0) = x^0 \tag{2}$$

having state variable $x \in R^n$ and control variable $u \in R^p$ with real coefficient matrices A and B prescribed or the discrete analogue

$$x_{k+1} = Ax_k + Bu \tag{1'}$$

$$x_{k_0} = x^0. \tag{2'}$$

Throughout the paper B is assumed to be of full positive rank $p \le n$.

The piecewise constant avoidance investigated takes the P,Q and K of Figure 2 to be manipulatable matrix multiplication operators held fixed during system operation or switched so as to be constant on each interval of any sequence of consecutive intervals. (In the discrete parameter case switching can occur at any discrete time point. The more complicated problem wherein P,Q and K vary continuously with time is dealt with under section 6.

2. Piecewise constant avoidance.

The problem is to study those real matrices P,Q and K which in Figure 2 provide the invariance $O^a = O_{A,B}$ for $O^i_{A,B} = O_{A,B}$. Since the result is the same for discrete dynamics as for continuous dynamics it suffices to consider continuous time. Hence if $x(t)$ is the response to (1) - (2) for $u = u(t)$ any locally integrable control function on $t_0 \le t < \infty$ and $y(t)$ is the response to $\dot{y} = Ay + B[Kx(t) + Qu(t)]$ with initial $y(t_0) = Px^0$ then the invariance requirement asks that $x(t) = P^{-1}y(t)$ for $t_0 \le t < \infty$. In other words P and its associated real matrices Q and K are to alter the initial input to $O_{A,B}$ without disrupting the intended responce $x(t)$. One can readily verify that the invariance translates into the matrix equations

$$PA - AP = BK \tag{3}$$

$$PB = BQ \tag{4}$$

for nonsingular P. (Note that if an invertible P satisfies (4) then Q is automatically invertible.) In summary, the employment of feedback in the quest to maximize fault avoidance potential leads to the following parameter set of interest.

Definition.

$$\mathcal{G}_{A,B} = \{P \mid PA - AP = BK, PB = BQ, \det(P) \ne 0, \text{ some } Q, K\}.$$

The first theorem points out the combined algebraic and topological structure of this parameter set.

THEOREM 1. $\mathcal{G}_{A,B}$ is a closed Lie subgroup of $GL(n,R)$.

Proof. Showing that $\mathcal{G}_{A,B}$ is a group involves algebraic manipulation of the defining equations. It is a Lie subgroup by virtue of the fact that it can be shown to be a closed subgroup of $GL(n,R)$. The details are in Lukes, [2].

Remark.   Theorem 1 remains valid for $B = 0$.

In the present discussion the matrices A and B are fixed. Consider Figure 2 in which we require that $P \in \mathcal{G}_{A,B}$ and for each choice of P determine K and Q by the respective formulas,

$$K = (B^*B)^{-1}B^*(PA - AP) \tag{5}$$

$$Q = (B^*B)^{-1}B^*PB . \tag{6}$$

Fix $P = P_1 \in \mathcal{G}_{A,B}$ and denote the consequent operator of Figure 2 by $P\mathcal{O}_{A,B}^i$. Take $\mathcal{O}_{A,B}^i$ to be $\mathcal{O}_{A,B}$. It has been established that $P_1\mathcal{O}_{A,B}$ is the same operator irrespective of the choice of $P_1 \in \mathcal{G}_{A,B}$, namely, $\mathcal{O}_{A,B}$, although four of the constituent blocks in Figure 2 contain matrices depending on the choice of $P = P_1$. Each $P\mathcal{O}_{A,B}$, for $P \in \mathcal{G}_{A,B}$, is called a *fault tolerant realization (f.t.r.)* of $\mathcal{O}_{A,B}$. Now consider taking $\mathcal{O}^i = P_1\mathcal{O}_{A,B}$ in Figure 2. The previous statements apply to $P_2(P_1\mathcal{O}_{A,B})$ which we identify with $(P_1P_2)\mathcal{O}_{A,B}$ for P_1, P_2 in $\mathcal{G}_{A,B}$ and it is likewise called a f.t.r. of $\mathcal{O}_{A,B}$. (See Figures 3 and 4.)

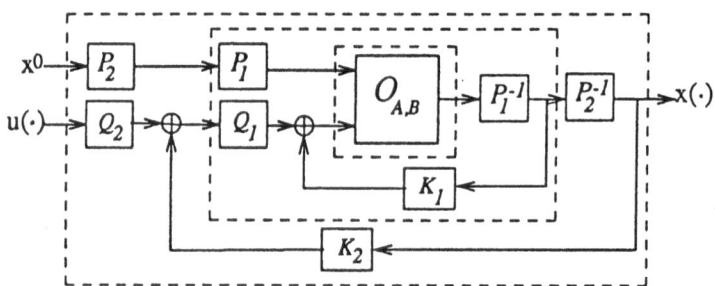

FIGURE 3. The block diagram for the fault tolerant
realization $P_2(P_1\mathcal{O}_{A,B})$ of $\mathcal{O}_{A,B}$.

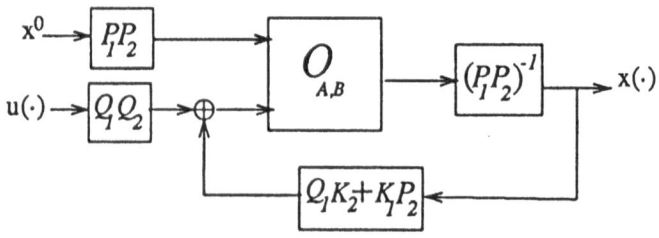

FIGURE 4. The equivalent block diagram for the fault tolerant
realization $(P_1P_2)\mathcal{O}_{A,B} = P_2(P_1\mathcal{O}_{A,B})$ of $\mathcal{O}_{A,B}$.

COROLLARY 1. *The collection of f.t.r.'s of $O_{A,B}$ is a Lie group isomorphic to $\mathcal{G}_{A,B}$ under the identification $P \rightarrow P O_{A,B}$.*

The above corollary emphasizes the fact that the Lie group $\mathcal{G}_{A,B}$ represents the algebraic and topological cohesion underlying the collection of all f.t.r.'s of $O_{A,B}$. The multiplication of two f.t.r.'s, say $P_2 O_{A,B}$ times $P_1 O_{A,B}$, has the natural system-theoretic interpretation of wrapping one layer of control structure from each about $O_{A,B}$ as illustrated in Figures 3 and 4. In particular, by adjunction of another layer, the control structure of $P O_{A,B}$ about $O_{A,B}$ imposed by P effectively can have its feedback loop broken ($K \rightarrow O$) and be erased away upon multiplication by the inverse, $P^{-1} O_{A,B}$. The manifold structure opens the door to doing global calculus and differential equations on the collection of all f.t.r.'s of $O_{A,B}$. Generally, every Lie group statement or concept bearing on $\mathcal{G}_{A,B}$ has its systems-theoretic interpretation in terms of f.t.r.'s.

The approach to fault avoidance under discussion selects the P in Figure 2 so as to avoid fault states at which $O'_{A,B}$ fails to respond in the same manner as $O_{A,B}$ and, by restricting $P \in \mathcal{G}_{A,B}$, this is done without disrupting the intended response $x(t)$. Moreover the initial choice of P can be switched to another value in $\mathcal{G}_{A,B}$ at a later time by taking the state x at that time to be the new initial state x^0. Of course the more extensive $\mathcal{G}_{A,B}$ is the more potential there is for doing the avoidance. One would expect, from the form of the equations (3) - (4) defining $\mathcal{G}_{A,B}$, that the feedback aspect of the control structure involved in the definition of f.t.r.'s would tend to provide a larger group of symmetries than would occur without it. Another factor expected to influence the potential for avoidance would be the coefficients A, B themselves. That is, some prescribed input-output operators would be expected to be naturally more fault tolerant than others. The extent to which these intuitions are correct and the answers to many related questions about fault tolerance that might be raised would revolve about $\mathcal{G}_{A,B}$ and the relationship between A, B and the group $\mathcal{G}_{A,B}$ that comes along with that pair. A closer examination of $\mathcal{G}_{A,B}$ seems to be in order.

3. **The detailed structure of $\mathcal{G}_{A,B}$.** It will be possible to describe the detailed structure of $\mathcal{G}_{A,B}$ in the general case where A, B is a controllable pair of matrices. Control theory defines the pair to be *controllable* if its controllability matrix has full rank; i.e., if

$$rank[B, AB, \cdots, A^{n-1}B] = n.\qquad(7)$$

LEMMA 1. *For (A, B) a controllable pair there exists a matrix K_0 and invertible matrices P_0 and Q_0 such that*

$$P_0(A + BK_0)P_0^{-1} = A^0 \qquad (8)$$

$$P_0 B Q_0 = B^0 \qquad (9)$$

in which (A^0, B^0) is the Brunovsky canonical form of (A, B).

(See Lukes, [1], for a discussion of a method for computing P_0, Q_0 and K_0.)

LEMMA 2. *For (A, B) a controllable pair of matrices the maps*

$$\mathcal{G}_{A^0,B^0} \rightarrow \mathcal{G}_{A,B}$$

$$\mathscr{G}_{A^0,B^0} \rightarrow \mathscr{G}_{A,B}$$

defined by $h(P) = P_0^{-1} P P_0$ are Lie group and Lie algebra isomorphisms, respectively.

Proof. Lemma 2 follows as a corollary to Theorem 1 and Lemma 1. The details are omitted.

The problem of computing $\mathcal{G}_{A,B}$ when (A, B) is controllable is solved by linking the problem to the feedback control theory of decoupling. The next theorem describes the structure of $\mathcal{G}_{A,B}$. In it the r_i are the dimensions of the decoupled subsystems occurring in the Brunovsky form of (A, B).

THEOREM 2. *For (A, B) a controllable pair and $\rho_1 > \rho_2 > \cdots > \rho_s$ the distinct values of its canonical dimension numbers $r_1 \geq r_2 \geq \cdots \geq r_p$, write $n = m_1\rho_1 + m_2\rho_2 + \cdots m_s\rho_s$ by denoting the multiplicity of ρ_k by m_k. There are semi-direct products $\times_\sigma$ and $\times_\tau$ for which there exists a Lie group isomorphism*

$$\mathcal{G}_{A,B} \cong \prod_{k=1}^{s} GL(m_k, R) \times_\sigma \left[R^{\alpha_*} \times_\tau R^{(n-q)q} \right]$$

in which

$$s_* = \text{the largest integer for which } \rho_{s_*} \geq 2$$

$$\alpha_* = \sum_{i=1}^{s_*-1} \alpha_i , \text{ where } \alpha_i = m_{i+1} \sum_{k=1}^{i} (\rho_i - \rho_{i+k} + 1)m_k$$

and

$$q = \begin{cases} 0 & \text{if } \rho_s > 1 \\ m_s & \text{if } \rho_s = 1 \end{cases}$$

Remarks about the proof of Theorem 2. Due to Lemma 2 it suffices to prove the theorem for (A,B) in its canonical form. The main task is solving the matrix equations defining $\mathcal{G}_{A,B}$ for P. The solution contains three families of parameters which accounts for the three Lie groups of the resultant decomposition established by the theorem. Those readers not familiar with the notion of a semi-direct product can consult Varadarajan, [1]. Formulas for $\times_\sigma$ and $\times_\tau$ as well as the extensive details of the proof appear in Lukes, [2].

Note. It is evident that $\mathcal{G}_{A,B}$ is not connected.

4. Piecewise constant avoidability and unavoidable events.

In the present context an *event E* for system (1) or (1') is defined to be any subset of the state space R^n. In specific applications a question of primary interest would be, is it possible that for some input function $u(\cdot)$ and some future time t_1 no piecewise constant switching of $P \in \mathcal{G}_{A,B}$ in the fault tolerant system $P O_{A,B}$ will prevent the occcurence of $y(t_1) \in E$ (or $y_k \in E$ in the discrete problem)? Here y denotes the output of $O_{A,B}$ in $P O_{A,B}$.

If the answer to the above question is affirmative then E is called *unavoidable*. Otherwise it is referred to as being an *avoidable* event. Certainly the existence of an unavoidable fault event E would be cause for concern in a critically designed system. In the definition of an unavoidable event the choice of the initial state x^0 is immaterial since as a consequence of the assumed controllability of (A, B) the unavoidability of an event from x^0 implies it unavoidability from every point in R^n. (Choose an input function $u(\cdot)$ that first steers $x(t)$ back to x^0.)

Recall that if $\mathcal{G}_{A,B}$ acts on R^n then for each $x \in R^n$ the set $\mathcal{G}_{A,B} x = \{Px \mid P \in \mathcal{G}_{A,B}\}$ is called the *orbit* of x in R^n. Theorem 3 indicates that $\mathcal{G}_{A,B}$ carries considerable information about which events are avoidable.

THEOREM 3. *Suppose that (A, B) is a controllable pair. Then an event $E \subseteq R^n$ is unavoidable if it contains an orbit of the action of $G_{A,B}$ on R^n . If the closure of an event contains no orbit of $G_{A,B}$ then the event is avoidable. The orbits are submanifolds of R^n.*

5. Computation of dimensions. Although the topic is not discussed here it is possible to compute the orbit of a point. However it is possible to compute the dimension of the orbit of a point $x \in R^n$ without actually finding the orbit. By computing the canonical dimension numbers $r_1 \geq r_2 \geq \cdots \geq r_p$ Theorem 2 can be applied to obtain an expression $P(\lambda)$ for an arbitrary element of G_{A^0,B^0} in terms of coordinates λ and then the appropriate formula is

$$dim(G_{A,B} x) = rank \frac{\partial}{\partial \lambda}[P(\lambda)] \tag{10}$$

in which $\frac{\partial}{\partial \lambda}[P(\lambda)]$ is the Jacobian matrix. (The Jacobian matrix will be independent of λ.)

A related group is

$$G_{A,B}^x = \{ P \mid P \in G_{A,B} , Px = x \} \tag{11}$$

which is a Lie subgroup of $G_{A,B}$ called the *isotropy* subgroup . A standard theorem of Lie theory gives the dimension relation

$$dim(G_{A,B}^x) = dim(G_{A,B}) - dim(G_{A,B} x) . \tag{12}$$

(See Varadarajan, [1].) Once the dimension numbers are found Theorem 2 can be applied to compute $dim(G_{A,B})$ and then that result can be combined with the one from (10) to determine the isotropy dimension by means of (12).

6. Continuous fault avoidance for differential systems. The analysis here parallels that of the previous sections with $O_{A,B}$ restricted to differential equations (1) - (2). A and B are as before but now P is an absolutely continuous matrix function defined on an interval I , $P(\cdot) : I \rightarrow GL(n,R)$ and $Q(\cdot)$ and $K(\cdot)$ are likewise allowed to be time dependent. (See Fig.5.)

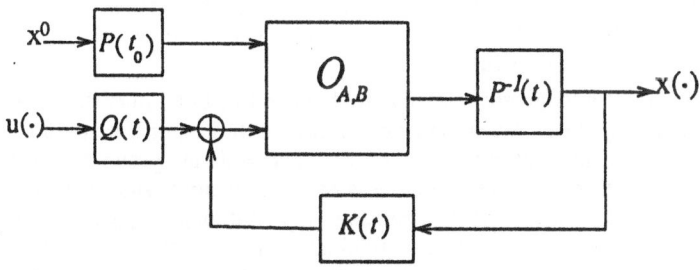

FIGURE 5. A continuous-avoidance
fault tolerant realization.

The interval I can be of either type, $I = [t_0 , \infty)$ or $[t_0 , t_1]$, with t_0 and t_1 real. Where the type makes a difference the choice will be indicated.

The invariance requirement that for all x^0 and $u(\cdot)$ the solution to $\dot{y} = Ay + B[K(t)x(t) + Q(t)u(t)]$, $y(t_0) = y^0$ satisfy $y(t) = P(t)x(t)$ where $x(t)$ is the solution to $\dot{x} = Ax + Bu(t)$, $x(t_0) = x^0$ translates into the equivalent conditions

$$\dot{P} + PA - AP = BK(t) \tag{13}$$

$$PB = BQ(t) , \tag{14}$$

generalizing (3) - (4). (Equality in differential equations is to be interpreted as a.e. $t \in I$ throughout the remainder of this article.) This leads to defining

$$\tilde{g}_{A,B}^{\infty} = \{ P(\cdot) | \dot{P} + PA - AP = BK(t) , PB = BQ(t) , t \in I , some\ Q(\cdot) , K(\cdot) \}$$

and

$$G_{A,B}^{\infty} = \{ P | P \in \tilde{g}_{A,B}^{\infty} , \det P(t) \neq 0 , all\ t \in I \}$$

in which $P(\cdot)$ and $Q(\cdot)$ are absolutely continuous and $K(\cdot)$ is locally integrable on I . (The $Q(\cdot)$ associated with $P \in G_{A,B}^{\infty}$ will have $\det Q(t) \neq 0$ since B is assumed to be of full rank.)

THEOREM 4. $\qquad\qquad\qquad\qquad G_{A,B}^{\infty}$ *is a group.*

$\qquad\quad \tilde{g}_{A,B}^{\infty}$ *is a Lie algebra relative to the commutator product.*

Proof. $G_{A,B}^{\infty}$ is closed under pointwise multiplication and inversion. The formulas for the $K's$ and $Q's$ encountered are the same as in the piecewise constant formulation. Everything stays in the required space. A routine calculation shows that $\tilde{g}_{A,B}^{\infty}$ is closed under the commutator product. Details are left to the reader.

An alternative characterization of the elements of $G_{A,B}^{\infty}$, similar to that found in the proof of Theorem 1 for the piecewise constant problem, is $G_{A,B}^{\infty} = f^{-1}(O)$ where we define

$$f_1(P) = [I - B(B^*B)^{-1}B^*][\dot{P} + PA - AP] \tag{15}$$

$$f_2(P) = [I - B(B^*B)^{-1}B^*]PB \tag{16}$$

and $f = f_1 \otimes f_2$ on the group of absolutely continuous matrix functions $P(\cdot) : I \rightarrow GL(n,R)$. The question of manifold structure is more complicated than in finite dimensions. Our approach is the constructive one which solves $f(P) = 0$.

The axioms of a finite dimensional manifold generalize in a rather direct way to manifolds modelled on Banach spaces. (An introduction to infinite dimensional manifolds can be found in Lange, [1].) Having made that generalization one then adds the group structure and, as in the finite dimensional setting, requires smoothness of multiplication and inversion to arrive at an axiomatic definition of infinite dimensional Lie group. Although $G_{A,B}$ is not itself made into a Lie group it does produce a Lie group in a useful and rather natural way. This is done by selecting any compact interval $I = [t_0 , t_1]$ of positive length and letting $G_{A,B}^{\infty}(I)$ be the group obtained by restricting the elements of $G_{A,B}^{\infty}$ to that interval. For (A , B) controllable I can prove that such a group can be given the structure of an infinite dimensional Lie group. To do this it is necessary to single out appropriate Banach spaces on which to base the coordinate systems. Due to the space limitation of this article I give a sketch of only one main result. The proofs are found in Lukes, [2], along with other results.

Recall from real analysis that $BV(I)$, the linear space of real-valued functions of bounded variation on I , is a Banach space relative to the norm $|\lambda| = |\lambda(t_0)| + V[\lambda]$ in which $V[\lambda]$ is the total variation of λ on $I = [t_0, t_1]$. In fact it is a Banach algebra relative to pointwise multiplication and thus $|\lambda_1 \lambda_2| \leq |\lambda_1| |\lambda_2|$ for $\lambda_i \in BV(I)$, $(i = 1, 2)$. (See Dunford and Schwartz, [1].) It proves useful to note that $|\lambda|^s \leq |\lambda|$ where $|\lambda|^s = \sup_t |\lambda(t)|$ and consequently any sequence convergent in $BV(I)$ is uniformly convergent on I. I work primarily within $AC(I)$, the absolutely continuous functions on I which is easily shown to be a closed linear subspace of $BV(I)$ and hence is a Banach subalgebra. A nested sequence of subspaces plays an important role. The notations $AC^0(I)$ for $AC(I)$ and $|\lambda|_0$ for $|\lambda|$ are adopted.

Definition.

$AC^r(I) = \{\lambda | \lambda \in AC^{r-1}(I), D\lambda \in AC^0(I)\}$ with norm $|\lambda|_r = |\lambda(t_0)| + |D\lambda|_{r-1}$, $(r = 1, 2, \cdots)$.

I can prove the following theorem.

THEOREM 5. $AC^r(I)$ *is a Banach space,* $(r = 0, 1, 2, \cdots)$. *For any integers* $r_1 \geq r_2 \geq 0$,

$$|h_2 D^s h_1|_{r_2} \leq [3 \cdot 2^{r_2} - 2]|h_1|_{r_1} |h_2|_{r_2}$$

for $s = 0, 1, 2, \cdots, r_1 - r_2$ *and all* $h_i \in AC^{r_i}(I)$, $(i = 1, 2)$.

COROLLARY 2. $AC^r(I)$ *is a Banach algebra relative to the (equivalent) norm* $|\lambda|_r = \dfrac{1}{c_r}|\lambda|$ *in which* $c_r = 3 \cdot 2^r - 2$, $(r = 0, 1, 2, \cdots)$. *Note:* $c_0 = 1$.

Theorem 5 and Corollary 2 are used to investigate $\mathcal{G}^\infty_{A,B}(I)$ and its applications to fault avoidance. The results include the following infinite dimensional version of Theorem 2.

THEOREM 6. *For* (A, B) *controllable* $\mathcal{G}^\infty_{A,B}(I)$ *is a Lie group. Moreover there are semi-direct products* $\times_\sigma$ *and* $\times_\tau$ *for which there is a Lie group isomorphism*

$$\mathcal{G}^\infty_{A,B}(I) \cong \prod_{k=1}^{s} GL(m_k, AC^{\rho_k}) \times_\sigma \left[\sum_{i=1}^{s_*-1} \sum_{j=1}^{\alpha_i} \oplus AC^{\rho_i} \times_\tau \sum_{i=1}^{(n-q)q} \oplus AC^0 \right].$$

REFERENCES

Ammann, P. and Knight, J.
   1. An approach to software fault tolerance,
   IEEE Transactions on Computers, 37, No.4, 1988
Dunford, N. and Schwartz, J.
   1. *Linear Operators, Part I*, Interscience, New York, 1985
Lange, S.
   1. *Introduction to Differentiable Manifolds*, Interscience, New York, 1967
Lukes, D. L.
   1. *Differential Equations: Classical to Controlled*,
   Academic Press, New York, 1982
   2. Lie groups and feedback control as a basis for fault tolerance, to appear
Varadarajan, V. S.
   1. *Lie Groups, Lie Algebras and their Representations*,
   Springer-Verlag, New York, 1984

ADDRESS: Dahlard L. Lukes, Department of Applied Mathematics, Thornton Hall, University of Virginia, Charlottesville, Virginia 22903

Computation and Implementation of Precision Control for Flexible Space Structures[1]

William H. Bennett[2]

TECHNO–SCIENCES, INC.

I. Introduction   Available methods for design of complex, multiloop control systems for flexible structures are currently quite limited. Current methods are largely based on state space models obtained by modal truncation and controllers are designed by LQG optimization [1]. Achieving *precision control* will depend on the ability to optimally assess design tradeoffs for models which capture the essential *distributed parameter* nature of the dynamic phenomenon involved [2].

We suggest that the use of transfer function models offer significant alternatives for control system design for the following reasons.

1. The interplay between 'state-space' and 'input-output' methods—evident in the theory of finite-dimensional systems [3,4]—is even more important for distributed parameter systems with potential for profound results in basic engineering methods.

2. Precise engineering specifications for distributed control system designs are extremely difficult to formulate in terms of state-space models (i.e., in terms of PDE's and their coefficients). These specifications can be rather easily characterized in terms of the system frequency response.

3. By working with the transfer function model directly, alternate control law implementations can be discovered which may be simple to implement and very effective.

4. It is convenient to characterize model uncertainty in a frequency domain setting and this approach provides insight useful for control system design [5].

Remarks on Implementation of Distributed Control   For a variety of reasons distributed state feedback control laws may be difficult to implement and state observer-based control will require the on-line realization of a distributed parameter system. In the usual approach to design the control realization is linked with the original state space model for the system. We assert that this paradigm can seriously limit available options for control realization for distributed parameter systems and we prefer to think in general terms of dynamic, output feedback control as *specified* by a general convolution equation, $u(t) = C(t) * z(t)$, where $z(t)$ are avaliable sensor measurements and $u(t)$ are control actuation. For linear, time-invariant system models the computation of the controller $C(t)$ can proceed in the frequency domain and result in a transfer function $\hat{C}(s)$ which may be irrational (or an appropriate approximation). It is our view point

[1] Work reported herein is partially supported by SDIO and AFWAL through contract F33615-88-C-3215.
[2] member IEEE, AIAA

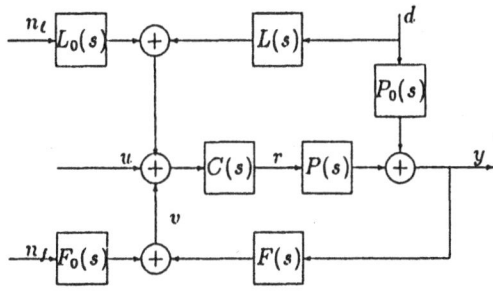

Figure 1: Multiloop Control Configuration for Wiener-Hopf Design

that control system design for distributed parameter systems can benefit from decoupling the computation and realization design steps.

In this paper we review the basis for a frequency domain method for design of control systems for flexible space structures based on Wiener-Hopf methods. We highlight the extension of available methods for rational transfer functions to an important class of distributed system models arising in the control of flexible structures. We consider the computational requirements and the advantages of frequency domain methods. In section III an example of boundary control of the wave equation is discussed. In section IV we consider options for controller realization based on frequency sampled computations.

II. Control Synthesis via Wiener-Hopf Methods

An approach to Wiener-Hopf design using transfer function models is given by Youla et al [3]. Youla's approach focuses on a specific control architecture displayed in Figure 1 where $P(s)$ is an $n \times m$ plant transfer function and $F(s)$ is $n \times n$ and models sensor measurement dynamics. The incorporation of $n \times n$ feedforward transfer function $L(s)$ is included to permit compensation for load disturbances which can be sensed prior to their effect being observed on the plant. In cases where plant delays are appreciable this feature may add substantially to the ability to compensate for output loads. In principal, the design model proposed offers several significant alternatives for control of large space structure control where a variety of sensors may be available which can predict dynamic loading.

Exogenous disturbances d, sensor noises n_ℓ, n_f, and desired set points u are modeled as zero mean, wide sense stationary random processes and are therefore characterized by their respective power spectral densities; $G_d(s)$, $G_{n_\ell}(s)$, $G_{n_f}(s)$, $G_u(s)$. Notice that no assumption is made about the form of the controller $C(s)$ except the obvious dimensions: $m \times n$.

A significant feature of the Youla method is the algebraic parametrization of all stabilizing compensators for a given plant and sensor combination. Following [3] we will call the combination of P and F admissible [3] if each transfer function is free of hidden modes and if the combination does not introduce cancellation of poles and zeros in the Closed Right Half Plane (C_+). Given that P and F are admissible let

$$F(s)P(s) = D_\ell^{-1}(s)N_\ell(s) = N_r(s)D_r^{-1}(s)$$

where N_ℓ, D_ℓ (resp. N_r, D_r) are left (right) coprime factorizations. As a result there will always exist a pair of polynomial matrices $X(s)$ and $Y(s)$ such that [6,4]

$$D_\ell(s)X(s) + N_\ell(s)Y(s) = I.$$

Then the closed loop system is asymptotically stable if and only if the controller is of the form

$$C(s) = [Y(s) + D_r(s)K(s)][X(s) - N_r(s)K(s)]^{-1}. \tag{1}$$

where K is any $m \times n$ real rational matrix, analytic in $\mathbf{C}_+$ and $\det[X(s) - N_r(s)K(s)] \neq 0$. Thus *all* stabilizing controllers can be expressed in terms of a "stable parameter", $K(s)$.

With the above parameterization in hand synthesis can be based on an optimal control problem which attempts to minimize tracking error $e = u - y$ subject to a power-like constraint on the control r. Thus let[3]

$$J_t = \frac{1}{2\pi i}E\left\{Tr\int_{-i\infty}^{i\infty} \epsilon(s)e_*(s)ds\right\}$$

subject to a constraint

$$J_s = \frac{1}{2\pi i}E\left\{Tr\int_{-i\infty}^{i\infty} P_s(s)r(s)r_*(s)P_{s*}(s)ds\right\},$$

where $P_s(s)$ is the transfer function from the control signal r to those sensitive plant modes which must be protected against saturation. The resulting optimal control problem is to find the controller (i.e. the parameter $K(s)$) such that the cost

$$J = J_t + kJ_s \tag{2}$$

is minimized where $K(s)$ is allowed to vary over all stable transfer functions of appropriate dimensions and $k > 0$ is a real parameter which plays the role of a Lagrange multiplier and effectively permits tradeoff between tracking and saturation. This is a standard Wiener-Hopf problem whose solution is based on spectral factorization.

A complete solution can be obtained by the following procedure:

Wiener-Hopf Design Procedure (Youla's Method)

Step 1 Obtain coprime factorizations

$$F(s)P(s) = D_\ell^{-1}(s)N_\ell(s) = N_r(s)D_r^{-1}(s) \tag{3}$$

Step 2 Compute spectral factorizations

$$D_{r*}[P_*P + kQ]D_r = \Lambda_*\Lambda \tag{4}$$

$$D_\ell[G_u + FP_0G_dP_{0*}F_* + F_0G_nF_{0*}]D_{\ell*} = \Omega\Omega_* \tag{5}$$

where Ω, Λ are analytic together with their inverses in CRHP.

Step 3 Find a solution $X(s)Y(s)$ to the Diophantine equation;

$$D_\ell(s)X(s) + N_\ell(s)Y(s) = I_n. \tag{6}$$

---

[3]We use the notation $u_*(s) = u^T(-s)$, E denotes expectation, and Tr denotes trace.

Step 4 Compute the transfer function

$$Z(s) = D_{r*}P_*[G_u + P_0 G_d P_{0*}F_*]D_{\ell*}. \tag{7}$$

and the stable parameter[4]

$$K(s) = \Lambda^{-1}\left(\{\Lambda_-^{-1}Z\Omega_-^{-1}\}_+ + \{\Lambda D_r^{-1}Y\Omega\}_-\right)\Omega^{-1} - D_r^{-1}Y. \tag{8}$$

Then the optimal controller has transfer function given by the parametric formula (1).

Step 4′ Alternately, compute the transfer function

$$H_o = D_r\Lambda^{-1}\left(\{\Lambda_-^{-1}Z\Omega_-^{-1}\}_+ + \{\Lambda D_r^{-1}Y\Omega\}_-\right) \tag{9}$$

and the optimal controller transfer function as

$$C = H_o[D_\ell^{-1}\Omega - FPH_o]^{-1}. \tag{10}$$

Wiener-Hopf Design for A Class of Irrational Transfer Functions The formal extension of the algebraic computations for stabilizing and optimal control synthesis in the frequency domain is an area of active research [6, pp. 357]. In this section we summarize available results for a limited class of 'pseudeo-meromorphic' irrational transfer functions which include most linear, time-invariant models arising in control of flexible space structures [2]. We refer the reader to recent work of Baras for proofs and details [7,8].

The results of Baras extend the critical constructions required for Wiener-Hopf design to the class of 'pseudo-meromorphic' transfer functions, denoted as $\mathcal{A}_m^\infty$, containing transfer functions

$$G \in H^\infty(\mathbf{C}_+)$$

where $\mathbf{C}_+$ denotes the closed right half plane and H^∞ is the Hardy space of bounded analytic functions on $\mathbf{C}_+$. Such transfer functions have (weak) coprime factorizations over $M(\mathcal{A}_m^\infty)$[5]

$$T(i\omega) = N_r(i\omega)D_r^{-1}(i\omega) = D_\ell^{-1}(i\omega)N_\ell(i\omega)$$

in the sense that N_r, D_r, D_ℓ, N_ℓ are in $H^\infty(\mathbf{C}_+)$ provided that D_ℓ, D_r are inner [4, pp. 635].

Alternately, we say that $N(s)(p \times m)$, $D(s)(m \times m)$ are *strongly coprime* if there exist $\delta > 0$ such that

$$\|N(s)x\| + \|D(s)x\| \geq \delta > 0$$

for all $x \in C^n$, $s \in \mathbf{C}_+$. Then the Carleson Corona theorem asserts that N, D are strongly coprime if and only if there exist X, Y in $H_{m\times m}^\infty$ such that

$$NY + DX = I_m.$$

Finally, the algebraic constructions required for Wiener-Hopf control are extended to the case $P, F, L, C \in \mathcal{A}_m^\infty$ by the following result [9,10].

---

[4]A rational (matrix) function has a (partial fraction) expansion $A(s) = \{A(s)\}_+ + \{A(s)\}_- + \{A(s)\}_\infty$ where $\{.\}_+$ (resp. $\{.\}_-$) is the part analytic in $\Re e s > 0$–the causal part ($\Re e s < 0$–the anti-causal part) and $\{.\}_\infty$ is the part associated with poles at infinity.

[5]Let $M(\mathcal{A}_m^\infty)$ be the set of matrix transfer functions (of dimensions determined by the context) with elements in $\mathcal{A}_m^\infty$.

Theorem 1 Assume that $P, F, L, C \in \mathcal{A}_m^\infty$ and $FP \in \mathcal{A}_m^\infty$. Then all of the following hold.

1. The singularities of $[I + CFP]^{-1}FP$ are the zeros of $\det[D_\ell X + N_\ell Y]$.

2. There exist stabilizing controllers provided that N_ℓ, D_ℓ are *strong coprime*.

3. If N_r, D_r are strongly coprime then the closed loop system is BIBO (or exponentially) stable if and only if

$$C = (Y + D_r K)(X - N_r K)^{-1}$$

where K is any element of $\mathcal{A}_m^\infty$, such that $\det(X - N_r K) \neq 0$, analytic in $\Re e s > 0$ ($\Re e s \geq 0$), and X and Y are solutions of the Diophantine identities

$$D_\ell X + N_\ell Y = I.$$

Computational Requirements for Wiener-Hopf Methods For rational transfer functions the required computational steps outlined by Youla and his colleagues for Wiener-Hopf control can all be carried out over the ring of polynomials. More recently, the approach outlined by Vidyassagar [6] suggests that by performing computations over a *ring of stable transfer functions* (denoted S and including all functions rational, proper, and analytic in $\mathbf{C}_+ \cup \{\infty\}$) the required computations can be directly supported in terms of state space realizations for the individual transfer functions. The key observation is that the collection of all proper, rational transfer functions with poles in $\mathbf{C}_+$ form an algebraic ring which has the structure of a principal ideal domain (S). In this section, we briefly outline several options for extending the required computations to $\mathcal{A}_m^\infty$. Just as in the rational transfer function case, the nonuniqueness of state space realizations for transfer functions may offer computational alternatives.

Computational requirements for Wiener-Hopf synthesis include the following.

1. **Coprime Factorization.** For rational transfer functions the coprimeness condition for a pair (N, D) essentially amounts to a requirement for noncollocation of zeros for N and D. In general irrational transfer functions may not have coprime factorizations[6]. In our studies the class of pseudo-meromorphic transfer functions includes most models for structural vibration control. Thus we have the existence of (strong) coprime factorizations in $\mathcal{A}_m^\infty$. Computation of the coprime factors can often be obtained using exponential polynomials (see the next section).

2. **Solution of Diophantine Relations.** Restricting attention to pseudo-meromorphic transfer functions gives the required existence of solutions to (6). Despite the potential lack of convergence of Euclidean division in $\mathcal{A}_m^\infty$ Berenstein and Struppa [11] have obtained explicit solutions to equations of the form (6) for (N, D) strongly coprime. The formulae obtained in [11] are nonalgebraic and extremely complex. Approximate solutions to the Bezout relations can also be obtained by sampling the Fourier response of the individual terms D_ℓ, N_ℓ, X, Y on a finite set of frequency samples. This reduces the computational problem to that of solving a set of symultaneous linear equations.

The insight from the algebraic approach to synthesis of stabilizing control [6] indicates that a particular solution X, Y to (6) is a *nominal stabilizing feedback control*; $K = YX^{-1}$. For most models of flexible structures arising in space applications the plant is nominally stable except for the existence of certain poles at the origin associated with rigid body inertias and it may be possible to obtain solutions to the Bezout relations by simple, classical output feedback methods.

3. **Spectral Factorization and Causal Projection.** In many cases arising in flexible structure control one can easily recognize the irrational transfer function is meromorphic and can be written as a ratio of a pair of exponential polynomials. In such cases the computational problems can often be reduced by the identification of an alternate state realization (see the Example). A more general approach is to use again frequency sampling. Methods for matrix spectral factorization have been studied as alternatives to solving infinite dimensional Riccati equations by Davis [12] and Bennett [13] where a numerical algorithm is given.

4. **Linear Fractional Combinations.** All remaining constructions involve linear fractional transformations which can be readily supported via either explicit transfer function computations or frequency sampling.

Our conclusion from this survey is that frequency sampling offers an attractive alternative for the computational requirements of Wiener-Hopf synthesis. It obtains a frequency sampled approximation to the ideal optimal controller for the given design problem and essentially provides a specification for such control. We remark that in contrast to state-space computations this approach decouples the realization of the controller from the computation of its response.

III. Example: Boundary Control of Wave Equation   Consider the wave equation,

$$\frac{\partial^2 y(t,x)}{\partial t^2} = \frac{\kappa G}{\rho} \frac{\partial^2 y(t,x)}{\partial x^2} \tag{11}$$

with boundary conditions

$$\text{at } x = 0 \qquad \kappa G \left.\frac{\partial y}{\partial x}\right|_{x=0} = d(t) \tag{12}$$

$$\text{at } x = L \qquad \kappa G \left.\frac{\partial y}{\partial x}\right|_{x=L} = u(t) \tag{13}$$

where u is a control force applied at one end $x = L$ and d is a disturbance force at the opposite end, $x = 0$[6]. Let $\alpha^2 = \kappa G/\rho$.

For the purposes of feedback control a sensor is mounted at the end $x = L$ which measures rate and is modeled as

$$v(t) = \dot{y}(t, L) + n(t). \tag{14}$$

The disturbance d and sensor noise n are modeled as zero mean, wide sense stationary, random processes with

$$E\{d(t)d(\tau)\} = \sigma_d^2 \delta(t - \tau)$$
$$E\{n(t)n(\tau)\} = \sigma_n^2 \delta(t - \tau).$$

---

[6] $y(t, x)$ can be thought of as the lateral deflection of a long, thin beam at time t and position x, $0 \le x \le L$, with parameters as given in [2]

A frequency domain model in the desired input-output form can be readily obtained via Laplace transform of the time variable, t, obtaining the equation,

$$\hat{y}(s,x) = \frac{e^{\frac{x}{a}s} + e^{-\frac{x}{a}s}}{\rho a s(e^{\frac{k}{a}s} - e^{-\frac{k}{a}s})}\hat{u}(s) - \frac{e^{\frac{x-L}{a}s} + e^{-\frac{x-L}{a}s}}{\rho a s(e^{\frac{k}{a}s} - e^{-\frac{k}{a}s})}\hat{d}(s). \tag{15}$$

With the sensor location as given the measurement equation becomes

$$\hat{v}(s) = P(s)\hat{u}(s) + P_0(s)\hat{d}(s) + \hat{n}(s), \tag{16}$$

$$P(s) = \frac{\coth(\frac{L}{a}s)}{\rho a}, \tag{17}$$

$$P_0(s) = -\frac{2}{\rho a}\operatorname{csch}(\frac{L}{a}s). \tag{18}$$

Modeling the exogenous disturbances and sensor noise terms as wide-band, uncorrelated signals we let the power spectral densities be of the form $G_d(\omega) = \sigma_d^2$, $G_{n_f}(\omega) = \sigma_n^2$, and $G_u = 0$ (c.f. Figure 1).

The optimal control problem is to regulate the rate $\dot{y}$ at the end $x = L$ subject to a constraint on the control energy. This is a standard Wiener-Hopf control problem for a distributed system with effective control objective,

$$J(u) = E\left\{\frac{1}{2\pi i}\int_{-i\infty}^{i\infty}|s\hat{y}(s,L)|^2 + k|\hat{u}(s)|^2 d\omega\right\}, \tag{19}$$

where $k > 0$ is a real parameter which plays the role of a Lagrange multiplier with respect to control constraint on $r(s)$.

The crucial observation which opens new options for control computation and realization comes when we write the transfer function as a ratio of exponential polynomials [11],

$$P(s) = \frac{1}{\rho a}\coth\frac{L}{a}s = \frac{e^{\frac{k}{a}s} + e^{-\frac{k}{a}s}}{\rho a(e^{\frac{k}{a}s} - e^{-\frac{k}{a}s})}. \tag{20}$$

The introduction of the map $e^{\frac{k}{a}s} \to z$ reduces the computational problem from one of infinite dimensions (c.f. [1,14]) to rational functions in the frequency domain. Indeed, the transfer functions required are now expressed in the z variable as,

$$P(z) = \frac{1 + z^{-2}}{\rho a(1 - z^{-2})} = \frac{z^2 + 1}{\rho a(z^2 - 1)}, \qquad P_0(z) = \frac{-2z}{\rho a(z^2 - 1)}. \tag{21}$$

In this form it is easy to identify (strong) coprime factorizations, to obtain the required solutions to the Bezout equation by Euclidean division, and to compute the required spectral factorizations all over the ring of polynomials.

At this point, all required computations may be carried out using state space constructions for the transformed, discrete time model. Recognition of an effective alternate state coordinates in this case is relatively easy from the transfer function model although it can be obtained from the orginal PDE with some additional physical insight.

The obvious choice for coprime factors for P are $D_\ell = D_r = d_p(z) = \rho a(z^2 - 1)$ and $N_\ell = N_r = n_p(z) = z^2 + 1$. Similarly write $P_0 = n_0(z)/d_p(z)$ where $n_0(z) = -2z$. A particular solution to (6) is readily obtained by Euclidean division over the ring of polynomials in z;

$$x(z) = \frac{-1}{2\rho a}, \qquad y(z) = \frac{1}{2}.$$

Following Youla's algebraic recipe we next compute causal spectral factors for the control problem (which in this case takes the form)

$$n_p(1/z)n_p(z) + kd_p(1/z)d_p(z) = \Lambda(1/z)\Lambda(z) \tag{22}$$

and the filtering problem

$$n_0(z)\sigma_d^2 n_0(1/z) + d_p(z)\sigma_n^2 d_p(1/z) = \Omega(1/z)\Omega(z), \tag{23}$$

where $\Lambda(z), \Omega(z)$ are the causal spectral factors.

Finally, the optimal controller can be obtained by a linear fractional transformation of the frequency domain data (in this case computed over the ring of polynomials). Here we compute from (7)

$$Z(z) = d_p(1/z)P(1/z)P_0(z)\sigma_d^2 P_0(1/z)d_p(1/z),$$

the optimal parameter H_o is obtained from (9)

$$H_o(z) = d_p\Lambda^{-1}(\{\Lambda_-^{-1}Z\Omega_-^{-1}\}_+ + \{\Lambda d_p^{-1}\Omega\}_-), \tag{24}$$

and the optimal controller C_o from (10), $C_o = H_o[d_p^{-1}\Omega - PH_o]^{-1}$. The causal and acausal projections required in (24) can be obtained in this case, by partial fraction expansion in the z coordinate.

The tradeoff between optimal wide-band disturbance rejection and control energy can now be addressed either in terms of the optimal cost E_t versus the saturation constraint E_s or directly in terms of the optimal sensitivity function frequency responses $S(i\omega)$ and $I - S(i\omega)$ as shown in Figure 2. Choosing $k = 2$ for implementation, the optimal transfer functions are given as

$$H_o(z) = \frac{a_0 + a_1 z + a_2 z^2}{b_0 + b_1 z},$$

$$C(z) = \frac{\sum_{k=0}^6 c_k z^k}{\sum_{k=0}^6 d_k z^k},$$

where the coefficients are listed in Table 1 were obtained by direct numerical computation[7]. Once a desirable value for the Lagrange multiplier is obtained by evaluation of the resulting optimal frequency response characteristics effecting saturation, sensitivity and robustness the optimal controller is then *specified* in terms of its frequency response as shown in Figure 4 on the unit disk.

Realization for Optimal Control The transformation $z = e^{\frac{k}{\alpha}s}$ suggests a realization for the optimal (infinite dimensional) controller in terms of a delay equation of the form,

$$u(t) = \sum_{k=0}^6 \frac{c_k}{d_6} y(t - (6 - k)\tau) - \sum_{k=1}^6 \frac{d_k}{d_6} u(t - (6 - k)\tau) \tag{25}$$

where $1/\tau = \alpha/L$ is the characteristic wave speed in the medium. Although this is an exact realization of the optimal controller considered here it may suffer from inherent sensitivity and robustness problems due to unmodeled delays in the sensor, actuator, or control computer as discussed in [15].

---

[7]This computation corrects some computational errors in the report [10].

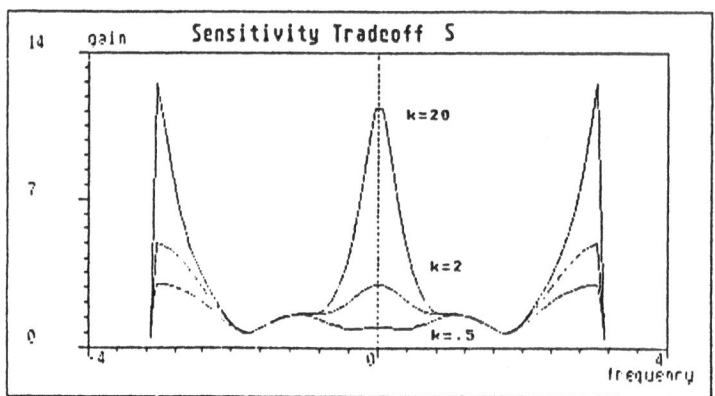

Figure 2: Sensitivity Tradeoff for S

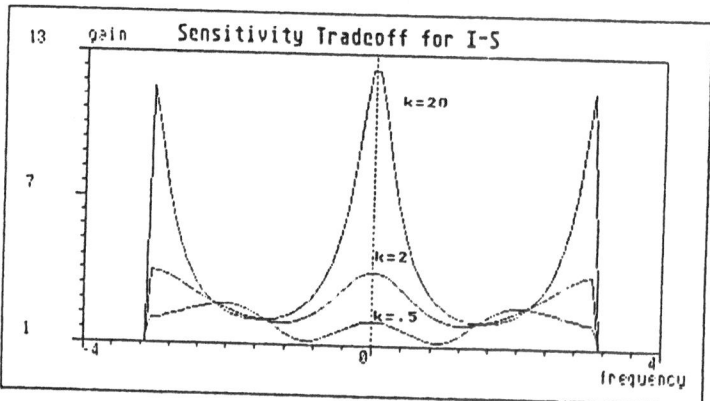

Figure 3: Sensitivity Tradeoff for $1 - S$

k	a_k	b_k	c_k	d_k
0	0.6862915	- 0.4142136	0.00	- 0.0517767
1	-1.6568542	1.0	0.00	0.125
2	3.3137085		0.2071068	0.5088835
3			- 0.5	- 1.2285534
4			0.7928932	1.2071068
5			0.5	- 0.5
6			- 1.0	1.0

Table 1: Coefficients of rational expressions for H_o and C_o

IV. Realization of Frequency Sampled Controllers by FIR Filters   Currently available technology offers several attractive alternatives for high speed finite impulse response filter implementations. Single chip 16/32-bit digital signal processors are now available which combine the flexibility of a high-speed control computer with the numerical capability of an array processor. Availability of such devices opens new opportunities for control system implementations. Before realistic designs can be attempted it is necessary to understand the limitations of the available theory for realizing feedback control based on high speed FIR filters. For the particular frequency sampled synthesis procedure discussed above the natural choice for FIR implementation is via FFT type processing.

The potential application of FIR filtering for realtime feedback control is limited by the extent to which the FIR processing can approximate the impulse response of optimal (IIR) controllers. Factors effecting feasibility of FIR control include:

1. whether or not the ideal IIR controller is stable,

2. whether the sampling requirements are commensurate with the available technology,

3. available FIR processing speed with respect to the required phase dependent stability margins for robust stabilization and performance.

With the exception of the first factor these requirements are largely technology dependent and depend on time–bandwidth product of available processors relative to the plant dynamics. It is well known that techniques for feedback compensation arising from Wiener-Hopf synthesis often employ unstable components to achieve required phase stabilization of plant dynamics [4]. The appearance of nonminimum phase zeros in transfer function models for flexible structures is almost unavoidable and this will ultimately lead to requirements for certain controller loops to be open loop unstable.

Consider a general feedback compensator in terms of the parametrization of *all* stabilizing controllers (1) written in the form,

$$u(s) = [Y(s) + D_r(s)K(s)]\xi(s) \tag{26}$$

$$X(s)\xi(s) = y(s) + N_r(s)K(s)\xi(s), \tag{27}$$

where we take $X, Y, N, D, K \in M(\mathcal{A}_m^\infty)$. Clearly, if X is not *unimodular* in $M(\mathcal{A}_m^\infty)$ in the sense that $\det X^{-1} \notin \mathcal{A}_m^\infty$ then even the 'nominal' controller, $C = YX^{-1}$ is not "stable" and direct

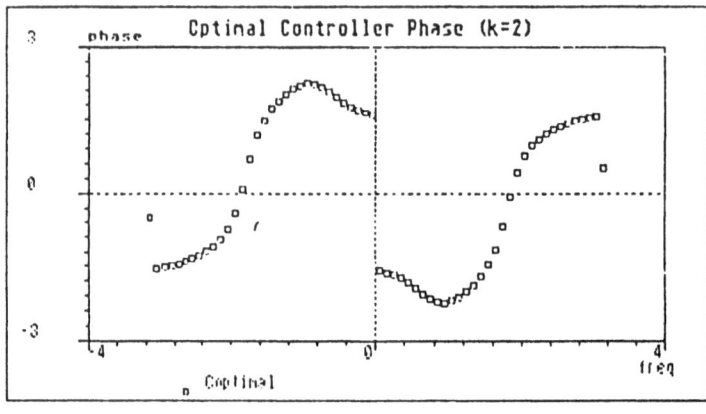

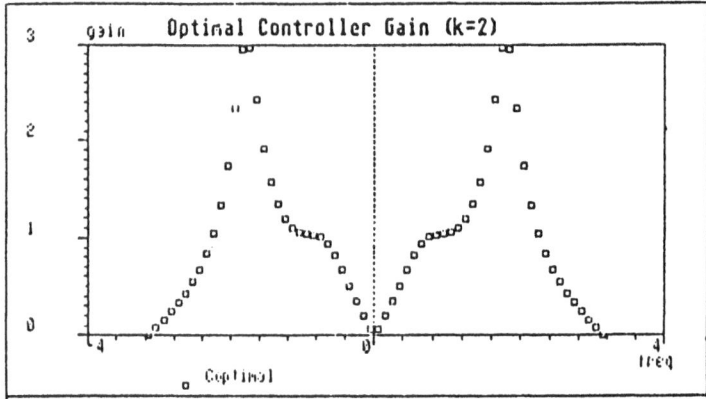

Figure 4: Frequency Response of Optimal Controller for $k = 2$

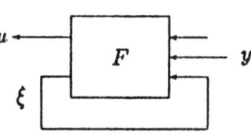

Figure 5: Option for Stable FIR controller implementation

implementation of the optimal feedback control as a convolution, $u(t) = C_{nom}(t) * y(t)$, via FIR filtering is impractical at best. In this case we prefer to construct a controller architecture by extracting a unimodular part from X as $X = U - \tilde{X}$ where U is unimodular in $M(\mathcal{A}_m^\infty)$ (i.e. has a stable inverse). Then writing the controller equations in the form,

$$\begin{pmatrix} u \\ \xi \end{pmatrix} = \begin{bmatrix} 0 & [Y + D_r K] \\ U^{-1} & [U^{-1} N_r K - I] \end{bmatrix} \begin{pmatrix} y \\ \xi \end{pmatrix} = F \begin{pmatrix} y \\ \xi \end{pmatrix}$$

which can be implemented as a stable filter with external feedbacks as shown in Figure 5. We remark that this approach differs from that proposed by Boyd et al [16].

V. Conclusion In this paper we have outlined a methodology for Wiener-Hopf design of control systems for problems arising in flexible space structures. The underlying distributed parameter dynamics are captured in a class of pseudo-meromorphic transfer functions for which the relevant constructions of coprime factorization are known to exist. It is shown that direct sampling of the system frequency response can be used to approximate the required optimal controller directly without recourse to modal truncation or other forms of rational approximation.

We indicate in a simple example the role of exponential polynomials in the required computations for a class of distributed parameter systems arising in flexible structure control. We also show how alternate realizations can be uncovered from the irrational transfer functions. Finally, for the general case, we suggest that the frequency sampled specification for the ideal, optimal controller can be implemented by one or more FIR filters.

Acknowledgements We wish to acknowledge the inspiration and guidance of Dr. J.S. Baras and Dr. C. Berenstein.

References

[1] M. Balas, "Trends in large space structure control theory: fondest hopes, wildest dreams," *IEEE Trans. Auto. Cntrl.*, vol. AC-27, pp. 522–535, 1984.

[2] W. Bennett and H. Kwatny, "Continuum modeling of flexible structures with application to vibration control," *AIAA J.*, January 1989. to appear.

[3] D. Youla, H. Jabr, and J. J.J. Bongiorno, "Modern Wiener-Hopf design of optimal controllers–parts I and II," *IEEE Trans. on Auto. Cntrl.*, vol. AC-30, no. 7, pp. 3–13 and 319–338, 1976.

[4] T. Kailath, *Linear Systems*. Prentice-Hall, 1980.

[5] J. C. Doyle, "Robustness of multiloop linear feedback systems," in *IEEE Conf on Dec. and Control*, (San Diego, CA), pp. 12–18, 1979.

[6] M. Vidyasagar, *Control System Synthesis: A Factorization Approach*. MIT Press, 1985.

[7] J. Baras, "Frequency domain design of linear distributed systems," in *Proc. of 19th IEEE CDC*, Dec. 1980.

[8] J. S. Baras, "Complex variable methods in the control of distributed systems," 1985. Presented at NSF, SIAM, AFOSR Workshop on Control of Distributed Systems.

[9] J. S. Baras, "Complex variable methods in distributed parameter control systems," in *Frontiers of Applied Mathematics*, (H. Banks, ed.), 1987. to appear.

[10] W. Bennett, H. Kwatny, and J. Baras, "Frequency domain design of robust controllers for large space structures," Tech. Rep. SEI TR-86-11, SEI, November 1986. Final Report, NASA Contract No. NAS1-18209.

[11] C. Berenstein and D. Struppa, "On explicit solutions to the Bezout equation," *Systems and Control Letters*, 1985.

[12] J. Davis and R. Dickinson, "Spectral factorization by optimal gain iteration," *J. Appl. Math.*, vol. 43, pp. 389–301, 1983.

[13] W. Bennett and I. Yan, "A computer algorithm for causal spectral factorization," in *Proc. 1988 NAECON*, May 1988.

[14] J. Gibson, "An analysis of optimal modal regulation: convergence and stability," *SIAM J. Cntrl and Optim.*, no. 5, pp. 686–707, 1981.

[15] W. Bennett and H. Kwatny, "Robustness issues in control of flexible systems," in *Proc. 1987 ACC*, June 1987.

[16] S. Boyd and et al, "A new CAD method and associated architectures for linear controllers," Tech. Rep. L-104-86-1, Stanford Electronic Labortories, December 1986.

ON PI AND PID/SLIDING MODE CONTROLLERS

Romano M. DeSantis

E E Department, Ecole Polytechnique de Montreal

Montreal, Canada

1. Introduction

Under ordinary operating conditions, the classical PI and PID
controllers currently employed in high accuracy speed drives and position servos usually offer satisfactory transient response and good
steady state disturbance rejection. They may prove inadequate,
however, to provide a satisfactory performance in the presence of
transient disturbances and/or plant parameter variations. This
problem, together with ever stricter industrial performance requirements and recent technological improvements in the available hardware
and software components, have generated renewed interest in the
development of more advanced controllers [1 - 4].

A natural approach in this direction is to embed the classical
controller in an adaptive control scheme [3 - 6]. Though possibly
effective in reducing the negative influence of slow variations in
plant parameters and low frequency disturbances, this approach is less
effective, however, in compensating for the presence of fast parameter
variations and strong transient disturbances. In an alternative
approach, the classical controller is replaced by a controller of the
sliding mode type which is characterized by a higher robustness to
plant parameter variations and external perturbations [7, 10]. The
difficulty, here, is in a tendency to high control levels and oscilla-
tions often leading to unacceptable trade-offs between quality of
transient performance and steady state disturbance rejection
capabilities.

In what follows, ideas and techniques from these references will be applied to consider sliding mode modified PI and PID directed towards alleviating this latter difficulty. These modified controllers will be developed with a particular emphasis at improving by reinforcing, rather than by replacing, the action of their classical counterpart. Their interest is in their potential to combine the good transient performance and the steady state disturbance rejection properties of the classical PI and PIDs with the robustness to plant parameters perturbations of the sliding mode controller. This reduces the level of trade-off which must usually be made between transient and steady state performance on the one hand, and disturbance rejection capabilities on the other.

2. The Main Idea

Given a system

$$\dot{x} = A_0 x + B_0 u_1 + B_1 u_2 + B_2 W$$

where:

$x(t)$: is a real n-dimensional vector representing the state;

$u_1(t)$, $u_2(t)$, $w(t)$, : are real p-dimensional vectors;

u_1: is generated by a control law in such a way that, in the absence of u_2 and W, the ensuing state trajectory $x(.)$ corresponds to a desired state trajectory $x_D(.)$;

u_2: is a supplementary control intended to reduce (neutralize) the influence of W over the state trajectory $x(.)$;

W: represents the effect of neglected nonlinearities and unmodeled dynamics, plus the influence of external perturbations and internal parameter variations;

A_0: an n^2 dimensional stable real matrix;

B_0, B_1, B_2: real matrices with an appropriate dimension;

Consider the problem of determining an effective u_2.

In line with the sliding mode philosophy [9], this problem can be solved by proceeding as follows: i) consider the (sliding mode) vector

$$slm(t) = \int_0^t \overset{\bullet}{x} - A_0 x - B_0 u_1$$

ii) determine a u_2 capable of attaining the (sliding mode) condition

$$slm(t) = 0$$

To justify this procedure, note that satisfaction of this condition implies

$$\overset{\bullet}{x} = A_0 x + B_0 u_1$$

which in turn implies

$$x(.) = x_D(.)$$

This means that the influence of W over the state trajectory of the system is neutralized.

For the determination of an adequate u_2, let

$$slm(t) = 0, \text{ for } t = 0,$$

and observe that to attain

$$slm(t) = 0, \text{ for } t > 0$$

it is sufficient to satisfy

$$d|slm(t)|^2/dt < 0, \text{ for } t > 0.$$

From

$$d|slm(t)|^2/dt = < slm(t), B_1 u_2 + B_2 W>,$$

assuming $B_1 = B_2$, it follows

$$d|slm(t)|^2/dt = <B_1{}^T slm(t), u_2 + W>.$$

Hence, to satisfy

$$d|slm(t)|^2/dt < 0 \text{ for } t > 0,$$

it is sufficient to determine a u_2 such that

$$SIGN[u_{2i} + W_i] = - SIGN[(B_1{}^T slm(t))_i]$$

with u_{2i}, W_i and $(B_1{}^T slm(t))_i$ representing the i-th component of the respective vector. This may be done by letting

$$u_{2i} = -M_{1i} (B_1{}^T slm(t))_i -M_{2i} SIGN[(B_1{}^T slm(t))_i]$$

with positive constants M_{1i} and M_{2i} chosen so that

$$-M_{2i}| (B_1{}^T slm(t))_i | < -(B_1{}^T slm(t))_i W_i + M_{1i}|(B_1{}^T slm(t))_i|^2$$

Note that u_2 is given by the sum of a linear plus a nonlinear on-off switching component. The relative strength of these components will depend on stability considerations, admissable level of chattering, the nature of W, the desired amount of sensitivity reduction.

In what follows the above procedure will be applied in the realm of speed drive and position servo controllers.

3. PI/Sliding Mode Controller for a Speed Drive

The dynamics of a speed drive is usually described by [11,12]

$$\dot{x}_1 = -x_1/\tau + (K_M/\tau)*(u_1 + u_2 + W)$$

where: x_1 represents speed; K_M and τ, represent the estimated values of the speed drive static gain and time constant; u_1, u_2, and W represent, respectively, control and perturbation voltages.

Following commun practice, the desired state trajectory of the speed drive is described in terms of the (Laplace transform) relation

$$x_1(s) = F_1(s)*REF(s)$$

where: REF(s) represent the set-point signal and

$$F_1(s) = 1/(1+\alpha_1 *s)$$

represents the specified input-output transfer function.
This implies that

$$\dot{x}_{1D}(t) = -x_{1D}(t)/\alpha_1 + REF(t)/\alpha_1,$$

and that the control law generating $u_1(t)$ is given by

$$u_1 = -K_{01}*(x_1 - REF) + K_{03}*REF)$$

with

$$K_{01} = -(p_1*\tau+1)/K_M$$

$$K_{03} = 1/K_M$$

where

$$p_1 = -1/\alpha_1.$$

In the absence of a supplementary control u_2, the sensitivity of the state trajectory x(.) to the influence of the perturbation W(.) is described, (in Laplace transform notations), by the relation

$$x_1(s) = F_2 * W(s)$$

where

$$F_2(s) = K_M / (r*s - r*p_1).$$

This sensitivity may be reduced by implementing the procedure suggested in the previous section. Observe that the sliding mode vector defined in the previous section corresponds to

$$slm(t) = \int_0^t [\dot{x}_1 + x_1/r - (K_M/r)(K_{01}*(x_1-REF) + K_{03}*REF)]$$

that is, to

$$slm(t) = \alpha_1 * x_1 + x_2$$

where

$$x_2 = \int_0^t (x_1 - REF) dt.$$

The ensuing expression for the control u_2 is

$$u_2 = -M_1(K_M/r)[\alpha 1 * x_1 + x_2] - M_2 SIGN[\alpha 1 * x_1 + x_2]$$

that is,

$$u_2 = -\Delta K_1 * (x_1 - REF) + \Delta K_2 x_2 + \Delta K_3 * REF - M_2 SIGN[\alpha 1 * x_1 + x_2]$$

with

$$\Delta K_1 = -p_2 * r / K_M$$

$$\Delta K_2 = p_1 * p_2 * r / K_M$$

$$\Delta K_3 = p_2 * r / K_M$$

where

$$p_2 = - M_1 * K_M^2 / (r^2 * p_1).$$

The overall control, $u = u_1 + u_2$, is then given by

$$u = - K_1 * (x_1 - REF) - K_2 * (x_2 - K_3 * REF) - M_2 SIGN[\alpha 1 * x_1 + x_2]$$

with

$$K_1 = -((p_1 + p_2) * r + 1) / K_M$$

$$K_2 = p_1 * p_2 * r / K_M$$

$$K_3 = (1 + 1/(r * p_2)) / p_1$$

It is of interest to observe that for $M_1 > 0$, $M_2 = 0$, this control is identical to that generated by the classical PI controller (see figure 1). In this case, the influence of the perturbation W(.) over the state trajectory x(.) is described, by

$$x_1(s) = F_2 * W(s),$$

with

$$F_2(s) = s*K_M/[\tau*(s - p_1)*(s - p_2)]$$

Clearly, such an influence can be attenuated as much as desirable by simply increasing the value of M_1. However, an infinite value of M_1 (hence infinite values for K_1, K_2, and K_3) would be required before a complete neutralization is obtained.

For $M_1 > 0$, $M_2 > 0$ the control, $u = u_1 + u_2$, corresponds to the control generated by the PI/sliding mode controller indicated in figure 1; the effect of M_2 is to make $F_2(s) = 0$, hence to completely neutralize the influence of the perturbation.

4. PID/Sliding Mode Controller for a Position Servo

The dynamics of a position servo actuator is usually described in terms of [11, 12]

$$\dot{x}_1 = x_2$$
$$\dot{x}_2 = -x_2/\tau + (K_M/\tau)*(u_1 + u_2 + W)$$

where: x_1 and x_2 denote respectively angular position and speed of the servo actuator; K_M and τ, represent the estimated values of the actuator static gain and time constant; u_1, u_2, and W are, respectively control and perturbation voltages.

Following once again common practice, the desired state trajectory of the position servo is given by

$$x_1(s) = F_1(s)*REF(s)$$

where: REF(s) represents the set-point signal and

$$F_1(s) = 1/(1 + \alpha_1*s + \alpha_2*s^2)$$

represents the specified input-output transfer function.
It follows that the desired state trajectory is determined by

$$\dot{x}_{2D}(t) = x_{1D}(t)$$

$$\dot{x}_{2D}(t) = -p_1 * p_2 * x_{1D}(t) + (p_1 + p_2) * x_{2D}(t) + p_1 * p_2 * REF(t)$$

and that the control law generating $u_1(t)$ is given by

$$u_1 = -K_{01} * (x_1 - REF) - K_{02} * x_2$$

with

$$K_{01} = p_1 * p_2 * \tau / K_M$$

$$K_{02} = -((p + p_2)\tau + 1)/K_M$$

and

$$p_1 + p_2 = -\alpha_1/\alpha_2 \quad , \quad p_1 * p_2 = 1/\alpha_2$$

In the absence of a supplementary control u_2, the sensitivity of the state trajectory $x(.)$ to the influence of the perturbation $W(.)$ is described, (in Laplace transform notations), by the relation

$$x_1(s) = F_2 * W(s)$$

where

$$F_2(s) = K_M/\tau * (1 + \alpha_1 * s + \alpha_2 * s^2)$$

The sliding mode vector to be considered for the determination of an effective sensitivity reducing u_2, is now given by

$$slm(t) = \begin{bmatrix} 0 \\ \int_0^t \dot{x}_2 + x_2/\tau - (K_M/\tau)u_1 \end{bmatrix}$$

that is by

$$slm = \begin{bmatrix} 0 \\ \alpha_1 * x_1 + \alpha_2 * x_2 + x_3 \end{bmatrix}$$

where

$$x_3 = \int_0^t \dot{x}_1 - REF)dt$$

In line with the procedure in section 2, this leads to

$$u_2 = -M_1(K_M/\tau) [\alpha_1 * x_1 + \alpha_2 * x_2 + x_3] - M_2 SIGN [\alpha_1 * x_1 + \alpha_2 * x_2 + x_3]$$

that is,

$$u_2 = -\Delta K_1 * (x_1 - REF) - \Delta K_2 * x_2 - \Delta K_3 (x_3 - \Delta K_4 * REF) - M_2 SIGN[\alpha_1 * x_1 + \alpha_2 * x_2 + x_3]$$

with

$$\Delta K_1 = (p_1 p_3 + p_2 p_3) * r / K_M ;$$

$$\Delta K_2 = -p_3 * r / K_M ;$$

$$\Delta K_3 = -p_1 * p_2 * p_3 * r / K_M ;$$

$$\Delta K_4 = (p_1 + p_2) / p_1 * p_2$$

where

$$p_3 = - M_1 * K_M^2 / (r^2 * p_1 * p_2)$$

The overall control, $u = u_1 + u_2$, is then given by

$$u = - K_1 * (x_1 - REF) - K_2 * x_2 - K_3 * (x_3 - K_4 * REF) - M_2 SIGN[\alpha_1 * x_1 + \alpha_2 * x_2 + x_3]$$

with

$$K_1 = (p_1 * p_2 + p_3 * p_1 + p_2 * p_3)) * r / K_M ;$$

$$K_2 = -((p_1 + p_2 + p_3) * r - 1) / K_M ;$$

$$K_3 = -p_1 * p_2 * p_3 * r / K_M ;$$

$$K_4 = (p_1 + p_2) / p_1 * p_2$$

Note that for $M_1 > 0$, $M_2 = 0$ this control corresponds to that generated by the classical PID servo controller (see figure 2). The influence of the perturbation $W(.)$ over the state trajectory $x(.)$ is then described, by

$$x_1(s) = F_2 * W(s),$$

where

$$F_2(s) = s * K_M / [r * p_1 * p_2 * (1 + \alpha_1 * s + \alpha_2 * s^2) * (s - p_3)]$$

As in the case of the PI speed controller , while not completely neutralized, the influence of the perturbation can be once again attenuated as much as desirable by simply increasing the value of M_1. Once again, however, an infinite value of M_1 (hence infinite values for the classical PID gains) would be required before a complete neutralization is attained.

For $M_1 > 0$, $M_2 > 0$ the control, $u = u_1 + u_2$, corresponds to the control generated by the PID/sliding mode controller indicated in figure 2; the effect of M_2 is to make $F_2(s) = 0$, which implies a complete neutralization of the perturbation.

Closure

The addition of an on-off switching element to classical PI and PID controllers used in high accuracy speed drive and position servo appears to represent a natural way to improve the sensitivity of currently available controllers. This improvement results from the capability to combine the wellknown transient and steady state properties of the classical controllers with the robustness to parameter variations and external perturbations of a sliding mode controller. This, without giving up the benefits of the long acquired familiarity with classical PI and PIDs: most of the design, tuning procedures or adaptive control techniques which are helpful to improve performance in the original controllers, do remain helpful in the context of the new controllers.

Because of the controller linear component, the high frequency chattering which typically overcasts the performance of sliding mode controllers can be substantially reduced without altering these basic properties. Under reasonable disturbances and plant parameters variations, this can be done by introducing a small threshold on the input of the switching component. Under larger parameter variations, an additional corrective action may be provided by equipping the controller linear component with a self-tuning loop [15, 16].

These theoretical results must of course be corroborated by extensive experimental tests before the practical implications of the PID/slm controllers may be fully assessed. While preliminary results in this direction are somewhat encouraging, they also appear to suggest that in actual industrial applications an improved performance may not be guaranteed by the sole utilization of a more advanced controller. The concurrent usage of higher quality open loop components such as a speed transducer free from magnetic linkages [15, 16], a small dead time in the loop, and a high rigidity motor axis to load axis transmission may well represent an equally essential requirement.

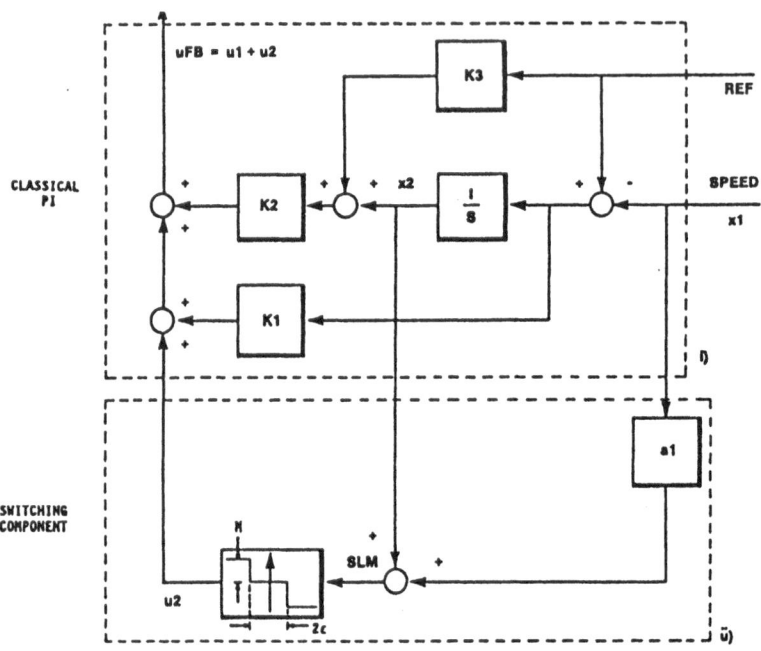

Figure 1 : The PI/Sliding mode controller: i) classical PI component;
ii) additional nonlinear component.

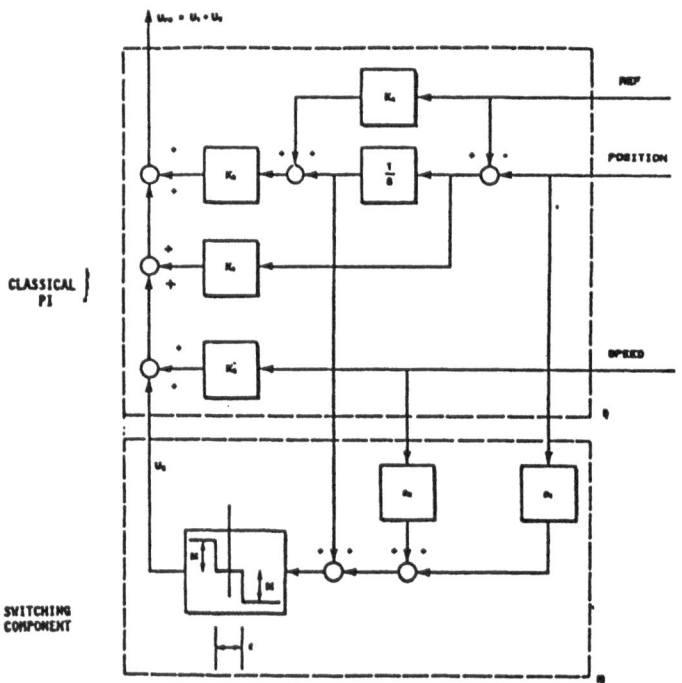

Figure 2 : The PID/Sliding mode controller: i) classical PID
component; ii) additional nonlinear component.

Acknowledgement

The results of this paper are from [15, 16]. Their presentation has benefited from the help of the reviewers of that ASME journal. Support from the Quebec Ministry of Education, grant FCAR-CRP-378-76, and from the Canadian Government, NSERC grant A-8244 is acknowledged.

References

[1] Mutoh, N., Nagase, H., Sakai, K., Ninomiya, H., High Response Digital Speed Control System for Induction Motors, IEEE Trans. on Industrial Applications, Vol IE-33, N.1, February 1986.

[2] Sule, R.R., Balakrishna, J.V., Krishnan, T., Kumar, M., Microprocessor-Based Speed Control System for High Accuracy Drives, IEEE Transactions on Industrial Electronics, Vol. IE-32, N.3, August 1985.

[3] Dubowsky, S., Kornbluh, R., On the Development of High Performance Adaptive Control Algorithms for Robotic Manipulators, The Robotic Research: The 2nd International Symposium, MIT Press 1985.

[4] Koivo, A.J., Guo,T.H, Adaptive Linear Controller for Robotic Manipulators, IEEE Transactions on Automatic Control, AC-28, pp 162-171, 1983.

[5] Astrom,K.J., Self-Tuning Regulators:Design Principles and Applications, Applications of Adaptive Control, Academic Press 1980.

[6] Landau, Y.D. , Adaptive Control: The Model Reference Approach, Dekker, 1979.

[7] Slotine, J.J.E., Coetsee, J.A., Adaptive Sliding Controller Synthesis for Nonlinear Systems, Int. J. Control, 1986.

[8] Balestrino, A., De Maria, G., Zinober, A.S.I., Nonlinear Adaptive Model-following Control, Automatica, Vol. 20, N.5, pp 559-568, 1984.

[9] Utkin, V.I., Sliding Modes and their Application to Variable Structure Systems, MIR Publishers, MOscow, 1978.

[10] Bellini, A., Figalli,G., Ulivi, G., Robotic Decoupling Control of an Industrial Manipulator Using Sliding Mode, International Journal of Robotics and Automation, Vol.3, N.1, 1988.

[11] Kuo, B.C., Automatic Control Systems, Prentice Hall, 1987.

[12] DC MOTORS SPEED CONTROLS SERVO SYSTEMS, an engineering handbook by ELECTRO-CRAFT CORPORATION, Pergamon Press 1977.

[13] DeSantis, R.M., Hurteau, R., Authie, G., A Real-Time Computer Control Demonstrator for Use in a University Control Laboratory, IEEE Transactions on Education, Vol. E-25, NO.1, Feb 1982.

[14] DeSantis, R.M., A PID Sliding Mode Controller for a Robotic Link, 2nd International Conference on Robotics and Factories of the Future, San Diego, California, July 1987.

[15] DeSantis,R.M., An Adaptive PI/Sliding Mode Controller for a Speed Drive, Journal of Dynamic Systems, Measurement and Control, (submitted).

[16] DeSantis,R.M., An Adaptive PID/Sliding Mode Controller for a Position Servo, Journal of Dynamic Systems, Measurement and Control, (submitted).

Tuned Values in Adaptive Control

Marc Bodson

Carnegie Mellon University

Introduction. The ability of a control system to maintain stability in the presence of unmodeled dynamics and measurement noise is a very important consideration for practical applications. In Rohrs *et al.* [1982], it was shown that adaptive algorithms discussed in the literature could be destabilized by relatively minor perturbations. The paper initiated a significant research effort to study the robustness of adaptive control systems.

Results to date may be classified into two groups: update law modifications and averaging analyses. In the first set (including Ioannou & Kokotovic [1984], Ioannou & Tsakalis [1986], Kreisselmeier and Anderson [1986], Narendra & Annaswamy [1987], Ortega, Praly & Landau [1985], and Sastry [1984]), one finds useful modifications of standard algorithms such as the deadzone and relative deadzone. These modifications are mostly designed to prevent the slow drift instability which occurs when inputs are not sufficiently rich, and outputs are perturbed by measurement noise.

The present paper falls in the second category of results (including Anderson *et al.* [1986], Astrom [1984], Bodson [1988], Fu & Sastry [1987], Mason *et al.* [1988], Riedle & Kokotovic [1985] and [1986], and Sastry & Bodson [1988]) which analyze the robustness of adaptive systems using averaging methods. The approach is aimed at obtaining an understanding of instability mechanisms based on frequency-domain relationships. Indeed, even in the presence of sufficiently rich inputs, adaptive algorithms may be destabilized if the frequency spectrum of the input is inadequately distributed. Although stability properties derived through averaging depend on the frequency spectrum of the input, results show that some algorithms are consistently more robust than others. Therefore, we expect to derive, from the averaging approach, methods to improve the robustness of adaptive algorithms that are of very different nature than existing update law modifications.

Tuned Values in Adaptive Control. To analyze the stability of adaptive algorithms in the presence of unmodeled dynamics, Kosut & Friedlander [1985] introduced the notion of tuned value and tuned system. A *tuned value* is a value of the controller parameter (vector) that would be used if the plant were known exactly (for example, a value minimizing some performance index). It is assumed that such value at least stabilizes the plant. The *tuned system* response is the response of the control system with parameters frozen at the tuned value.

Riedle & Kokotovic [1985] analyzed the stability of an adaptive control system by linearizing the system around an arbitrary tuned value and by applying averaging. The resulting linearized-averaged system was a linear time invariant system so that its stability properties were determined by the eigenvalues of some matrix. Depending on the frequency spectrum of the input, the adaptive system was stable, or unstable. Through simulations, Riedle & Kokotovic showed that the analysis provided a tight stability/instability boundary for the *original* system.

A drawback of the linearization approach is that it may be inconclusive depending on which tuned value is chosen. Indeed, an important question is to determine *where* the parameters converge in the presence of unmodeled dynamics. Although the choice of a tuned value minimizing a cost criterion (for example quadratic) is intuitive, analysis and simulations show that parameters do not necessarily converge to such a value *in the presence of unmodeled dynamics*.

To provide a well-defined and more appropriate notion of tuned value, Riedle & Kokotovic [1986] proposed to replace its definition by letting the tuned value be the equilibrium of the averaged system. Subsequently, in

Mason *et al.* [1988], an equation error adaptive identifier was analyzed using averaging methods, and it was shown that the adaptive parameters converged to neighborhoods of the equilibrium of the averaged system. The averaged system was linear and it was concluded that the tuned value was *unique* under sufficient richness conditions. Further, it was shown to be *stable*, no matter what unmodeled dynamics were present. The value depended on the input spectrum, but a closed-form expression could be obtained.

In Bodson [1988], an example similar to the Riedle & Kokotovic example was studied. It was shown that the averaged system was linear so that a tuned value could again be defined as the unique equilibrium of the averaged system. The tuned value was *not* always stable (as observed by Riedle & Kokotovic), but a remarkable result was that some algorithms would remain stable when others would not. Again, the tuned value expressions and the stability conditions involved frequency-domain relations on the input spectrum and on the plant transfer function in a computable closed-form expression.

In this paper, we pursue the analysis on an example where the averaged system is *nonlinear*. The emphasis is on deriving frequency domain expressions allowing to determine the location and stability of the equilibrium points of the averaged system. The nonlinearity of the averaged system presents new challenges which the simplicity of the example allows to resolve. It is shown that the equilibrium is *not unique* anymore. How many equilibria there are depends on the spectral content of the input, but bounds on their location may be obtained. The stability of the averaged system is also discussed. Simulations illustrate the validity of the averaging approximation on a specific example.

A Model Reference Adaptive System. We consider a simple *model reference adaptive control* system. For the design, a first-order linear time-invariant plant P^* is assumed, that is,

$$y_p = P^* [u] \qquad P^*(s) = \frac{k}{s+a_p} \qquad\qquad [3.1]$$

where a_p is unknown, and $k > 0$ is known. The control objective is to track the output y_m of a reference model given by

$$y_m = M [r] \qquad M(s) = \frac{k}{s+a_m} \qquad\qquad [3.2]$$

where $a_m > 0$ is arbitrary. The *reference input r* is arbitrary but bounded. If a_p was known, the control objective would be achieved by choosing the control input as in Figure 1, that is,

$$u = r + d_0 y_p \qquad\qquad [3.3]$$

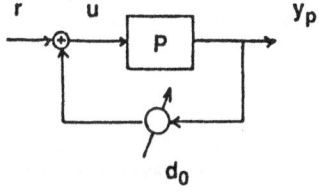

Figure 1: Plant and Controller

Indeed, transforming [3.1] and [3.2] in the time domain

$$\dot{y}_p = -a_p y_p + k u$$

leads, with [3.3], to

$$\dot{y}_p = (-a_p + k\, d_0)\, y_p + k\, r \qquad\qquad [3.5]$$

The *nominal value* of the parameter d_0 is the unique value

$$d_0^* = \frac{a_p - a_m}{k} \qquad\qquad [3.6]$$

such that the transfer function from r to y_p is the reference model transfer function.

Since the parameter a_p is unknown, an *adaptive* law is designed so that the *output error*

$$e_0 = y_p - y_m \qquad\qquad [3.7]$$

tends to zero as $t \to \infty$, with all internal signals remaining bounded. It may be shown (Narendra & Valavani [1978]) that the update law

$$\dot{d}_0 = -g\, e_0\, y_p \qquad\qquad g > 0 \qquad\qquad [3.8]$$

satisfies these requirements (taking a Lyapunov function $v(e_0, d_0) = e_0^2/2 + k/2g\,(d_0 - d_0^*)^2$).

In reality, the plant is only *approximately* a first-order plant $P^*(s)$. We assume that the true plant $P(s)$ is a linear time-invariant plant of arbitrary order. We may write, for example,

$$P(s) = P^*(s)\,(1 + L_m(s)) \qquad\qquad [3.9]$$

where $L_m(s)$ is a multiplicative uncertainty, describing the high-frequency dynamics that were neglected to obtain a reduced-order controller.

Introduction to Averaging . Averaging (see, for example, Bogoliuboff & Mitropolskii [1961], Hale [1980], and Volosov [1962]) is a method of analysis for differential equations of the form

$$\dot{x} = \varepsilon f(t, x) \qquad\qquad x(0) = x_0 \qquad\qquad [4.1]$$

The main assumptions are that $\varepsilon > 0$ is small and that the limit

$$f_{av}(x) = \lim_{T \to \infty} \frac{1}{T} \int_{t_0}^{t_0+T} f(\tau, x)\, d\tau \qquad\qquad [4.2]$$

exists uniformly in t_0 and x. The *averaged system* is then defined by

$$\dot{x}_{av} = \varepsilon f_{av}(x_{av}) \qquad\qquad x_{av}(0) = x_0 \qquad\qquad [4.3]$$

Given some smoothness assumptions on f and f_{av}, the following facts follow for ε sufficiently small:
- The trajectories $x(t)$ and $x_{av}(t)$ are arbitrarily close as $\varepsilon \to 0$, on intervals $[0, T/\varepsilon]$ where T is arbitrary.
- If $x_{av} = 0$ is an exponentially stable equilibrium of the averaged system, then trajectories of the original system converge to a neighborhood of the origin that shrinks to zero when $\varepsilon \to 0$.
- If $x_{av} = 0$ is an exponentially stable equilibrium of the averaged system and $x = 0$ is an equilibrium of

the original system, then $x = 0$ is exponentially stable.

The theory may be extended to two-time scale systems of the form

$$\dot{x} = \varepsilon f(t, x, y)$$ [4.4]

$$\dot{y} = A(x) y + h(t, x) + \varepsilon g(t, x, y)$$ [4.5]

The state x is varying slowly as compared to y when ε tends to zero. Therefore, x is called the *slow state* and y the *fast state*. This type of equation naturally arises in adaptive control, where x is the adaptive parameter vector, and y the vector of plant and controller states. The averaged system is then obtained by finding the steady-state solution of [4.5] with $\varepsilon = 0$, *assuming that x is fixed*, i.e.,

$$y(t, x) = \int_0^t e^{A(x)(t-\tau)} h(\tau, x) \, d\tau$$ [4.6]

and letting

$$f_{av}(x) = \lim_{T \to \infty} \frac{1}{T} \int_{t_0}^{t_0+T} f(\tau, x, y(\tau, x)) \, d\tau$$ [4.7]

It is assumed that $A(x)$ is a (uniformly) stable matrix for all values of x considered, and that the average [4.7] exists uniformly in t_0, x.

Stability and instability theorems for averaging, with applications to adaptive systems, may be found in several places, including Anderson *et al.* [1986], Bodson *et al.* [1986], Fu, Bodson & Sastry [1986], Fu & Sastry [1987], Mason *et al.* [1988], Riedle & Kokotovic [1986], and Sastry & Bodson [1988]. We do not derive averaging theorems in this paper. Rather, we concentrate on obtaining closed-form expressions for the averaged system in the frequency domain, and without linearization. We are especially interested in determining the number and location of the equilibrium points of the averaged system, and their stability.

Averaging Analysis of the Adaptive System. The adaptive system is completely described by

$$\dot{d}_0 = -g(y_p - y_m) y_p$$ [4.8]

$$y_p = P[r + d_0 y_p] \qquad y_m = M[r]$$ [4.9]

By writing a state-space representation for P, the system [4.8]-[4.9] may be represented in the form [4.4]-[4.5] where x replaces d_0, ε replaces g, and y replaces the states of P and M. We will assume that
- r is *periodic*, of the form

$$r = \sum_{i=1}^n r_i \sin(\omega_i t)$$ [4.10]

- the parameter g is sufficiently small.

To derive an expression for the averaged system, we must first solve [4.9] (i.e. [4.5]) assuming that d_0 (i.e. x) is constant. For brevity, we will keep a compact notation and write the solution simply as

$$y_p = \frac{P}{1 - d_0 P} [r]$$ [4.11]

The averaged system is then obtained by calculating the averaged value

$$AVG[(y_p - y_m) y_p] = AVG [\frac{P}{1-d_0 P} [r] \frac{P}{1-d_0 P} [r]] - AVG [\frac{P}{1-d_0 P} [r] M [r]]$$

$$= AVG [\frac{P}{1-d_0 P} [r] \frac{P}{1-d_0 P} [r]] - AVG [\frac{P}{1-d_0 P} [r] \frac{M}{1-d_0 P} [r]]$$

$$+ d_0 AVG [\frac{P}{1-d_0 P} [r] \frac{P M}{1-d_0 P} [r]] \qquad [4.12]$$

Since r is given by [4.10], the average in [4.12] may be found to be

$$AVG[(y_p - y_m) y_p] = \sum_{i=1}^{n} \frac{r_i^2}{2} (\frac{|P(j\omega_i)|^2 - \text{Re}[P(j\omega_i) M^*(j\omega_i)]}{|1 - d_0 P(j\omega_i)|^2}$$

$$+ d_0 \frac{|P(j\omega_i)|^2 \text{Re}[M(j\omega_i)]}{|1-d_0 P(j\omega_i)|^2}) \qquad [4.13]$$

If we define

$$d_i = \frac{\text{Re}[P(j\omega_i) M^*(j\omega_i)] - |P(j\omega_i)|^2}{|P(j\omega_i)|^2 \text{Re}[M(j\omega_i)]} \qquad [4.14]$$

$$\alpha_i (d_0) = \frac{r_i^2}{2} \frac{|P(j\omega_i)|^2 \text{Re}[M(j\omega_i)]}{(1 - d_0 \text{Re}[P(j\omega_i)])^2 + (d_0 \text{Im}[P(j\omega_i)])^2} \qquad [4.15]$$

the averaged system is simply given by

$$\dot{d}_{av} = -g \sum_{i=1}^{n} \alpha_i (d_{av}) (d_{av} - d_i) \qquad [4.16]$$

Properties of the Averaged System. The averaged system is time invariant (autonomous) and is therefore considerably simpler than the original system. However, averaging was applied without linearization, so that the nonlinear behavior of the original system is reflected in the averaged system. The right-hand side of [4.16] is a rational function of d_{av}, due to the presence of the term $\alpha_i (d_{av})$. By [4.15], this term satisfies

$$\alpha_i (d_{av}) > 0 \qquad \text{for all } d_{av} \qquad [4.17]$$

Two underlying assumptions of the analysis are that
  • along the trajectories of the averaged system, d_{av} stabilizes P (this usually restricts the applicability to some interval of the real axis)
  • for all i, $|P(j\omega_i)| \neq 0$.

Assuming that these additional assumptions are satisfied, we proceed with the analysis of the averaged system.

Case 1: n=1. Consider the case when the input contains only one sinusoid of amplitude $r_1 \neq 0$. The averaged system is then given by

$$\dot{d}_{av} = -g \alpha_1(d_{av}) (d_{av} - d_1) \qquad [4.18]$$

Since $\alpha_1(d_{av}) > 0$, we conclude that

- the system has a unique equilibrium $d_{av} = d_1$,
- the equilibrium is exponentially stable (take a Lyapunov function $v = (d_{av} - d_1)^2$).

Case 2: n=2. When two sinusoids are present, the averaged system becomes

$$\dot{d}_{av} = - g \left(\alpha_1(d_{av}) \, (d_{av} - d_1) + \alpha_2(d_{av}) \, (d_{av} - d_2) \right) \tag{4.19}$$

The equilibrium points d^+ of the averaged system satisfy

$$\alpha_1(d^+) \, (d^+ - d_1) + \alpha_2(d^+) \, (d^+ - d_2) = 0 \tag{4.20}$$

or

$$\frac{d^+ - d_1}{\alpha_2(d^+)} + \frac{d^+ - d_2}{\alpha_1(d^+)} = 0 \tag{4.21}$$

Using [4.15], equation [4.21] defines a 3rd order polynomial with real coefficients. Therefore, there exists at least one equilibrium, and at most three. Further, any equilibrium satisfies [4.20], or

$$d^+ = \beta(d^+) \, d_1 + (1 - \beta(d^+)) \, d_2 \tag{4.22}$$

where

$$\beta(d^+) = \frac{\alpha_1(d^+)}{\alpha_1(d^+) + \alpha_2(d^+)} \tag{4.23}$$

By [4.17], the equilibrium points satisfy the *convex* relationship [4.22] with $0 < \beta < 1$, and therefore

$$d^+ \in [d_1, d_2] \quad \text{if} \quad d_1 \le d_2 \qquad \text{or} \qquad d^+ \in [d_2, d_1] \quad \text{if} \quad d_2 \le d_1$$

where d_1, d_2 are the equilibrium points for single sinusoidal inputs.

When three equilibrium points d_1^+, d_2^+, d_3^+ exist, the general form of the right-hand side of [4.19] is shown on Figure 2. Consequently, the equilibrium d_2^+ in the model is unstable, while the equilibria d_1^+, d_3^+ are stable (in the

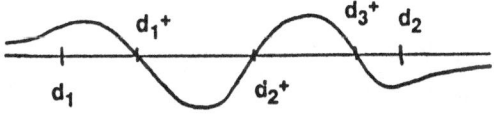

Figure 2: Sketch of the Right-Hand Side of [4.19]

sense of Lyapunov). In the case when two roots are colocated, the equilibrium point is unstable. Finally, if only one equilibrium point exists, it is stable.

Case n. The analysis for $n = 2$ may be repeated, leading to a polynomial of order $2n-1$. Since the order is odd, there exists *at least one* equilibrium point (this is a conclusion also found in Pomet, Coron & Praly [1987]). There may be up to $2n-1$ equilibrium points, and up to n stable equilibrium points. Convergence to one or the other equilibrium points is dependent on the initial conditions in the adaptive system. However, in all cases, there exists *at least one stable equilibrium point.*

Illustrative Example. For illustration, we present simulations with values inspired from the Rohrs examples (Rohrs *et al.* [1982]). We let

$$P^*(s) = \frac{2}{s+1} \qquad \text{i.e., } k = 2, a_p = 1$$

$$M(s) = \frac{2}{s+3} \qquad \text{i.e., } d_0^* = -1$$

$$P(s) = \frac{2}{s+1} \frac{229}{s^2+30s+229} \qquad\qquad\qquad [5.1]$$

The unmodeled dynamics consist of a pair of stable, well-damped complex poles located more than a decade away from the main plant pole. We let

$$r(t) = r_1 \sin(t) + r_2 \sin(10t)$$

that is, $\omega_1 = 1$, $\omega_2 = 10$. The equilibrium points for single sinusoids are calculated to be

$$d_1 = -1.05 \qquad\qquad d_2 = -12.86$$

Figure 3 shows the locus of the equilibrium points of the averaged system [4.19] for varying r_2 and $r_1 = 10 - r_2$. For a range of values between $r_2 \approx 6$ and $r_2 = 7$, there are three equilibrium points, but for most of the range, there exists only one equilibrium. The lack of symmetry of the graph may be attributed to the low-pass filtering of the

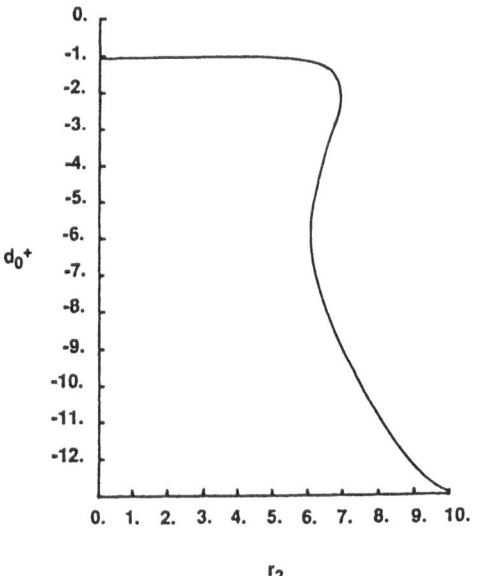

Figure 3: Locus of Equilibrium Points

plant. Also, the equilibrium remains close to d_1 for a large range of r_2, suggesting that the adaptive system "tunes" to the dominant low frequency $\omega_1 = 1$ despite the high-frequency component.

The bifurcation of equilibrium is reminiscent of Bai & Salam [1987]. In their example, bifurcation occured as the amplitude of a constant output disturbance was varied, while the reference input was zero. Here, a time-varying adaptive system is originally considered. A bifurcation in the averaged system is observed when the relative amplitude of the frequency components of the input is varied.

Figure 4 shows simulations for the case when $r_1 = 3.5$, $r_2 = 6.5$. The adaptation gain g is equal to 1. The three

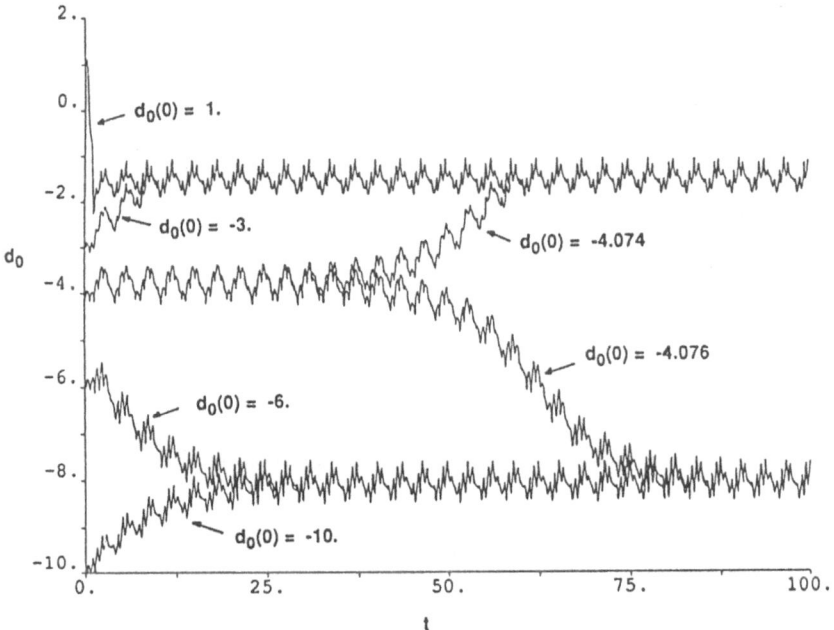

Figure 4: Adaptive Parameter Responses

equilibrium points are then

$$d_1{}^+ = -1.44 \ \text{(stable)} \qquad d_2{}^+ = -3.63 \ \text{(unstable)} \qquad d_3{}^+ = -8.14 \ \text{(stable)}$$

The simulations with different initial conditions illustrate the convergence to neighborhoods of two different equilibrium points of the averaged system. Close to the unstable equilibrium, trajectories slowly drift toward one of the stable equilibrium points. For example, two very close initial conditions lead to close trajectories for 40 seconds, but eventually diverging. Figures 5 and 6 show the output error for $d_0(0) = -4.074$ and $d_0(0) = -4.076$ respectively. Consistently with Figure 4, the initial responses are very similar, while the steady-state responses are quite different (different frequency content is observed, reflecting the different tuned values).

Referring to Figure 3, it may be expected that a slow transition from sin(t) to sin(10t) would lead to a *jump* phenomenon. Figure 7 illustrates the parameter response when $r = r_1(t) \sin(t) + r_2(t) \sin(10t)$, with $r_1(t) = 3 (1 - t / 1000)$ and $r_2(t) = 3 t / 1000$. The adaptation gain g is equal to 5. For a long period of time, the adaptive parameter remains close to the original value around -1, then suddenly jumps toward -13.

Conclusions. This paper illustrated the application of *nonlinear* averaging methods for the analysis of an adaptive control system with unmodeled dynamics.

Usually, the approach to such averaging analysis has been to define tuned values arbitrarily, as parameter

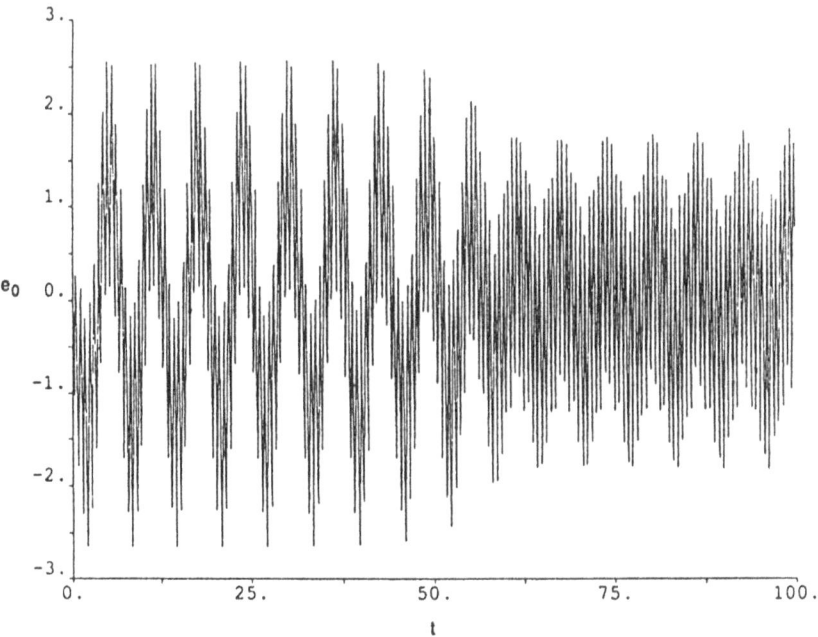

Figure 5: Output Error Response ($d_0(0)$=-4.074)

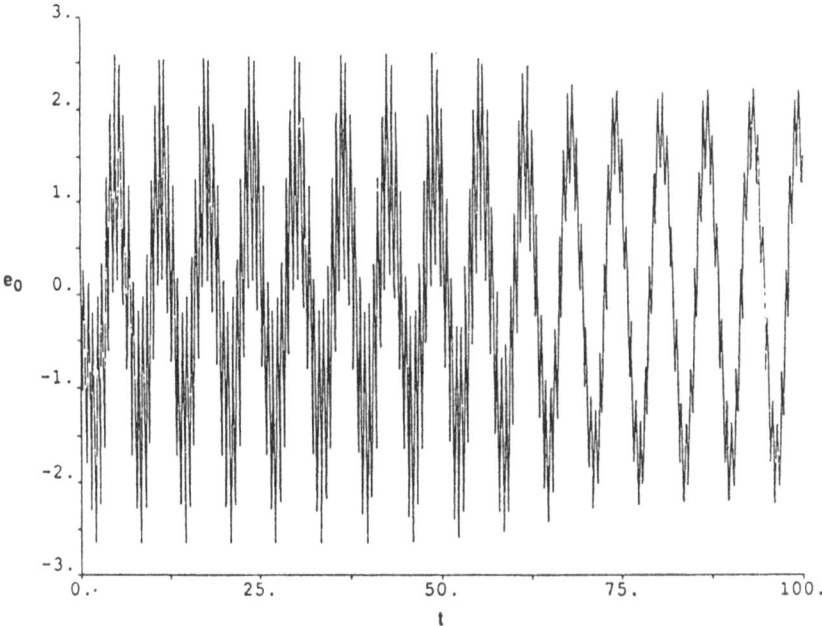

Figure 6: Output Error Response ($d_0(0)$=-4.076)

values which stabilize the inner loop (plant and controller). The adaptive system may then be studied by linearization around the trajectories of the tuned system. A main drawback of such an analysis is that the

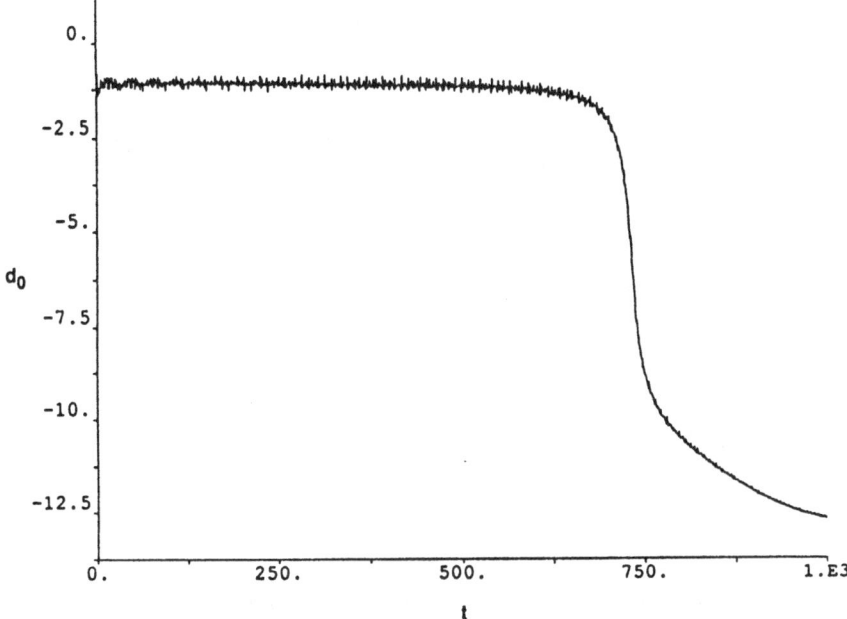

Figure 7: Adaptive Parameter Response (Jump)

analysis may be inconclusive depending on the choice of tuned value, and that it fails to determine *where* the adaptive parameters actually converge.

In this paper, we restricted tuned values to be the equilibrium points of the nonlinear averaged system. The stable equilibrium points determine where the adaptive parameters converge. Due to the nonlinearity of the equations, several equilibria may exist, and steady-state responses may depend on initial conditions. It was shown that there always existed at least one stable equilibrium. Bounds on the location of the tuned values were also determined. The tuned values were calculated from frequency-domain expressions of the input spectrum and of the plant transfer function by solving a nonlinear polynomial equation. The main assumptions were that the adaptation gain was sufficiently small and that the input consisted of a finite number of sinusoids.

Acknowledgements. The support of the National Science Foundation under grant ECS-8810145 is gratefully acknowledged. The author is Assistant Professor in the Department of Electrical and Computer Engineering, Carnegie-Mellon University, Pittsburgh, PA 15213. Simulations were carried out using the simulation package SIMNON at the University of California, Berkeley. The author would like to thank Professor S. Sastry for his hospitality, and Professor K. Astrom for the use of the software package.

References

Anderson, B.D.O., R.R. Bitmead, C.R. Johnson, P.V. Kokotovic, R.L. Kosut, I.M.Y. Mareels, L. Praly, & B.D. Riedle, *Stability of Adaptive Systems, Passivity and Averaging Analysis*, MIT Press, Cambridge, Massachusetts, 1986.

Astrom, K.J., "Interactions Between Excitation and Unmodeled Dynamics in Adaptive Control," *Proc. of the 23rd IEEE Conf. on Decision and Control*, pp. 1276-1281, Las Vegas, Nevada, 1984.

Bai, S., & F.M.A. Salam, "Disturbance-Generated Bifurcation in an Adaptive System with e_1-Modification Law," *Proc. of the 1988 Automatic Control Conference*, pp. 772-773, Atlanta, Georgia, 1988.

Bodson, M., "Effect of the Choice of Error Equation on the Robustness Properties of Adaptive Control Systems," *Int. J. of Adaptive Control and Signal Processing*, to appear, 1988.

Bodson, M., S. Sastry, B.D.O. Anderson, I. Mareels, & R.R. Bitmead, "Nonlinear Averaging Theorems, and the Determination of Parameter Convergence Rates in Adaptive Control," *Systems & Control Letters,* Vol. 7, no. 3, pp. 145-157, 1986.

Bogoliuboff, N.N., & Y.A. Mitropolskii, *Asymptotic Methods in the Theory of Nonlinear Oscillators,* Gordon & Breach, New York, 1961.

Fu, L.-C., M. Bodson, & S. Sastry, "New Stability Theorems for Averaging and Their Application to the Convergence Analysis of Adaptive Identification and Control Schemes," *Singular Perturbations and Asymptotic Analysis in Control Systems,* Lecture Notes in Control and Information Sciences, P. Kokotovic, A. Bensoussan, & G. Blankenship (Eds.), Springer-Verlag, New York, 1986.

Fu, L.-C., & S. Sastry, "Slow Drift Instability in Model Reference Adaptive Systems - An Averaging Analysis," *Int. J. Control,* Vol. 45, no. 2, pp. 503-527, 1987.

Hale, J.K., *Ordinary Differential Equations,* Krieger, Huntington, New York, 1980.

Ioannou, P.A., & P.V. Kokotovic, "Robust Redesign of Adaptive Control," *IEEE Trans. on Automatic Control,* Vol. AC-29, no. 3, pp. 202-211, 1984.

Ioannou, P.A., & K. Tsakalis, "Robust Direct Adaptive Controller," *IEEE Trans. on Automatic Control,* Vol. AC-31, No. 11, pp. 1033-1043, 1986.

Kosut, R.L., and B. Friedlander, "Robust Adaptive Control: Conditions for Global Stability," *IEEE Trans. on Automatic Control,* Vol. AC-30, no. 7, pp. 610-624, 1985.

Kreisselmeier, G., & B.D.O. Anderson, "Robust Model Reference Adaptive Control," *IEEE Trans. on Automatic Control,* Vol. AC-31, no. 2, pp. 127-133, 1986.

Mason, J.E., E.W. Bai, L.-C. Fu, M. Bodson, & S. Sastry, "Analysis of Adaptive Identifiers in the Presence of Unmodeled Dynamics: Averaging and Tuned Parameters," *IEEE Trans. on Automatic Control,* to appear, 1988.

Narendra, K.S., & A.M. Annaswamy, "A New Adaptive Law for Robust Adaptation without Persistent Excitation, " *IEEE Trans. on Automatic Control,* Vol. AC-32, no. 2, pp. 134-145, 1987.

Narendra, K.S., & L.S. Valavani, "Stable Adaptive Controller Design - Direct Control," *IEEE Trans. on Automatic Control,* Vol. AC-23, no. 4, pp. 570-583, 1978.

Ortega, R., L. Praly, & I.D. Landau, "Robustness of Discrete-Time Direct Adaptive Controllers," *IEEE Trans. on Automatic Control,* Vol. AC-30, no. 12, pp. 1179-1187, 1985.

Pomet, J.-B., J.M. Coron, & L. Praly, "About Periodic Solutions of Adaptive Systems in Presence of Periodic Forcing Terms," Research Paper, 1987.

Riedle, B.D., & P.V. Kokotovic, "Stability-Instability Boundary for Disturbance-Free Slow Adaptation with Unmodeled Dynamics," *IEEE Trans. on Automatic Control,* Vol. AC-30, no. 10, pp. 1027-1030, 1985.

Riedle, B.D., & P.V. Kokotovic, "Integral Manifolds of Slow Adaptation," *IEEE Trans. on Automatic Control,* Vol. AC-31, no. 4, pp. 316-324, 1986.

Rohrs, C.E., L. Valavani, M. Athans, & G. Stein, "Robustness of Adaptive Control Algorithms in the Presence of Unmodeled Dynamics," *Proc. of the 21st IEEE Conference on Decision and Control,* Florida, pp. 3-11, 1982.

Sastry, S., "Model-Reference Adaptive Control - Stability, Parameter Convergence, and Robustness," *IMA Journal of Mathematical Control & Information,* Vol. 1, pp. 27-66, 1984.

Sastry, S., & M. Bodson, *Adaptive Control: Stability, Convergence and Robustness,* Prentice Hall, Englewood-Cliffs, New Jersey, to appear, 1988.

Volosov, V.M., "Averaging in Systems of Ordinary Differential Equations," *Russian Mathematical Surveys,* Vol. 17, no. 6, pp. 1-126, 1962.

ON THE ADAPTIVE CONTROL OF A PARTIALLY
OBSERVABLE BINARY MARKOV DECISION PROCESS

Emmanuel Fernández-Gaucherand, Aristotle Arapostathis, and Steven I. Marcus

Department of Electrical and Computer Engineering
The University of Texas at Austin
Austin, Texas 78712-1084

1. Introduction. As noted in [AM], despite the considerable amount of work in adaptive stochastic control (see [KUM1] for a survey), there has been little work on problems with incomplete or noisy state observations, aside from the linear case. A first step in this direction was taken in [AM], in which the adaptive estimation of the state of a finite state Markov chain with incomplete state observations and in which the state transition probabilities depend on unknown parameters is studied. In this context, the adaptive estimation problem is that of computing recursive estimates of the conditional probability vector of the state at time t given the observations up to time t, when the transition matrix P is not completely known (i.e., it depends on a vector of unknown parameters θ — this dependence is expressed as $P(\theta)$).

The approach to this problem which we adopted in [AM] has been widely used in linear filtering: we use the previously derived recursive filter for the conditional probabilities, and we simultaneously recursively estimate the parameters, plugging the parameter estimates into the filter. This adaptive estimation algorithm is then analyzed via the Ordinary Differential Equation (ODE) Method. That is, it is shown that the convergence of the parameter estimation algorithm can be analyzed by studying an "averaged" ordinary differential equation. The most crucial and difficult aspect of the proof is that of showing that, for each value of the unknown parameter, an augmented Markov process has a unique invariant measure, and that the expectations of certain functions of the augmented state converge geometrically to the same expectations under the invariant measure. The convergence of the recursive parameter estimates is studied, and optimality of the adaptive state estimator is proved, for a given criterion.

In this paper, we take some initial steps in the direction of applying similar techniques to adaptive finite state Markov decision (control) problems with incomplete observations. One intriguing set of problems for which some results are available when the parameters are *known* are those involving quality control and machine replacement, and we study the properties of these problems in this paper, with the eventual aim of developing optimal adaptive stochastic controllers. However, the presence of feedback makes this stochastic adaptive control problem much more difficult than the adaptive estimation problem of [AM]. We consider here only problems involving two states and two actions; for some more general analysis, see [FAM].

2. The Two-State Binary Replacement Problem. Consider the situation in which a machine/production process deteriorates over time under operation. The ("core") state of the machine is such that 0 is the more desirable ("as new" or "good") state, and 1 represents the failed (or "bad") state. The efficiency of the machine, or quality of the items produced is a function of the core state. Items are produced at the beginning of each time epoch t, for $t \in N_0 := \{0\} \cup N$,

and at that moment one of two decisions must be made: continue to produce (0) or repair/replace (1); the word "binary" refers to the fact that there are only two actions. Under production, the machine may deteriorate to the "bad" state 1, and once in that state it stays there if no repair/replace action is taken. If the decision to repair/replace is taken, then the machine will be in the "good" state by the beginning of the next time epoch w.p.1. Imperfect observations of the state of the system are available while producing.

Putting this problem in the standard framework of partially observed Markov decision processes (POMDP) (see, e.g., [BE], [MO], [HLM]), we have the following model. Let (Ω, F, P) be a probability space and let $X = Y = \{0, 1\}$, $U = \{0, 1\}$. The system's state process will be modelled as a finite state controlled Markov chain with ("core") state space X and action space U with 2×2 transition matrices $\{P(u)\}_{u \in U}$. Thus, the core process is given by a random process $\{x_t\}$ on (Ω, F, P), $t \in \mathbf{N}_0$, where for a sequence of U-valued random variables $\{u_k\}_{k=0}^{t}$ on (Ω, F, P), the controls (or decisions),

$$P\{x_{t+1} = j | x_t = i, x_{t-1}, \ldots, x_0; u_t, u_{t-1}, \ldots, u_0\} = [P(u_t)]_{i,j} =: p_{i,j}(u_t), \quad t \in \mathbf{N}_0$$

Only partial observations of $\{x_t\}_{t \in \mathbf{N}_0}$ are available in the form of a random process $\{y_t\}_{t \in \mathbf{N}_0}$ taking values in the ("message") space Y. The sequence of events is assumed as follows: at time epoch t the system is in state x_t, observation y_t becomes available, and action u_t is taken; transition to a state x_{t+1} has occurred by the beginning of time epoch $t+1$, another observation y_{t+1} becomes available, and then a new decision u_{t+1} is made; and so on. The core and observation processes are related probabilistically as

$$P\{y_{t+1} = k | y_t, \ldots, y_0; x_{t+1} = i, \ldots, x_0; u_t, \ldots, u_0\}$$

$$= P\{y_{t+1} = k | x_{t+1} = i, u_t\} =: q_{i,k}(u_t), \quad t \in \mathbf{N}_0$$

$$P\{y_0 = k | x_0 = i\} =: q_{i,k}^{(0)}$$

which leads to the definition of a collection of 2×2 observation matrices Q_0 and $\{Q(u)\}_{u \in U}$ such that

$$Q(u) := [q_{i,k}(u)]_{i \in X, k \in Y}; \quad Q_0 := [q_{i,k}^{(0)}]_{i \in X, k \in Y}$$

It is assumed that the probability distribution of the initial state, $p_0 := [P\{x_0 = i\}]_{i \in X} \in \Delta$, is available for decision making, where $\Delta := \{p \in \Re^2 : p^{(i)} \geq 0, 1'p = 1\}$; here, $1 = [1, 1]'$, "prime" denotes transposition, and $p^{(i)}$ denotes the ith component of p. We endow Δ with the metric topology induced by the norm $\| \cdot \|_1$ given by $\|p\|_1 := \frac{1}{2}1'p$. Define recursively the *information spaces*

$$H_0 := \Delta \times Y; \quad H_t := H_{t-1} \times U \times Y, \quad t \in \mathbf{N}; \quad H_\infty := H_0 \times (U \times Y)^\infty,$$

each equipped with its respective product topology. An element $h_t \in H_t, t \in \mathbf{N}_0$, is called an *observable history* and represents the information available for decision making at time epoch t. It is straightforward to show that X, Y, U, and Δ are Borel spaces, and hence so is $H_t, t \in \mathbf{N}_0$ (see [BS, p. 119]); this leads to well-defined probabilistic structure [BS].

Let $\mu_t : H_t \to U$ be a measurable map. An admissible control law, policy, or strategy μ is a sequence of maps $\{\mu_t(\cdot)\}_{t \in \mathbf{N}_0}$, or $\{\mu_0(\cdot), \ldots, \mu_n(\cdot)\}$ for a finite horizon. When $\mu_t(\cdot) = \mu(\cdot)$ for all values of t, then the policy is said to be *stationary*. Let $c : U \times X \to [0, M]$ be a given measurable map, where M is a positive scalar; $c(u, x)$ is interpreted as the cost incurred given that the system was in state x and control action u was selected. To each admissible strategy μ and probability distribution p_0 of the initial state, the following expected costs are associated.

Finite Horizon:

$$J_\beta(\mu, p_0, n) := E^\mu_{p_0}\left[\sum_{t=0}^n \beta^t c(u_t, x_t)\right], \quad n \in N_0, \quad 0 < \beta \tag{FH}$$

Discounted Cost:

$$J_\beta(\mu, p_0) := \lim_{n \to \infty} E^\mu_{p_0}\left[\sum_{t=0}^n \beta^t c(u_t, x_t)\right], \quad 0 < \beta < 1 \tag{DC}$$

Average Cost:

$$J(\mu, p_0) := \lim_{n \to \infty} \sup E^\mu_{p_0}\left[\frac{1}{n+1}\sum_{t=0}^n c(u_t, x_t)\right] \tag{AC}$$

where $E^\mu_{p_0}$ is the expectation with respect to the (unique) probability measure on H_∞ induced by p_0 and the strategy μ, or an appropriate marginal (see [BS, pp. 140-144 and 249]). For a strategy μ, $\{u_t = \mu_t(h_t)\}_{t=0}^n$ is the control process governing state transitions. The *optimal control (or decision) problem* is that of selecting an (optimal) admissible stragegy such that one of the above criteria is minimized over all admissible strategies. The optimal (DC) cost function is obtained as $\Gamma_\beta(p_0) := \inf_\mu \{J_\beta(\mu, p_0) : \mu \text{ is an admissible strategy}\}$, for each $p_0 \in \Delta$. Similarly denote by $\Gamma_\beta(\cdot, n)$ and $\Gamma(\cdot)$ the optimal cost functions for the horizon n (FH) and (AC) cases, respectively.

It is well known that a separation principle holds for the problems listed above (see [BS], [KV]). Specifically, let $\{p_{t|t}\}_{t\in N_0}$ denote the conditional probability distribution process, whose i^{th} component is given by

$$p^{(i)}_{t|t} := P\{x_t = i|y_t, \ldots, y_0; u_{t-1}, \ldots u_0\}, \quad t \in N$$
$$p^{(i)}_{0|0} := p^{(i)}_0$$

Then, assuming that $P\{h_t, u_t, y_{t+1} = k\} \neq 0$ for $t \in N_0$ and for each $k \in Y$, and using Bayes' rule it is easily shown that (see also [KV, Sect. 6.6], [AS])

$$p_{t+1|t+1} = \sum_{k \in Y} \frac{\overline{Q}_k(u_t)P'(u_t)p_{t|t}}{1'\overline{Q}_k(u_t)P'(u_t)p_{t|t}} \cdot I[y_{t+1} = k] \tag{2.1}$$

for each sample path, where $I[A]$ denotes the indicator function of the event A and the 2×2 matrices $\overline{Q}_k(u)$ are given by $\overline{Q}_k(u) := \text{diag} \{q_{i,k}(u)\}$. Note that $p_{t|t}$ is a function of $(y_t, p_{t-1|t-1}, u_{t-1})$. Define a *separated* admissible law for time epoch t as measurable map $\mu_t : \Delta \to U$; separated admissible strategies are defined similarly as before. It is straightforward to show that the process $\{p_{t|t}\}$ obtained via feedback using a separated strategy is a Markov process [AS]. A separated admissible law μ can be regarded as an admissible law via $h_t \mapsto \mu_t(p_{t|t})$, where $p_{t|t}$ is obtained from h_t by applying (2.1) recursively. Then the partially observed, finite horizon problem (FH) is equivalent (i.e., equal minimum costs for each $p_0 \in \Delta$) to the *completely observed* finite horizon problem, with state space Δ, of finding a separated admissible strategy which minimizes

$$J_\beta'(\mu, p_0, n) := E_{p_0}^\mu \left[\sum_{t=0}^n \beta^t p_{t|t}' c(u_t) \right] \qquad (FH')$$

where $u_t := \mu_t(p_{t|t})$, $c(u) = [c(u,i)]_{i \in X}$. Similarly, (DC') and (AC') are defined.

Returning to the binary replacement problem with two states, the model of the POMDP takes the form

$$P(0) = \begin{bmatrix} 1-\theta & \theta \\ 0 & 1 \end{bmatrix}; \quad P(1) = \begin{bmatrix} 1 & 0 \\ 1 & 0 \end{bmatrix} \qquad (2.2)$$

$$Q := Q(0) = Q(1) = \begin{bmatrix} q & 1-q \\ 1-q & q \end{bmatrix} \qquad (2.3)$$

where $\theta \in [0,1]$ and $q \in (0.5, 1)$ (i.e., the core state is strictly partially observed). Actually, it is natural to expect that θ is some small positive number. Note that if the decision to replace has been taken at time epoch t, then the state of the machine is known at the beginning of time epoch $t+1$, hence the observation at this time is irrelevant and thus $Q(1)$ can be chosen arbitrarily. For the observation quality structure under decision $u = 0$, the probability of making a correct observation of the state is q. The conditional probability vector can be written as $p_{t|t} = [1 - \rho_{t|t}, \rho_{t|t}]'$, where $\rho_{t|t}$ is the conditional probability of the process being in the bad state. From (2.1)-(2.3), with $\rho_{0|0} = \rho_0 := [0 \ 1]p_0$, we get

$$\rho_{t+1|t+1} = T(1, \rho_{t|t}, u_t)y_{t+1} + T(0, \rho_{t|t}, u_t)(1 - y_{t+1}) \qquad (2.4)$$

where u_t is the decision made at time epoch t, and for $\rho \in [0,1]$

$$V(0, \rho, 0) := q(1-\rho)(1-\theta) + (1-q)[\rho(1-\theta) + \theta] \qquad (2.5)$$

$$V(1, \rho, 0) := (1-q)(1-\rho)(1-\theta) + q[\rho(1-\theta) + \theta] \qquad (2.6)$$

$$V(0, \rho, 1) := 1 - q; \quad V(1, \rho, 1) := q \qquad (2.7)$$

$$T(0, \rho, 0) := \frac{(1-q)[\rho(1-\theta) + \theta]}{V(0, \rho, 0)}; \quad T(1, \rho, 0) := \frac{q[\rho(1-\theta) + \theta]}{V(1, \rho, 0)} \qquad (2.8)$$

$$T(k, \rho, 1) := 0; \quad k = 0, 1. \qquad (2.9)$$

Here, given an a priori probability ρ of the system being in the bad state, $V(k, \rho, u)$ is interpreted as the (one-step ahead) conditional probability of the observation being k, under decision u. Similarly, $T(k, \rho, u)$ is interpreted as the a posteriori conditional probability of the system being in the bad state, given that decision u was made, observation k obtained, and an a priori probability ρ. Note that $\{\rho_{t|t}\} = \{\rho_{t|t}(\theta)\}$.

We now study some important properties of $T(k, \cdot, 0), k = 0, 1$. Define

$$f_0(\rho) := T(0, \rho, 0) - \rho, \qquad f_1(\rho) := T(1, \rho, 0) - \rho.$$

Thus, the roots of $f_0(\cdot)$ and $f_1(\cdot)$ are the fixed points of $T(0, \cdot, 0)$ and $T(1, \cdot, 0)$, respectively. The pertinent quadratic equation for $f_0(\cdot)$ is

$$\xi_0^2 [(2q-1)(1-\theta)] - \xi_0 [(2q-1)(1-\theta) + (1-q)\theta] + (1-q)\theta = 0$$

and for $\theta \neq 1$ its roots are

$$\xi_0' = 1; \qquad \xi_0 = \frac{(1-q)\theta}{(2q-1)(1-\theta)}$$

For $\theta = 1$, both roots are equal to 1. Replacing q by $(1-q)$ above, the expressions corresponding to $f_1(\cdot)$ are obtained. Then, the following can be shown (see [FAM]).

Lemma 2.1: For $q \in (0.5, 1)$ and $\theta \in (0, 1)$, the following holds:

(a) $T(k, \cdot, u)$ is monotone increasing in $[0, 1)$, for each $u = 0, 1$ and $k = 0, 1$;

(b) $T(0, \cdot, 0) < T(1, \cdot, 0)$, in $[0, 1)$;

(c) $\rho < T(1, \rho, 0)$, for $\rho \in [0, 1)$;

(d) if $\frac{1}{2-\theta} < q$, then $\xi_0 \in (0, 1)$ and $\rho_1 < T(0, \rho_1, 0)$ and $T(0, \rho_2, 0) < \rho_2$, for $\rho_1 \in [0, \xi_0)$ and $\rho_2 \in (\xi_0, 1)$;

(e) if $q \leq \frac{1}{2-\theta}$, then $\xi_0 \geq 1$ and $\rho < T(0, \rho, 0)$ for $\rho \in [0, 1)$.

Remark 2.1: When $\theta = 1$ (not a very interesting situation), it follows that $T(0, \cdot, 0) = T(1, \cdot, 0) = 1$ on $[0, 1]$. Also, parts (c) and (d) above can be interpreted as meaning that an observation of the process being in the bad state is trusted more than an observation of the process being in the good state, in general.

For an optimization problem within some class of decision processes, it is of great importance to establish qualitative properties on the structure of the optimal policies. Such insight may aid in accelerating the computation of optimal policies for specific problems, and also hints at what "good," possibly easily implementable, control laws should be for the physical system being modelled [HS]. Results of this nature can be found, e.g., in machine replacement problems [W2], [RO], [BE], competing queues schemes [SM], etc. Also, when an adaptive control scheme based on a Certainty Equivalence Principle is to be studied, structural specificiation of optimal (or "good") policies is of critical importance in order to determine the policies over which adaptation must take place. This is well known for the self-tuning regulator [KV]; for other applications, see also [MMS], [KUM2].

For the cost structure of our problem, we choose for production $c(0, i) = C_i$, with $C_0 = 0$, $C_1 = C$, and for replacement $c(1, i) = c(1) = R$, $0 < C < R < \infty$. Considering a (DC') criterion, it is straightforward to show that it is always optimal to produce at $\rho = 0$ [AKO], [W1]. Thus, since the region $\Delta_r \subset \Delta$ in which it is optimal to repair is convex [LO1], then optimal strategies will be monotone in $\rho \in [0, 1]$, and hence to each such optimization problem there corresponds an $\alpha \in (0, 1]$ such that the structure of an optimal stationary strategy is as shown in Figure 1. Furthermore, it is easily shown that $\alpha = 1$, i.e., it is optimal to produce for all $\rho \in [0, 1]$, if and only if $R \geq \frac{C(1+\beta\theta)}{[1-\beta(1-\theta)]}$; [W1]. Such separated policies are said to be of the control-limit type, with the corresponding α being the "control-limit."

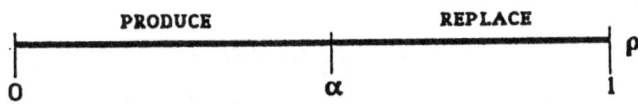

Figure 1: Control-limit policy structure

Under a control-limit stationary feedback strategy and with nonvoid replacement region, i.e. $\alpha \in (0,1)$, it is clear from Lemma 2.1 that the time to replace will be finite, and uniformly bounded with respect to the initial probability distribution, if $1/2 < q < \frac{1}{2-\theta}$. However, since θ is to be expected to be "small," this condition only says that the system is close to being completely unobservable, and thus periodic replacement should be scheduled. In general, when the quality of the observations may be more significant, e.g. $\frac{1}{2-\theta} < q$, then we have the following.

Lemma 2.2: Let $q \in (0.5, 1)$, and let μ be a control-limit stationary policy with nonvoid replacement region. Then there exists an $M \in \mathbb{N}$ such that, under feedback by μ, M consecutive observations of the core process being in the bad state result in at least one replace action, independently of the initial probability of being in the bad state.

Proof: Define recursively $T^{n+1}(1,\rho,0) := T(1,T^n(1,\rho,0),0)$, for $n \in \mathbb{N}$ and $T^1(1,\rho,0) = T(1,\rho,0)$ for $\rho \in [0,1]$. Now, $\xi_1 = 1$ is the only stable fixed point of $T(1,\cdot,0)$: for any $\rho \in [0,1]$ $T^n(1,\rho,0) \xrightarrow[n\to\infty]{} 1$ since by (c) in Lemma 2.1, $T^{n-1}(1,\rho,0) < T^n(1,\rho,0)$. Then, for fixed $\alpha \in (0,1)$, there is a minimum $M' \in \mathbb{N}$ such that $\alpha < T^{M'}(1,0,0)$. By part (a) in Lemma 2.1, $\alpha < T^{M'}(1,\rho,0)$ for any $\rho \in [0,1]$. Taking $M = M' + 1$ to account for a starting probability of being in the bad state within the replace region, the result follows. Q.E.D.

3. The Adaptive Two-State Binary Replacement Problem. If the parameter θ in $P(0)$ (see (2.2)) is unknown, we cannot compute $\rho_{t|t}$, nor can we determine optimal strategies. The "certainty equivalence" approach which we will adopt involves recursively computing estimates of the unknown parameter $\hat{\theta}_t$, at each decision epoch t, use the latest available estimate in the filtering equation (2.4) to compute $\rho_{t+1|t+1}(\hat{\theta}_t) =: \hat{\rho}_{t+1|t+1}$, where the decision u_t is made taking $\hat{\theta}_t$ as if it were the true parameter. For decision-making, it is assumed that a set of stationary separated strategies $CL = \{\mu(\cdot,\theta)\}_{\theta \in [0,1]}$, parameterized by θ, is available; these policies will be restricted to be of the control-limit type with $0 < \alpha(\theta) < 1$ (not necessarily optimal for each value of θ).

We denote by $\Theta = [0,1]$ the parameter space in which θ takes its values. The dependence of the transition matrix, one-step ahead probabilities, and aposteriori bad state probabilities on θ will be expressed in the form $P(u,\theta)$, $V(k,\rho,u,\theta)$, and $T(k,\rho,u,\theta)$, respectively; of course, from (2.2) and (2.5)-(2.9), we see that only $P(0,\theta)$, $V(k,\rho,0,\theta)$, and $T(k,\rho,0,\theta)$ depend on θ. Also, the dependence of a policy (or of the "control-limit") in Figure 1 on θ is denoted by $\mu(\rho,\theta)$ (or $\alpha(\theta)$). We also let θ_0 denote the (unknown) true value of the parameter, which we assume to be constant.

The certainty equivalence approach to this control problem is similar to that used for adaptive estimation in [AM], with the additional complication of feedback here. For a policy $\mu(\cdot,\theta) \in CL$, the adaptive algorithm takes the form:

$$\epsilon_t = y_t - V(1,\hat{\rho}_{t-1|t-1},\mu(\hat{\rho}_{t-1|t-1},\hat{\theta}_t),\hat{\theta}_t) \tag{3.1}$$

$$\hat{\theta}_{t+1} = \pi_\Theta(\hat{\theta}_t + \frac{1}{t+1}R_{t+1}^{-1}\Psi_t\epsilon_t) \tag{3.2}$$

$$\hat{\rho}_{t+1|t+1} = T(1,\hat{\rho}_{t|t},\mu(\hat{\rho}_{t|t},\hat{\theta}_{t+1}),\hat{\theta}_{t+1}) \cdot y_{t+1} + T(0,\hat{\rho}_{t|t},\mu(\hat{\rho}_{t|t},\hat{\theta}_{t+1}),\hat{\theta}_{t+1}) \cdot (1 - y_{t+1}) \tag{3.3}$$

where $\hat{\rho}_{0|0} = \rho_0$ and $\hat{\theta}_1 \in [0,1]$ is arbitrary. Here, ϵ_t is the prediction error, R_t is a positive definite matrix which modifies the search direction, Ψ_t is an approximation of the negative gradient of ϵ_t

with respect to θ (evaluated at $\hat{\theta}_t$), and $V(k,\rho,u,\theta)$ and $T(k,\rho,u,\theta)$ are given in (2.5)-(2.9). Note that the values that ϵ_t, $\hat{\rho}_{t+1|t+1}$, x_t and y_t take depend implicitly on the whole path $\{\hat{\theta}_t\}_{t=1}^{t+1}$. The map π_Θ is a projection onto the parameter space. Its inclusion is necessary since otherwise $\hat{\theta}_t$ will not necessarily be in Θ. The recursive estimate (3.1)-(3.3) is of the type analyzed by Kushner and Shwartz [KUS1], [KUS2], [KUS3], [KS], and Ljung and Söderström [LS]. The objective is first to prove convergence of $\hat{\theta}_t$ to θ_0 in an appropriate sense, and then to prove the long-run average cost (or asymptotically discounted cost, see [FAM], [HLM]) due to the adaptive policy is the same as would have been incurred if the true parameter had been known.

We use a Gauss-Newton search direction computed via

$$R_{t+1} = R_t + \frac{1}{t+1}[\Psi_t\Psi_t' - R_t], \quad R_1 = I. \tag{3.4}$$

It is useful to first write (3.1) and (3.3) for the constant sequence $\{\hat{\theta}_t = \theta\}_{t=1}^{t+1}$ [KUS2, p. 225]. That is, for each θ and $\mu(\cdot,\theta) \in CL$, we define the process $\{x_t(\theta), y_t(\theta), \rho_t(\theta)\}$, where $\{x_t(\theta)\}$ is governed by $P(\mu(\rho_t(\theta),\theta),\theta_0)$, $\{y_t(\theta)\}$ is related to $\{x_t(\theta)\}$ by Q defined in (2.3), $\{\rho_t(\theta)\}$ is defined recursively by

$$\rho_{t+1}(\theta) = T[1, \rho_t(\theta), \mu(\rho_t(\theta),\theta),\theta] \cdot y_{t+1}(\theta) + T[0, \rho_t(\theta), \mu(\rho_t(\theta),\theta),\theta] \cdot (1 - y_{t+1}(\theta)), \tag{3.5}$$

with $\rho_0(\theta) = \rho_0$, and

$$\epsilon_t(\theta) = y_t(\theta) - V[1, \rho_{t-1}(\theta), \mu(\rho_{t-1}(\theta),\theta),\theta]. \tag{3.6}$$

In [AM], the approximate gradient Ψ_t is obtained by deriving an equation for $\partial\epsilon_t(\theta)/\partial\theta$ (for fixed θ) and then evaluating at $\theta = \hat{\theta}_t$. Thus,

$$\partial\epsilon_t/\partial\theta = \partial y_t/\partial\theta - \partial V/\partial\theta$$

Contrary to the situation considered in [AM], the above derivatives are not well defined for all sample paths. This problem arises due to the discontinuity of $\mu(\rho,\theta)$ in its first argument at $\rho = \alpha(\theta)$, and the dependence on θ of both $\alpha(\theta)$ and $\rho_t(\theta)$. Preliminary simulation results suggest a smooth dependence of $\alpha(\theta)$ on θ. This, combined with the fact that both $\{y_t(\theta)\}$ and $\{\mu(\rho_t(\theta),\theta)\}$ take only two values make the following statement seem plausible: using an approximation that treats the derivatives as if they exist everywhere, and are equal to zero, then the algorithm will exhibit a satisfactory performance, as measured by the expectation with respect to an invariant measure (shown to exist uniquely, for each value of θ, in the sequel). The rationale behind the above is the fact that the expectation operator results in additional "smoothing," which may enable us to show that $\frac{\partial}{\partial\theta}E^\theta[\rho_t(\theta)] = E^\theta[\frac{\partial}{\partial\theta}\rho_t(\theta)]$, a fundamental result in the developments in [AM], where E^θ is the expectation operator with respect to the invariant measure. Then, using (2.6) and (2.8), we have that

$$\partial\epsilon_t(\theta)/\partial\theta = -(1-2q)[1 - \rho_{t-1}(\theta) + (1-\theta)\partial\rho_{t-1}(\theta)/\partial\theta], \quad \text{if } \mu(\rho_{t-1}(\theta),\theta) = 0, \tag{3.7}$$

$$\partial\epsilon_t(\theta)/\partial\theta = 0, \quad \text{if } \mu(\rho_{t-1}(\theta),\theta) = 1, \tag{3.8}$$

Using the aforementioned approximations, we obtain $\zeta_t(\theta) := \frac{\partial}{\partial\theta}\rho_t(\theta)$ from (3.5) as

$$\zeta_{\ell+1}(\theta) = \{\gamma[1, \rho_\ell(\theta), \mu(\rho_\ell(\theta), \theta), \theta]\zeta_\ell(\theta) + \delta[1, \rho_\ell(\theta), \mu(\rho_\ell(\theta), \theta), \theta]\} \cdot y_{\ell+1}(\theta)$$
$$+ \{\gamma[0, \rho_\ell(\theta), \mu(\rho_\ell(\theta), \theta), \theta]\zeta_\ell(\theta) + \delta[0, \rho_\ell(\theta), \mu(\rho_\ell(\theta), \theta), \theta]\} \cdot (1 - y_{\ell+1}(\theta)), \quad (3.9)$$

where $\zeta_0(\theta) := 0$, and

$$\gamma(k, \rho, 0, \theta) = \frac{q(1-q)(1-\theta)}{[V(k, \rho, 0, \theta)]^2}, \quad k = 0, 1$$

$$\delta(k, \rho, 0, \theta) = \frac{q(1-q)(1-\rho)}{[V(k, \rho, 0, \theta)]^2}, \quad k = 0, 1$$

$$\gamma(k, \rho, 1, \theta) = \delta(k, \rho, 1, \theta) = 0, \quad k = 0, 1$$

Ψ_ℓ in (3.2) is thus calculated as

$$\Psi_\ell = (1 - 2q)[1 - \hat{\rho}_{\ell-1|\ell-1} + (1 - \hat{\theta}_\ell)\hat{\zeta}_{\ell-1}], \quad \text{if } \mu(\hat{\rho}_{\ell-1|\ell-1}, \hat{\theta}_\ell) = 0 \qquad (3.10)$$

$$\Psi_\ell = 0, \quad \text{if } \mu(\hat{\rho}_{\ell-1|\ell-1}, \hat{\theta}_\ell) = 1 \qquad (3.11)$$

where

$$\hat{\zeta}_{\ell+1} = \{\gamma[1, \hat{\rho}_{\ell|\ell}, \mu(\hat{\rho}_{\ell|\ell}, \hat{\theta}_{\ell+1}), \hat{\theta}_{\ell+1}]\hat{\zeta}_\ell + \delta[1, \hat{\rho}_{\ell|\ell}, \mu(\hat{\rho}_{\ell|\ell}, \hat{\theta}_{\ell+1}), \hat{\theta}_{\ell+1}]\} \cdot y_{\ell+1}$$
$$+ \{\gamma[0, \hat{\rho}_{\ell|\ell}, \mu(\hat{\rho}_{\ell|\ell}, \hat{\theta}_{\ell+1}), \hat{\theta}_{\ell+1}]\hat{\zeta}_\ell + \delta[0, \hat{\rho}_{\ell|\ell}, \mu(\hat{\rho}_{\ell|\ell}, \hat{\theta}_{\ell+1}), \hat{\theta}_{\ell+1}]\} \cdot (1 - y_{\ell+1}), \quad (3.12)$$

and $\hat{\zeta}_0 := 0$.

Equations (3.1)-(3.4), (3.10)-(3.12) constitute the adaptive algorithm. To facilitate the analysis, we define the Markov chain $\hat{\xi}_\ell := (x_\ell, y_\ell, \hat{\rho}_{\ell|\ell}, \hat{\zeta}_\ell)$ and, for fixed θ, the Markov chain $\xi_\ell(\theta) = (x_\ell(\theta), y_\ell(\theta), \rho_\ell(\theta), \zeta_\ell(\theta))$. Then $\Sigma := \{0, 1\} \times \{0, 1\} \times [0, 1] \times \Re$ is the state space of either of these processes, and we let B_Σ denote its Borel σ-algebra. With this notation, we can write

$$\eta_{\ell+1} = \pi_C(\eta_\ell + a_\ell G(\eta_\ell, \hat{\xi}_\ell)) \qquad (3.13)$$

with C a convex, closed set, $a_\ell = 1/(\ell + 1)$, and $\eta_\ell = (\hat{\theta}_\ell, R_\ell)$. Notice that $\{\eta_\ell, \hat{\xi}_{\ell-1}\}$ is a Markov chain. The ODE approach proceeds as follows: Let $\tau_k = \sum_{i=1}^{k-1} a_i$ and denote by $\bar{\eta}^0(\cdot)$ the piecewise linear function with value η_k at τ_k. Define the shifted function $\bar{\eta}^k(\cdot)$ by $\bar{\eta}^k(\tau) = \bar{\eta}^0(\tau + \tau_k)$, $\tau \geq 0$, and observe that $\bar{\eta}^k(0) = \eta_k$. The idea is then to show either weak or w.p.1 convergence, as $k \to \infty$, of the sequence $\{\bar{\eta}^k(\cdot)\}$ to the solutions of an ordinary differential equation (ODE) associated with the algorithm (3.13). The study of the asymptotic behavior of $\{\eta_k\}$ is thus reduced to the analysis of the associated ODE.

In [AM], we utilized a theorem of Kushner [KUS2, Th. 3] in order to show almost sure convergence for the adaptive estimation algorithm. Define the "partial" transition function, for $A \in B_\Sigma$,

$$q^1(A|\xi, \eta) = P\{\hat{\xi}_{n+1} \in A | \hat{\xi}_n = \xi, \eta_{n+1} = \eta\},$$

and recursively, for $j > 1$,

$$q^j(A|\xi, \eta) = \int q^1(d\xi'|\xi, \eta)q^{j-1}(A|\xi', \eta).$$

This is thus the homogeneous probability transition function of the Markov process $\xi_\ell(\theta)$ for fixed θ, since $\hat{\xi}_\ell$ depends only on $\hat{\theta}_\ell$ and not on R_ℓ; because of this, we will sometimes use the notation $q^j(A|\xi, \theta)$ for the same transition function. One of the crucial hypotheses in Kushner's theorem is:

(H1)   There exists a function $\overline{G}$ on C such that, for each $n \in N_0$ and $\eta \in C$, the function

$$W_o(\eta, n) := \lim_{N \to \infty} \sum_{j=n+1}^{N} a_j [\int G(\eta, \xi') q^{j-n}(d\xi'|\xi_n, \eta) - \overline{G}(\eta)]$$

is well defined and bounded in norm by $K a_{n+1}$, for some $K < \infty$, w.p.1.

In [AM], (H1) is proved by a detailed analysis of the sample paths of $\{\xi_t(\theta)\}$, which turn out to be stable in a very strong sense. In the adaptive control problem considered here, this is not true, and a different approach toward verifying (H1) must be taken. In addition, it is much more difficult to show for this problem that a given candidate for $\overline{G}(\cdot)$ is continuously differentiable, and this is part of the hypothesis if one is to directly apply Kushner's theorem. However, (H1) can be shown via the *geometric ergodicity* of $\{\xi_t(\theta)\}$, which is defined as follows (c.f., [ME], [NUM], [OR]). For each θ, an *invariant probability measure* for the process $\{\xi_t(\theta)\}$ is a probability measure $\pi(\cdot, \theta)$ on B_Σ such that

$$\pi(A, \theta) = \int \pi(d\xi', \theta) q^1(A|\xi', \theta), \quad A \in B_\Sigma.$$

Define the total variation norm $\|\lambda_1 - \lambda_2\|$ for λ_1, λ_2 probability measures on B_Σ by

$$\|\lambda_1 - \lambda_2\| := \sup \left| \int f \, d\lambda_1 - \int f \, d\lambda_2 \right|$$

where the supremum is taken over all Borel functions $f : \Sigma \to [-1, 1]$.

<u>Definition 3.1</u>: Fix $\theta \in [0, 1]$. The Markov chain $\{\xi_t(\theta)\}$ is *uniformly geometrically ergodic* if there exist $r < 1$, $\Lambda < \infty$, and a measure $\pi(\cdot, \theta)$ on B_Σ such that, for every probability measure λ_0 on B_Σ and $k \in N$,

$$\left\| \int \lambda_0(d\xi') q^k(\cdot|\xi', \theta) - \pi(\cdot, \theta) \right\| \leq \Lambda r^k$$

If $\overline{G}(\eta) := \int G(\eta, \xi') \pi(d\xi', \theta)$, then hypothesis (H1) is clearly a consequence of uniform geometric ergodicity, which we will concentrate on proving. The following is a consequence of [OR, Proposition 6.1 and Theorem 7.1].

<u>Proposition 3.1</u>: Fix $\theta \in [0, 1]$. Suppose that the Markov chain $\{\xi_t(\theta)\}$ is aperiodic and that there exists a σ-finite measure φ on B_Σ such that *Doeblin's condition* is satisfied: for each $A \in B_\Sigma$ such that $\varphi(A) > 0$, there exist $n > 0$ and $\epsilon > 0$ such that

$$\sum_{k=1}^{n} q^k(A|\xi, \theta) > \epsilon, \quad \text{for all } \xi \in \Sigma \tag{3.14}$$

Then the Markov chain $\{\xi_t(\theta)\}$ has an invariant probability measure $\pi(\cdot, \theta)$ and is *uniformly geometrically ergodic*.

In the terminology of Orey [OR], Doeblin's condition implies that the Markov chain is uniformly φ-recurrent. In the next section, we verify the hypotheses of Proposition 3.1, thus proving uniform geometric ergodicity of $\{\xi_t(\theta)\}$ for each $\theta \in [0, 1]$, and hence verifying (H1).

4. Uniform Geometric Ergodicity of $\{\xi_t(\theta)\}$. In this section, we consider, for each fixed $\theta \in [0,1]$ and fixed policy $\mu(\cdot, \theta) \in CL$, the Markov chain $\{\xi_t(\theta) = (x_t(\theta), y_t(\theta), \rho_t(\theta), \zeta_t(\theta))\}$ defined above. In order to simplify the notation, let $z_t(\theta) = (\rho_t(\theta), \zeta_t(\theta))$; then $z_t(\theta)$ takes values in $S := [0,1] \times \Re$ ($\mathbf{B}_S$ denoting its Borel σ-algebra) and satisfies an equation of the form

$$z_{t+1}(\theta) = \chi[1, z_t(\theta), \mu(\rho_t(\theta), \theta), \theta] \cdot y_{t+1}(\theta) + \chi[0, z_t(\theta), \mu(\rho_t(\theta), \theta), \theta] \cdot (1 - y_{t+1}(\theta)) \qquad (4.1)$$

with the obvious definitions of χ. We also define, for $A \in \mathbf{B}_S$, and $i,j,k,m = 0,1$,

$$[\Gamma(A|z)]_{i+2k,j+2m} = P\{\xi_{t+1}(\theta) \in (j,m,A) | \xi_t(\theta) = (i,k,z)\}. \qquad (4.2)$$

The transition function can then be written in the form, for $A \in \mathbf{B}_S$ and $z = (\rho, \zeta)$,

$$\Gamma(A|z) = \begin{bmatrix} P(\mu(\rho,\theta),\theta_0)\bar{Q} & P(\mu(\rho,\theta),\theta_0)[I - \bar{Q}] \\ P(\mu(\rho,\theta),\theta_0)\bar{Q} & P(\mu(\rho,\theta),\theta_0)[I - \bar{Q}] \end{bmatrix} \cdot \underline{I}(A,z), \qquad (4.3)$$

where

$$\underline{I}(A,z) = \begin{bmatrix} I(\chi[0,z,\mu(\rho,\theta),\theta] \in A) \cdot I_2 & 0 \\ 0 & I(\chi[1,z,\mu(\rho,\theta),\theta] \in A) \cdot I_2 \end{bmatrix}; \qquad (4.4)$$

here, $\bar{Q} := diag(q, 1-q)$ and I_2 denotes the 2×2 identity matrix. Using (2.2), we thus have

$$\Gamma(A|z) = \begin{bmatrix} (1-\theta_0)q & \theta_0(1-q) & (1-\theta_0)(1-q) & \theta_0 q \\ 0 & 1-q & 0 & q \\ (1-\theta_0)q & \theta_0(1-q) & (1-\theta_0)(1-q) & \theta_0 q \\ 0 & 1-q & 0 & q \end{bmatrix} \cdot \underline{I}(A,z), \qquad (4.5a)$$

for $\mu(\rho,\theta) = 0$, while if $\mu(\rho,\theta) = 1$,

$$\Gamma(A|z) = \begin{bmatrix} q & 0 & 1-q & 0 \\ q & 0 & 1-q & 0 \\ q & 0 & 1-q & 0 \\ q & 0 & 1-q & 0 \end{bmatrix} \cdot \underline{I}(A,z). \qquad (4.5b)$$

Proposition 4.1: Let $q \in (0.5, 1)$, $\theta \in [0,1]$, and a policy $\mu(\cdot, \theta) \in CL$ be fixed. Then for every $t \in \mathbf{N}$, there exists $\epsilon_t > 0$ such that for all $\xi_0(\theta)$,

$$P\{y_0(\theta) = 1, y_1(\theta) = 1, \ldots, y_t(\theta) = 1 | \xi_0(\theta)\} = \epsilon_t \qquad (4.6)$$

Proof: Notice first that (omitting, for simplicity, the explicit dependence on θ) for $k = 0,1$,

$$P\{y_{t+1} = k | y_t, x_t, \rho_t, \xi_0\}$$
$$= P\{y_{t+1} = k, x_{t+1} = 0 | y_t, x_t, \rho_t, \xi_0\} \cdot P\{x_{t+1} = 0 | y_t, x_t, \rho_t, \xi_0\}$$
$$+ P\{y_{t+1} = k, x_{t+1} = 1 | y_t, x_t, \rho_t, \xi_0\} \cdot P\{x_{t+1} = 1 | y_t, x_t, \rho_t, \xi_0\} \qquad (4.7)$$

An easy calculation from (4.5) shows that there is a number $\bar{\epsilon} > 0$ such that the quantity in (4.7) is greater than or equal to $\bar{\epsilon}$ for all ξ_0. Now, letting $\{\underline{y}_t := \underline{1}\} = \{y_0 = 1, y_1 = 1, \ldots, y_t = 1\}$, we have

$$P\{\underline{y}_{t+1} = \underline{1} | \xi_0\} = P\{y_{t+1} = 1 | \xi_0, \underline{y}_t = \underline{1}\} \cdot P\{\underline{y}_t = \underline{1} | \xi_0\} \qquad (4.8)$$

By the smoothing property of conditional expectations, the first quantity on the right hand side of (4.8) can be written via indicator functions as

$$E[E[I(y_{t+1} = 1)|\xi_0, \underline{y}_t = 1, x_t, \rho_t]|\xi_0, \underline{y}_t = 1] \geq \bar{\epsilon}.$$

Performing this calculation recursively, using the Markov property of $\{\xi_t(\theta)\}$, shows that the proposition follows if we let $\epsilon_t = (\bar{\epsilon})^t$. Q.E.D.

In order to verify the hypotheses of Proposition 3.1, we now define a measure φ on B_Σ. This is in terms of the "recurrent" state $(0,0,0,0) \in \Sigma$, which is interpreted as follows. Notice from (2.4), (3.5), and (3.9) that if $\mu(\rho_t(\theta), \theta) = 1$, then $x_{t+1}(\theta) = \rho_{t+1}(\theta) = \zeta_{t+1}(\theta) = 0$, while there is a positive probability that $y_{t+1}(\theta) = 0$. We then define φ to be a measure on B_Σ such that $\varphi(\{(0,0,0,0)\}) = 1$ and $\varphi(A) = 0$ if $(0,0,0,0) \notin A$.

<u>Theorem 4.1</u>: Let $q \in (0.5, 1)$, $\theta \in [0, 1]$, and a policy $\mu(\cdot, \theta) \in CL$ be fixed. Then $\{\xi_t(\theta)\}$ satisfies (H1).

<u>Proof</u>: As noted in Section 3, we need only verify the hypotheses of Proposition 3.1. With φ defined as above, the only sets $A \in B(\Sigma)$ with $\varphi(A) > 0$ are those containing $(0,0,0,0)$; hence, Doeblin's condition need only be verified for $A = \{(0,0,0,0)\}$. This follows easily if we take (a) n in Doeblin's condition to be $M(\theta) + 1$, where $M(\theta)$ is defined in Lemma 2.2; and (b) ϵ in Doeblin's condition to be ϵ_n, where ϵ_n is defined in Proposition 4.1 and n is defined in (a). The only remaining hypothesis of Proposition 3.1 is the aperiodicity of $\{\xi_t(\theta)\}$, which follows easily from the definition (see [OR, pp. 12-15]). Q.E.D.

5. <u>Conclusions</u>. This paper represents the beginning stages of a program to address the adaptive control of partially observable Markov decision processes (POMDP) with finite state, action, and observation spaces. We have reviewed the results on the control of POMDP with *known* parameters, and, in particular, the results on the control of quality control/machine replacement models. We have chosen to study the adaptive control of a problem with simple structure: the two-state binary replacement problem. An adaptive control algorithm was defined, and initial results in the direction of using the ODE method were obtained. The next steps, on which work is in progress, are (1) the verification of the other hypotheses of Kushner's theorem (or the proof of a modification of the theorem for this algorithm); (2) the analysis of the limit points of the ODE and of the convergence of the sequence of parameter estimates $\{\hat{\theta}_t\}$; (3) an analysis of the optimality of the adaptive control algorithm.

The principal contribution of this paper has been the derivation of some ergodic properties for the process $\{\xi_t(\theta)\}$, for policies $\mu(\cdot, \theta) \in CL$. It should be noted that such properties are to some degree independent of the $\{\zeta_t\}$ process used: the same would be true for any process with a "recurrent" state and which can be put in a form similar to (3.9).

<u>Acknowledgments</u>: The authors would like to thank Dr. Sean Meyn of the Australian National University for helping us to understand geometric ergodicity and to prove Theorem 4.1.

This research was supported in part by the Air Force Office of Scientific Research under grant AFOSR-86-0029, in part by the National Science Foundation under grant ECS-8617860, in part by the DoD Joint Services Electronics Program through the Air Force Office of Scientific Research (AFSC) Contract F49620-86-C-0045, and in part by the Texas Higher Education Advanced Technology Program.

References

[AKO] V.A. Andriyanov, I.A. Kogan and G.A. Umnov, "Optimal Control of a Partially Observable Discrete Markov Process," *Aut. Remot. C.*, 4, 1980, 555-561.

[AM] A. Arapostathis and S.I. Marcus, "Analysis of an Identification Algorithm Arising in the Adaptive Estimation of Markov Chains," *Mathematics of Control, Signals and Systems*, to appear.

[AS] K.J. Åström, "Optimal Control of Markov Processes with Incomplete State Information," *J. Math. Anal. Appl.*, 10, 1965, 174-205.

[BE] D.P. Bertsekas, *Dynamic Programming: Deterministic and Stochastic Models*, Prentice-Hall, Englewood Cliffs, New Jersey, 1987.

[BS] D.P. Bertsekas and S.E. Shreve, *Stochastic Optimal Control: The Discrete Time Case*, Academic Press, New York, NY, 1978.

[FAM] E. Fernández-Gaucherand, A. Arapostathis, and S. I. Marcus, "On the Adaptive Control of a Partially Observable Markov Decision Process," *Proc. 27th IEEE Conf. on Decision and Control*, Austin, Texas, 1988.

[HLM] O. Hernández-Lerma, "Adaptive Markov Control Processes," preprint, 1987.

[HS] D.P. Heyman and M.J. Sobel, *Stochastic Models in Operations Research, Vol. II: Stochastic Optimization*, McGraw-Hill, New York, 1984.

[KS] H.J. Kushner and A. Shwartz, "An Invariant Measure Approach to the Convergence of Stochastic Approximations with State Dependent Noise," *SIAM J. Control and Optim.*, 22, 1984, 13-27.

[KUM1] P.R. Kumar, "A Survey of Some Results in Stochastic Adaptive Control," *SIAM J. Control and Optim.*, 23, 1985, 329-380.

[KUM2] P.R. Kumar, "Optimal Adaptive Control of Linear Quadratic Gaussian Systems," *SIAM J. Control and Optimiz.*, 21, 1983, 163-178.

[KUS1] H.J. Kushner, "Stochastic Approximation with Discontinuous Dynamics and State Dependent Noise: w.p.1 and Weak Convergence," *J. Math. Anal. Appl.*, 82, 1981, 527-542.

[KUS2] H.J. Kushner, "An Averaging Method for Stochastic Approximations with Discontinuous Dynamics, Constraints, and State Dependent Noise," in *Recent Advances in Statistics*, Rizvi, Rustagi and Siegmund, Eds., Academic Press, New York, 1983, 211-235.

[KUS3] H.J. Kushner, *Approximation and Weak Convergence Methods for Random Processes*, MIT Press, Cambridge, MA, 1984.

[KV] P.R. Kumar and P. Varaiya, *Stochastic Systems: Estimation, Identification and Adaptive Control*, Prentice-Hall, Englewood Cliffs, New Jersey, 1986.

[LO1] W.S. Lovejoy, "On the Convexity of Policy Regions in Partially Observed Systems," *Operations Res.*, 35, 1987, 619-621.

[LS] L. Ljung and T. Söderström, *Theory and Practice of Recursive Identification*, MIT Press, Cambridge, MA, 1983.

[ME] S.P. Meyn, "Ergodic Theorems for Discrete Time Stochastic Systems Using a Generalized Stochastic Lyapunov Function," 1988, preprint.

[MMS] D.J. Ma, A.M. Makowski and A. Shwartz, "Estimation and Optimal Control for Constrained Markov Chains," *Proc. 25th IEEE Conf. on Decision and Control*, Athens, Greece, 1986, 994-999.

[MO] G.E. Monahan, "A Survey of Partially Observable Markov Decision Processes: Theory, Models, and Algorithms," *Management Sci.*, 28, 1982, 1-16.

[NUM] E. Nummelin, *General Irreducible Markov Chains and Non-Negative Operators*, Cambridge University Press, New York, 1984.

[OR] S. Orey, *Limit Theorems for Markov Chain Transition Probabilities*, Van Nostrand Reinhold Mathematical Studies 34, London, 1971.

[RO] S.M. Ross, *Introduction to Stochastic Dynamic Programming*, Academic Press, New York, NY, 1983.

[SM] A. Shwartz and A.M. Makowski, "An Optimal Adaptive Scheme for Two Competing Queues with Constraints," *7th Internat. Conf. on Analys. and Optim. of Systems*, Antibes, France, 1986, 515-532.

[W1] C.C. White, "A Markov Quality Control Process Subject to Partial Observation," *Management Sci.*, 23, 1977, 843-852.

[W2] C.C. White, "Optimal Control-Limit Strategies for a Partially Observed Replacement Problem," *Internat. J. Systems Science*, 10, 1979, 321-331.

ADAPTIVE CONTROL OF LINEAR SYSTEMS WITH RAPIDLY-VARYING PARAMETERS

E.W. Kamen   T.E. Bullock
University of Pittsburgh   University of Florida

                    C.-H. Song
                University of Florida

1. Introduction

Much of the existing work on adaptive control deals with the case
of linear time-invariant plants; however, in most applications the
plant parameters are time varying. Some results do exist on the
adaptive control of linear systems with slowly-varying parameters; for
example, see Anderson and Johnstone [1], Kreisselmeir [2], Middleton
and Goodwin [3], and Koh and Kamen [4]. In the case of systems with
rapidly-varying parameters, Xianya and Evans [5] and Zheng [6] have
developed adaptive control schemes for the class of systems possessing
a stable inverse. Unfort inately, many systems arising in practice are
not stably invertible. Examples include the linearized system
dynamics of air-to-air missiles, which are also known to contain
rapidly-varying coefficients due to missile acceleration and high
velocity.

In this paper we develop an indirect adaptive control scheme for
multi-input multi-output linear discrete-time systems with
rapidly-varying parameters. A system is specified by an nth-order
vector input/output difference equation which provides an appropriate
framework in which to carry out parameter estimation. The specific
problem of interest is the design of a feedback control which results
in global asymptotic stability of the closed-loop system (i.e., all
signals converge to zero for any initial conditions).

It is assumed that the time-varying parameters can be modeled in
terms of linear combinations of known bounded functions with unknown
constant coefficients. It is also assumed that the unknown
coefficients belong to known bounded intervals. This leads to the
definition of the parameter set Ω consisting of all possible values of
the unknown constant coefficients. The unknown constant coefficients
are estimated using a recursive least squares (RLS) scheme with a
projection operation that forces parameter estimates to belong to the
parameter set. The RLS scheme also includes a variable forgetting
factor that provides robustness to errors in the modeling of the time
variations in the system parameters. It is shown that the estimates

produced by the RLS scheme always converge to fixed values, although
in general these values are not equal to the true values of the
unknown constant coefficients. To guarantee convergence to the true
values, we would need to include a persistency of excitation (PE)
condition. However, it is possible to obtain global asymptotic
stability without requiring convergence to the true system parameters,
and thus the PE condition is not required. It is also shown that the
output prediction error satisfies a key property which plays a very
important role in proving global asymptotic stability of the control
scheme.

In addition to the above assumptions on the plant parameters, the
only additional assumptions are knowing an upper bound on the system
order and that the observable canonical realization of the given
system is stabilizable for all possible values of the unknown constant
coefficients. The stabilizability assumption on the given system is
necessary in order to be able to solve the adaptive control problem.
We emphasize that the given system does not have to be stably
invertible.

It is shown that the stabilizabilty assumption on the given
system implies the existence of a state feedback gain matrix that
stabilizes the estimated system model computed from the observable
canonical form by replacing the unknown constant coefficients by their
estimates. The adaptive controller is then constructed by applying
the output of an adaptive observer into the stabilizing feedback gain
matrix. The resulting closed-loop system is globally asymptotically
stable.

2. Preliminaries

Consider the r-input m-output linear time-varying discrete-time
system given by the nth-order vector input/output difference equation

$$y(k) = \sum_{i=1}^{n} A_i(k) y(k-i) + \sum_{i=1}^{n} B_i(k) u(k-i). \qquad (1)$$

In (1), $u(k)$ is the r-vector control, $y(k)$ is the m-vector
output, and n is a known upper bound on the order of the input/output
difference equation. It is assumed that the coefficient matrices
$A_i(k)$ and $B_i(k)$ in (1) can be expressed in the form

$$A_i(k) = \sum_{j=1}^{s} A_{ij} f_j(k), \quad i=1,2,\ldots,n \qquad (2)$$

$$B_i(k) = \sum_{j=1}^{s} B_{ij} f_j(k), \quad i=1,2,\ldots,n \tag{3}$$

where the A_{ij} are unknown $m \times m$ matrices, the B_{ij} are unknown $m \times r$ matrices, s is a known integer, and the $f_j(k)$ are known real-valued bounded functions of k. For example, the $f_j(k)$ could be equal to powers of k over some finite time interval; that is,

$$f_j(k) = k^j, \quad k_0 \le k \le k_1,$$

$$f_j(k) \text{ bounded for } k > k_1,$$

Note that the assumption that the $f_j(k)$ are bounded functions of k implies that the coefficient matrices $A_i(k)$ and $B_i(k)$ in (1) are bounded functions of k. The rate of time variation of the $A_i(k)$ and the $B_i(k)$ depends on the rate of time variation of the $f_j(k)$ which can be arbitrarily fast (as long as the $f_j(k)$ are bounded).

We are assuming that the given system is specified by the vector difference equation (1) due in part to the convenience of this form in carrying out parameter estimation. If the system is initially specified in some other form (e.g., a state model), it will be necessary to convert the system representation to the form (1) (assuming this is possible).

Now we want to consider the construction of a feedback control $u(k)$ so that the resulting closed-loop system is globally asymptotically stable; that is, all signals in the system converge to zero for any initial conditions. In the tracking problem, the objective is to select a feedback control $u(k)$ so that the closed-loop system is globally asymptotically stable and $y(k) - y_r(k)$ converges to zero where $y_r(k)$ is a given reference signal. Since the stabilization problem is a special case of the tracking problem (take $y_r(k) = 0$), we shall first consider the solution to the stabilization problem. The solution to this problem can then be applied to the tracking problem which is considered in a separate paper.

We would also like to insure that the adaptive control scheme is robust to model errors and disturbance inputs. Here robustness means maintaining stability in the presence of unmodelled dynamics, unknown bounded disturbances, and errors in the modeling of parameter time variations. Robustness to unmodelled dynamics and unknown bounded disturbances will not be considered in this paper (see Section 5). With respect to errors in the models (2) and (3) for the time-varying coefficient matrices, if the mismatch terms

$$A_i(k) - \sum_{j=1}^{s} A_{ij}f_j(k) \tag{4}$$

$$B_i(k) - \sum_{j=1}^{s} B_{ij}f_j(k) \tag{5}$$

are slowly varying, robustness to these modeling errors can be achieved by using a forgetting factor in the estimation of the unknown constant coefficients (the A_{ij} and B_{ij}). Slowly varying means that the percent change in the mismatch terms (4) and (5) is small (say less than 10 percent) over the transient response time of the given system. In the next section, we consider a recursive parameter estimation scheme with a variable forgetting factor.

3. The Parameter Estimator

Given the system specified by (1)-(3), let θ denote the $sn(m+r) \times m$ parameter matrix defined by

$$\theta^T = [A_{11} \ A_{12} \ \cdots \ A_{1s} \ A_{21} \ A_{22} \ \cdots \ A_{2s} \ \cdots \ A_{n1} \ A_{n2} \ \cdots \ A_{ns}$$
$$B_{11} \ B_{12} \ \cdots \ B_{1s} \ B_{21} \ B_{22} \ \cdots \ B_{2s} \ \cdots \ B_{n1} \ B_{n2} \ \cdots \ B_{ns}]$$

where "T" denotes the transpose operation. With θ_{ij} equal to the i,j entry of θ, it is assumed that

$$\theta_{ij} \in [\theta_{ij}^{min}, \theta_{ij}^{max}] \text{ for all } i,j, \tag{6}$$

where the θ_{ij}^{min} and θ_{ij}^{max} are known real numbers. Thus the unknown constant parameters take their values from known bounded intervals.

With $N = sn(m+r)$, let Ω denote the set of all $N \times m$ matrices with i,j entry belonging to the interval $[\theta_{ij}^{min}, \theta_{ij}^{max}]$. The set Ω consists of all possible matrix parameter values of the system defined by (1)-(3). Clearly, Ω is a closed bounded subset of the space of all $N \times m$ matrices over the field of real numbers.

Let $\phi(k-1)$ denote the N-element regression vector defined by

$$\phi^T(k-1) = [f_1(k)y^T(k-1) \quad f_2(k)y^T(k-1) \quad \cdots \quad f_s(k)y^T(k-1)$$
$$f_1(k)y^T(k-2) \quad f_2(k)y^T(k-2) \quad \cdots \quad f_s(k)y^T(k-2) \quad \cdots$$
$$f_1(k)y^T(k-n) \quad f_2(k)y^T(k-n) \quad \cdots \quad f_s(k)y^T(k-n)$$
$$f_1(k)u^T(k-1) \quad f_2(k)u^T(k-1) \quad \cdots \quad f_s(k)u^T(k-1) \quad \cdots$$
$$f_1(k)u^T(k-n) \quad f_2(k)u^T(k-n) \quad \cdots \quad f_s(k)u^T(k-n)].$$

Then in terms of the parameter matrix θ and the regression vector $\phi(k-1)$ we can rewrite the input/output difference equation (1) in the standard form

$$y(k) = \theta^T \phi(k-1).\qquad(7)$$

In the following development, we need to work with a normalized version of (7): Define the normalized signals

$$y_n(k) = \frac{y(k)}{\eta(k-1)}, \quad \phi_n(k-1) = \frac{\phi(k-1)}{\eta(k-1)},$$

where

$$\eta(k-1) = \max(1, ||\phi(k-1)||),$$

with $||\phi(k-1)|| =$ Euclidean norm of $\phi(k-1)$ defined by

$$||\phi(k-1)|| = (\phi^T(k-1)\phi(k-1))^{1/2}$$

Then (7) is equivalent to

$$y_n(k) = \theta^T \phi_n(k-1),\qquad(8)$$

where $\phi_n(k-1)$ is bounded (by definition).

Based on the model (8), we shall construct estimates of the parameter matrix θ using a recursive least squares (RLS) scheme with a projection operation into Ω and a variable forgetting factor. With $\theta(k)$ equal to the projected estimate of θ at time k and $\theta^o(k)$ equal to the unprojected estimate at time k, the RLS estimation scheme is given by

$$\theta^o(k) = \theta(k-1) + \frac{P(k-1)\phi_n(k-1)}{1+\phi_n^T(k-1)P(k-1)\phi_n(k-1)}[y_n^T(k) - \phi_n^T(k-1)\theta(k-1)]$$

$$(9)$$

$$P(k) = \frac{1}{\lambda(k)}\left[P(k-1) - \frac{P(k-1)\phi_n(k-1)\phi_n^T(k-1)P(k-1)}{1+\phi_n^T(k-1)P(k-1)\phi_n(k-1)}\right], \quad P(0)>0.$$

$$(10)$$

$$\lambda(k) = 1 - \frac{1}{TrP(0)}\left[\frac{\phi_n^T(k-1)P^2(k-1)\phi_n(k-1)}{1+\phi_n^T(k-1)P(k-1)\phi_n(k-1)}\right],\qquad(11)$$

where $TrP(0)$ is the trace of the positive definite matrix $P(0)$.

The estimate $\theta(k)$ is computed by projecting $\theta^o(k)$ into the parameter set Ω via the following procedure:

(1) If $\theta^o(k) \epsilon \Omega$, set $\theta(k) = \theta^o(k)$.

(2) If $\theta^o(k) \not\epsilon \Omega$, let $P^{-1/2}(k)$ denote the square root of $P^{-1}(k)$, which always exists since $P(k)$ is positive definite for all $k \geq 0$. Orthogonally project $P^{-1/2}(k)\theta^o(k)$ onto the boundary of the image of Ω under the transformation $P^{-1/2}(k)$. Let the result be denoted by $\rho(k)$. Then set $\theta(k) = P^{1/2}(k)\rho(k)$.

It follows from the results in Goodwin and Sin [7] that

$$\mathrm{Tr}[(\theta(k) - \theta)^T P^{-1}(k)(\theta(k) - \theta)] \leq \mathrm{Tr}[(\theta^o(k) - \theta)P^{-1}(k)(\theta^o(k) - \theta)],$$

(12)

where again "Tr" denotes the trace operation.

By the results of Lozano-Leal and Goodwin [8], the forgetting factor $\lambda(k)$ defined by (11) keeps the trace of $P(k)$ constant (equal to the trace of $P(0)$). Hence $P(k)$ does not converge to zero which is usually the case when no forgetting factor is used. As a result parameter updating will not eventually stop in the case when there are slowly-varying errors in the models (2)-(3) of the time-varying coefficients. Thus we have a type of robustness to errors in the modeling of the time-varying coefficients.

The RLS estimation process has the following properties which are crucial in proving global asymptotic stability of the adaptive control scheme to be defined later.

Theorem 1: Assuming that the system coefficient matrices $A_i(k)$, $B_i(k)$ are given by (2)-(3), the RLS parameter estimation scheme defined by (9)-(11) has the following properties:

(1) The output prediction error $e(k) = y(k) - \theta^T(k-1)\phi(k-1)$ satisfies the inequality

$$||e(k)|| \leq c_1(k)||\phi(k-1)|| + c_2(k),$$

where again $||.||$ denotes the Euclidean norm and $c_1(k)$, $c_2(k)$ are bounded functions with both $c_1(k)$ and $c_2(k)$ converging to zero as $k \to \infty$.

(2) As $k \to \infty$ the estimate $\theta(k)$ converges to some matrix $\theta^* \epsilon \Omega$ (which is not necessarily equal to θ).

Proof: Define the functionals

$$V(k) = \mathrm{Tr}[(\theta(k) - \theta)^T P^{-1}(k)(\theta(k) - \theta)]$$

$$V^o(k) = \mathrm{Tr}[(\theta^o(k) - \theta)^T P^{-1}(k)(\theta^o(k) - \theta)].$$

Using (9)-(10) and the property that $0 < \lambda(k) < 1$ for all k, we have that

$$V^o(k) \leq V(k-1) - \frac{||e_n(k)||^2}{1 + \phi_n^T(k-1)P(k-1)\phi_n(k-1)}$$

where $e_n(k)$ is the normalized output prediction error defined by $e_n(k) = y_n(k) - \theta^T(k-1)\phi_n(k-1)$. By (12), $V(k) \leq V^o(k)$, and thus

$$V(k) \leq V(k-1) - \frac{||e_n(k)||^2}{1 + \phi_n^T(k-1)P(k-1)\phi_n(k-1)}. \tag{13}$$

Hence $V(k)$ is monotone decreasing and since $V(k) \geq 0$ for all k, $V(k)$ must converge to some constant $V_\infty \geq 0$. Thus $V(k) - V(k-1) \to 0$ and defining

$$c_2(k) = \frac{||e_n(k)||}{(1 + \phi_n^T(k-1)P(k-1)\phi_n(k-1))^{1/2}}, \tag{14}$$

from (13), we have that $c_2(k) \to 0$ as $k \to \infty$. Now $e_n(k) = e(k)/\eta(k-1)$ and $\phi_n(k-1) = \phi(k-1)/\eta(k-1)$ where $\eta(k-1) = \max(1, ||\phi(k-1)||)$. Inserting these expressions for $e_n(k)$ and $\phi_n(k-1)$ into (14) yields

$$c_2(k) = \frac{||e(k)||}{(\eta^2(k-1) + \phi^T(k-1)P(k-1)\phi(k-1))^{1/2}},$$

and thus,

$$||e(k)|| = (\eta^2(k-1) + \phi^T(k-1)P(k-1)\phi(k-1))^{1/2} c_2(k). \tag{15}$$

But

$$\phi^T(k-1)P(k-1)\phi(k-1) \leq [\lambda_{max}P(k-1)] \, ||\phi(k-1)||^2, \tag{16}$$

where $\lambda_{max}P(k-1)$ is the maximum eigenvalue of $P(k-1)$, and

$$\eta^2(k-1) \leq 1 + ||\phi(k-1)||^2. \tag{17}$$

Combining (15)-(17) and defining

$$c_1(k) = [1 + \lambda_{max}P(k-1)]^{1/2} c_2(k),$$

we have that $c_1(k) \to 0$ as $k \to \infty$ and

$$||e(k)|| \leq c_1(k)||\phi(k-1)|| + c_2(k).$$

Hence, property (1) is verified. The proof of property (2) follows from (13) and the results in [8].∎

4. The Adaptive Control Scheme

For the system specified by the vector input/output difference equation (1) with the coefficients given by (2), (3), consider the observable canonical realization given by the nm-dimensional state model

$$x(k+1) = A(k)x(k) + B(k)u(k) \tag{18}$$

$$y(k) = Cx(k) \tag{19}$$

where

$$A(k) = \begin{bmatrix} \bar{A}_1(k+1) & I & 0 & \cdots & 0 \\ A_2(k+2) & 0 & I & & 0 \\ \vdots & \vdots & & \ddots & \\ A_n(k+n) & 0 & 0 & \cdots & 0 \end{bmatrix}, \quad B(k) = \begin{bmatrix} B_1(k+1) \\ B(k+2) \\ \vdots \\ B_n(k+n) \end{bmatrix}$$

and

$$C = [I_m \quad 0 \quad \cdots \quad 0].$$

In the following development, we shall denote A(k) and B(k) by A(k,θ) and B(k,θ), respectively, to denote the fact that A(k) and B(k) depend on the parameter matrix θ defined in terms of the constant coefficients in (2) and (3).

To be able to solve the adaptive control problem, we need to make the following key assumption on the stabilizability of the given system.

Assumption: There exists a rxmn matrix L(k,θ), which is a bounded function of k and a continuous function of θεΩ, such that the unforced system

$$x(k+1) = [A(k,θ) - B(k,θ)L(k,θ)]x(k)$$

is uniformly asymptotically stable for all θεΩ.

The above assumption implies that the given system can be stabilized for all possible values of the parameter matrix θ. This condition is necessary in order to be able to solve the adaptive control problem since we do not know what the true value of θ is, nor will we be able in general to identify the true value of θ. (Recall from Theorem 1 that the estimate θ(k) converges to some matrix θ* which in general is not equal to θ.) However, by also requiring persistency of excitation (PE), we can insure that θ(k) converges to θ in which case we would need stabilizability for only the true value of θ, not for all possible values of θ ranging over the parameter set Ω. Since we do not want to impose the PE condition (which cannot be

guaranteed to hold in practice), we are forced to assume
stabilizability for all possible parameter values.

A matrix $L(k,\theta)$ satisfying the above assumption will be called a
stabilizing feedback matrix. One can attempt to compute a priori a
stabilizing $L(k,\theta)$ as a function of θ. In particular, in some cases a
stabilizing $L(k,\theta)$ can be computed from the Q-step reachability

grammian defined by

$$Y(k,k-Q+1,\theta) = \sum_{i=k-Q+1}^{k} \Phi(k+1,i+1,\theta)B(i,\theta)B^T(i,\theta)\Phi^T(k+1,i+1,\theta)$$

where $\Phi(k,i,\theta)$ is the state transition matrix for the vector
difference equation $x(k+1)=A(k,\theta)x(k)$. If for every $\theta\epsilon\Omega$, there exists
a real number $a_\theta>0$ such that

$$Y(k,k-Q+1,\theta) \geq a_\theta I \text{ for all } k,$$

then a stabilizing feedback is

$$L(k,\theta) = B^T(k,\theta)\Phi^T(k+Q+2,k+1,\theta)Y^{-1}(k+Q+1,k,\theta)\Phi(k+Q+2,k+1,\theta)A(k,\theta).$$

The proof that $L(k,\theta)$ is stabilizing follows from the results of Moore
and Anderson [9].

Now given the estimate $\theta(k)$ of θ produced by the RLS scheme, we

define the estimated system model to be

$$x(k+1) = A(k,\theta(k))x(k) + B(k,\theta(k))u(k),$$

where $A(k,\theta(k))$ and $B(k,\theta(k))$ are equal to $A(k,\theta)$ and $B(k,\theta)$,
respectively, with θ replaced by the estimate $\theta(k)$. As shown in the
next result, a stabilizing feedback $L(k,\theta)$ leads to a stabilizing
feedback for the estimated system model.

Theorem 2: Let $L(k,\theta)$ be a stabilizing feedback matrix with $L(k,\theta)$
bounded in k and continuous in $\theta\epsilon\Omega$. Let $L(k,\theta(k))$ denote $L(k,\theta)$ with
θ replaced by $\theta(k)$. Then the unforced system

$$x(k+1) = [A(k,\theta(k)) - B(k,\theta(k))L(k,\theta(k))]x(k)$$

is uniformly asmptotically stable.

Proof: By property (2) of Theorem 1, $\theta(k)\rightarrow\theta^*\epsilon\Omega$ and by the

stabilizability assumption on the given system,

$$x(k+1) = [A(k,\theta^*) - B(k,\theta^*)L(k,\theta^*)]x(k)$$

is uniformly asymptotically stable. The result then follows by
compactness of the parameter set Ω and continuity of $A(k,\theta)$, $B(k,\theta)$,
and $L(k,\theta)$ as functions of θ.∎

If $L(k,\theta)$ is a stabilizing feedback, the adaptive control scheme (applied to the given system) is realized by setting

$$u(k) = -L(k,\theta(k))\hat{x}(k),$$

where $\hat{x}(k)$ is an estimate of the state $x(k)$ of the observer canonical realization. The estimate $\hat{x}(k)$ is generated by the adaptive observer

$$\hat{x}(k+1) = A(k,\theta(k))\hat{x}(k) + B(k,\hat{\theta}(k))u(k) + M(k,\theta(k))[y(k) - C\hat{x}(k)],$$

$$(20)$$

where

$$M^T(k,\theta(k)) = [-A_1^T(k+1,\theta(k)) - A_2^T(k+2,\theta(k)) \ldots -A_n^T(k+n,\theta(k))]$$

$$A_i(k+i,\theta(k)) = A_i(k+i) \text{ with } \theta=\theta(k).$$

By construction, $A(k,\theta(k))-M(k,\theta(k))C$ is a constant matrix with all eigenvalues located at the origin of the complex plane. Thus the observer (20) has a finite settling time; that is, $\hat{x}(k)=0$ after a finite number of steps when $y(k)=u(k)=0$.

Combining the constructions and results in this section and the previous section, we have the following main theorem.

Theorem 3: Let $L(k,\theta)$ be a stabilizing feedback matrix with $L(k,\theta)$ bounded in k and continuous in $\theta\epsilon\Omega$ and let $\hat{x}(k)$ denote the state of the adaptive observer given by (20). Then the adaptive control law $u(k)=-L(k,\theta(k))\hat{x}(k)$ results in a closed-loop system that is globally asymptotically stable.

Proof: The proof of the theorem follows from Theorems 1 and 2 and a generalization of the proof technique given in [10] and [11].∎

5. Discussion

The adaptive control scheme developed in this paper has been applied to the design of an autopilot for air-to-air missiles using bank-to-turn steering. Computer simulations have shown that the control scheme works very well, even though large errors have been observed in the estimation of some of the time-varying parameters (see [12]). These simulation results substantiate the analytical result (derived in this paper) that control can be achieved without having to identify the true values of the time-varying system parameters.

Current work is centering on the development of an adaptive control scheme that can be shown analytically to be robust to unmodeled dynamics, unknown bounded disturbances, and errors in the modeling of the time-varying parameters. Preliminary work indicates that the control scheme derived in this paper may be modified in such a way that one can guarantee robustness to model errors and bounded disturbances.

Acknowledgement: This work was supported by the US Air Force, AFATL, Eglin AFB, Florida, under Contract No. F08635-86-K-0265.

REFERENCES

[1] B.D.O. Anderson and R.M. Johnstone, "Adaptive Systems and Time Varying Plants," **International J. of Control**, Vol. 38, pp. 309-319, 1983.

[2] G. Kreisselmeier, "Adaptive Control of a Class of Slowly Time Varying Plants," **System and Control Letters**, Vol. 8, pp. 97-103, 1986.

[3] R.H. Middleton and G.C. Goodwin, "Adaptive Control of Time Varying Systems," **IEEE Trans. Automatic Control**, Vol. AC-33, No. 2, pp. 150-155, Feb. 1988.

[4] K. Koh and E.W. Kamen, "Robust Indirect Adaptive Control of Linear Time-varying Systems," American Control Conference, Atlanta, GA, pp. 777-778, June 1988.

[5] X. Xianya and R.J. Evans, Discrete-time Adaptive Control for Deterministic Time-Varying Systems," **Automatica**, Vol. 20, No. 3, pp. 309-319, 1984.

[6] L. Sheng, "Discrete-Time Adaptive Control of Deterministic Fast Time-Varying Systems," **IEEE Trans. on Automatic Control**, Vol. AC-32, No. 5, pp. 444-447, May 1987.

[7] G.C. Goodwin and K.S. Sin, **Adaptive Filtering, Prediction and Control**, Englewood Cliffs, NJ: Prentice-Hall, 1984.

[8] R. Lozano-Leal and G.C. Goodwin, "A Globally Convergent Adaptive Pole Placement Algorithm without a Persistency of Excitation Requirement," **IEEE Trans. Automatic Control**, Vol. AC-30, No. 8, pp. 795-798, Aug. 1985.

[9] J.B. Moore and B.D.O. Anderson, "Coping with Singular Transition Matrices in Estimation and Control Stability Theory," **Int. J. of Control**, pp. 571-586, 1980.

[10] K.A. Ossman and E.W. Kamen, "Adaptive Regulation of MIMO Linear Discrete-Time Systems without Requiring a Persistent Excitation," **IEEE Trans. Automatic Control**, Vol. AC-32, No. 5, pp. 397-404, May 1987.

[11] C. Samson and J.J. Fuchs, "Discrete Analysis Regulation of Not-Necessarily Minimum Phase Systems," **Proc. IEE**, Vol. 128, Pt. D, No. 3, pp. 102-108, May 1981.

[12] E.W. Kamen, T.E. Bullock, C-H. Song, "Adaptive Control Applied to Missile Autopilots," American Control Conference, Atlanta, GA, pp. 555-560, June 1988.

LEARNING CONTROL FOR ROBOTICS

Kevin L. Moore

Texas A&M University

Mohammed Dahleh

Texas A&M University

S. P. Bhattacharyya

Texas A&M University

I. **INTRODUCTION** *Learning control* is a new approach to the problem of improving transient behavior in the face of design or modelling uncertainty for processes that are repetitive in nature and operate over a fixed time interval. It is also well-suited to problems in which a system must be able to follow different types of inputs. An example of such a system is a robot arm in a flexible manufacturing assembly line which must execute a given trajectory many times but may be configured to perform different tasks, depending upon production needs. Learning control is an iterative approach to generating the optimal reference input for the system so that the system output is as close as possible to the desired output. The idea is illustrated in Figure 1. Each time the system operates the input to the system u_k is stored, along with the resulting system output y_k. The learning controller then evaluates the performance error as compared to some desired signal y_d and computes a new input u_{k+1} which is stored for use the next time the system operates. The new input is chosen in such a way as to guarantee that the performance error will be reduced on the next trial. The task in learning control is to specify the algorithm for generating the next input, given the current input and output, so that convergence to the desired output is attained (in the sense of some norm). Ideally, the convergence property of the learning control algorithm would require minimal knowlege of the system parameters and would be independent of the desired response y_d. Thus it would be possible to derive the best control signal for any of several desired responses, in the presence of parameter uncertainty, simply by using input-output data obtained from actual operation.

A wide variety of learning control algorithms have been proposed by Arimoto and his colleagues ([1–4] are representative). The most general algorithm presented is found in [3]. Here a PID-type of learning controller is used which updates the input by $u_{k+1} = u_k + \Phi e_k + \Gamma \dot{e}_k + \Psi \int e_k dt$. Although the use of derivative action is undesirable, the effectiveness of Arimoto's algorithms has been demonstrated with actual robots. Togai and Yamano [13] consider the problem of learning control for a discrete-time LTI system by using gradient methods to optimize the gain G in the input updating law $u_{k+1}(i) = u_k(i) + Ge_k(i+1)$. With this rulte the authors pick G to minimize the error between successive trials. The approach of Mita and Kato [12] is to specify the learning control algorithm in the frequency domain. Their algorithm updates the input by $U_{k+1}(s) = L(s)[U_k(s) + aE_k(s)]$. One feature of this is that there will always be a non-zero error, although they show that this error can be made arbitarily small and they present a specific design technique for finding the transfer matrix $L(s)$ and the parameter a. Furuta and Yamakita [8] present a system-theoretic treatment of the learning control problem. Their algorithm for updating the input is $u_{k+1} = u_k + \epsilon_k T_p^* e_k$, where T_p^* is the adjoint operator of the system and ϵ_k is a time-varying gain. A contribution of this paper is to provide some singular value conditions that ensure robustness of convergence when there is parameter uncertainty. Other learning control schemes found in the literature can be classified with one of the methods described above. Craig [7] has independently proposed learning and adaptive schemes similar to those of Arimoto's, as have Atkeson and McIntyre [5], Gu and Loh [9], and Hara [10]. Harokopos has considered learning from the viewpoint of minimizing a cost functional [11], similar to that of Togai and Yamano. Wang [14] has applied the idea of learning control to the problem of determining the inverse dynamics of an unknown nonlinear system. Bondi, et.al., [6] give a high-gain feedback, model-reference approach to learning control for nonlinear systems that are typical of robot manipulator models.

In the sequel we present a generalized formulation of the problem which includes the approaches given above as special cases. We analyze the problem for the class of linear, time-invariant learning controllers. Next, we consider a specific time-varying learning controller for a class of systems which includes typical robotic manipulator models. This method is based on an existence result by Bondi, et. al., [6] and is illustrated by a simulation example.

II. **PROBLEM FORMULATION** We will describe our problem in operator-theoretic notation. Let U and Y be normed linear spaces over R whose elements are time functions taking values in C^r and C^m, respectively. Assume the same norm is used in each space. Let $T : U \mapsto Y$ be a bounded linear operator mapping elements in U to those in Y. We write this as $y = Tu$ where $u \in U$ and $y \in Y$. The induced norm of an operator T is $\|T\|_i = \sup_{\|x\|=1} \|Tx\|$. The compositon of two operators T_1 and T_2 is denoted T_1T_2. The nonlinear operator $f : U \mapsto Y$ mapping elements in U to those in Y will be written as $y = f(u)$ where $u \in U$ and $y \in Y$. With this notation we may state the learning control problem as follows.

PROBLEM : Given a system N (possibly nonlinear, time-varying), with $y_k(t) = f_{n_k}(u_k(t))$, and given a desired response $y_d(t)$, defined on the interval (t_0, t_f), find a system L (possibly non-linear, time-varying) so that the sequence of inputs produced by the iteration

$$u_{k+1}(t) = f_{L_k}(u_k, y_k, y_d) = f_{L_k}(u_k, f_{n_k}(u_k), y_d),$$

converges to a fixed point u^* such that

$$\lim_{k \to \infty} \|y_d - y_k\| = \|y_d - T_s u^*\|$$

is minimum on the interval (t_0, t_f) over all possible systems L, for a specified norm. In addition, such a system L should require as little information about the system N as possible.

Of course, if the operator f_n is left invertible then we simply let $u^* = f_n^{-1}(y_d)$ and we are done. If the system is not invertible it may still be possible to obtain $y(t) = y_d(t)$ for some input u^*, as long as y_d is in the image of the operator f_n (i.e., the system is invertible from the input y_d). If y_d is not in the image of f_n then the best we can hope for is to find a $u^*(t)$ which minimizes $\|y_d - y\|$ over all possible inputs $u(t)$, for a given norm.

Note that we may propose learning control algorithm. in both the time domain and in the frequency domain. To implement an algorithm designed in the frequency domain will require storing signals that are defined on all of R^+. While in practice this is not possible, such schemes will still work provided we take measurements well past the settling time of the signals. Hence, analysis in the frequency domain will be valid even though our problem is stated over a finite interval.

The learning control problem as we have stated it is general enough to include all the problem formulations presented in the literature, including those that employ time-varying or non-causal learning controllers. However, this formulation is so general as to be almost intractable. In the next section we simplify the problem by assuming that both the system to be controlled and the learning controller are linear. These assumptions make it possible to gain some insight into the problem.

III. **LINEAR TIME-INVARIANT LEARNING CONTROL** Consider the configuration of Figure 1 where the system is a causal, LTI system S described by $y = T_s u$, with T_s known. Note that all the work presented in the literature assumes some degree of knowlege on the plant. In this section we assume full knowlege of the dynamics. This will be necessary to minimize the norm of the final error. If the iteration causes $\|c^*(t)\| \to 0$ then the question of plant knowledge would be less important. However, it is clear that without some plant knowlege our learning control scheme will be suboptimal when the final error is non-zero. Our goal at this point is to examine the nature of the best possible solution. Hence we suppose the plant is known. We also assume our desired response $y_d(t)$ is defined on the interval (t_0, t_f) and the learning controller L is restricted to the class of causal, LTI systems given by

$$u_{k+1} = T_u u_k + T_e(y_d - y_k),$$

where T_u and T_e are linear operators. This is the most general linear algorithm one might consider because it allows separate weighting of both the current input and the current error. For this scheme it is easy to show the following.

THEOREM 1 For the plant $y_k = T_s u_k$, the linear time-invariant learning control algorithm, $u_{k+1} = T_u u_k + T_e(y_d - y_k)$, converges to a fixed point $u^*(t)$ given by

$$u^*(t) = (I - T_u + T_e T_s)^{-1} T_e y_d(t)$$

with a final error $e^*(t) = \lim_{k \to \infty}(y_k - y_d) = (I - T_s(I - T_u + T_e T_s)^{-1} T_e) y_d(t)$ on the interval (t_0, t_f) if

$$\|T_u - T_e T_s\|_i < 1.$$

A. <u>Convergence with zero error</u> Notice that in general the error $e^*(t) \neq 0$ over the entire interval. However, if we choose $T_u = I$ then we can show that $e^*(t) = 0$ for all $t \in (t_0, t_f)$. So given a system T_s we can use the learning control algorithm $u_{k+1} = u_k + T_e(y_d - y_k)$ and obtain the desired convergence $\lim_{k \to \infty} y_k = y_d$ if the operator T_e is designed so that $\|I - T_e T_s\|_i < 1$.

Many of the learning control schemes in the literature require a condition of this form to achieve convergence with zero error. Unfortunately this condition is overly restrictive because for some problems it corresponds to a plant invertibility condition. For instance, consider the learning control problem with the popular H_∞ optimality criterion. Referring to Figure 1, let $u \in U = L_2^l(0, \infty)$ and $y \in Y = L_2^m(0, \infty)$. Recall that $L_2(0, \infty)$ is isomorphically isometric to the Hardy space H_2 so we may work in the frequency domain with operators that are transfer matrices whose elements are in the space H_∞. In this case our learning control algorithm is

$$U_{k+1}(s) = U_k(s) + H_e(s)[Y_d(s) - Y_k(s)]$$

and $y_k(t) \to y_d(t)$ in the sense of the L_2-norm whenever $\|I - H_e(s)H_s(s)\|_\infty < 1$. Here $H_s(s)$ is the transfer matrix of the system S and $\| \cdot \|_\infty$ is the H_∞-norm, defined for a transfer matrix $H(s)$ by

$$\|H(s)\|_\infty = \sup_\omega \lambda_{max}^{\frac{1}{2}}\{H^*(j\omega)H(j\omega)\}.$$

That is, the H_∞-norm is the maximum singular value of $H(s)$ over the $j\omega$-axis. This is a good way to formulate the problem because the class $L_2(0, \infty)$ includes many signals of interest and because there are results in H_∞ control theory for solving problems of this type [18]. However, a closer look at the basic definitions yields the following.

THEOREM 2 There exists a proper, stable, LTI system $H_e(s) \in H_\infty$ such that

$$\|I - H_e(s)H_s(s)\|_\infty < 1,$$

if and only if $H_s(s)$ is invertible over the space H_∞.

This result shows that to obtain converence in an L_2 sense for the class of proposed learning controllers we must have an invertible system, which is a very restrictive condition. (A less general version of this Theorem was given by Mita in [12] for $\|I - aH_s(s)\|_\infty$.)

Note that the condition $\|I - T_eT_s\|_i < 1$ is sufficient but not necessary. Although there may be no stable, proper, LTI system T_e that satisfies this condition this does not imply that there are no learning controllers which solve our original problem. We comment that all the schemes currently found in the literature, with the exception of Mita and Kato's, use a learning control update law where the current input is not weighted (i.e., $T_u = I$). In some of these schemes convergence to zero error is achieved without a requirement of system invertibility. This is done by using different norms and/or an enlarged class of admissible learning controllers (such as the use of derivative action by Arimoto, or time-varying controllers used by other researchers). However, for a useful class of problems (i.e., those with an H_∞-optimality criterion) convergence with zero error of the learning control scheme for the class of causal, LTI learning controllers is equivalent to a requirement of plant invertibility.

B. Convergence with non-zero error The problem of requiring plant invertibility for convergence of the learning control scheme given above can be avoided by using the more general form of the algorithm where $T_u \neq I$. That is, let $u_{k+1} = T_u u_k + T_e(y_d - y_k)$. In this case the condition for convergence is $\|T_u - T_eT_s\|_i < 1$, which is less restrictive than $\|I - T_eT_s\|_i < 1$. However, the error $e_k = (y_d - y_k)$ will not converge to zero. Thus the task in this approach is to choose T_u and T_e so that the norm of the final error, $e^*(t) = \lim_{k \to \infty}(y_d - y_k)$, is minimized. Let U and Y be normed linear spaces and let X be the space of causal LTI operators mapping elements from U into Y. Then our learning control problem can be cast as the following optimization problem.

OPT1 : Let $u_k \in U$, $y_d, y_k \in Y$ and $T_s, T_u, T_e \in X$. Then given y_d and T_s, find T_u^* and T_e^* that solve

$$\min_{T_u, T_e \in X} \|(I - T_s(I - T_u + T_eT_s)^{-1}T_e)y_d\|$$

subject to $\|T_u - T_eT_s\|_i < 1$.

To state the solution to OPT1 we first introduce another optimization problem.

OPT2 : Let $y_d \in Y$ and let $T_n, T_s \in X$. Then given y_d and T_s, find T_n^* that solve

$$\min_{T_n \in X} \|(I - T_sT_n)y_d\|$$

The following Theorem relates the solution of OPT1 and OPT2.

THEOREM 3 : Let T_n^* be the solution of OPT2. Factor $T_n^* = T_m^*T_e^*$ where $T_m^{*-1} \in X$ and $\|I - T_m^{*-1}\|_i < 1$. Define $T_u^* = I - T_m^{*-1} + T_e^*T_s$. Then T_u^* and T_e^* are the solution of OPT1.

PROOF : First note that if T_u^* and T_e^* are given as in the Theorem then

$$\|(I - T_s(I - T_u^* + T_e^*T_s)^{-1}T_e^*)y_d\| = \|(I - T_sT_n^*)y_d\| = \min_{T_n \in X} \|(I - T_sT_n)y_d\|.$$

Also note that $\|T_u^* - T_e^*T_s^*\|_i = \|I - T_m^{*-1}\|_i < 1$ so T_u^* and T_e^* are candidate solutions for OPT1. Now suppose $\bar{T}_u \neq T_u^*$ and $\bar{T}_e \neq T_e^*$ are solutions to OPT1. Define $\bar{T}_m = (I - \bar{T}_u + \bar{T}_eT_s)^{-1}$. Note that $\bar{T}_m$ exists and is invertible because if $\bar{T}_u$ and $\bar{T}_e$ solve OPT1 then $\|\bar{T}_u - \bar{T}_eT_s\|_i < 1$. Also note that as T_u and T_e range over X, subject to $\|T_u - T_eT_s\|_i < 1$, then $T_m = (I - T_u + T_eT_s)^{-1}$ ranges over all invertible operators in X. Thus we may write

$$\min_{T_e, T_u \in X} \|(I - T_s(I - T_u + T_eT_s)^{-1}T_e)y_d\| = \min_{T_e, T_m, T_m^{-1} \in X} \|(I - T_sT_mT_e)y_d\|,$$
$$= \|(I - T_s(I - \bar{T}_u + \bar{T}_eT_s)^{-1}\bar{T}_e)y_d\|,$$
$$= \|(I - T_s\bar{T}_m\bar{T}_e)y_d\|.$$

Further, let $\bar{T}_n = \bar{T}_m\bar{T}_e$ and note that as T_e and T_m range over X with T_m invertible then $T_n = T_mT_e$ ranges over X. So we may write

$$\min_{T_s,T_m,T_m^{-1} \in \mathbf{X}} \|(I - T_s T_m T_e)y_d\| = \min_{T_n \in \mathbf{X}} \|(I - T_s T_n)y_d\|,$$

$$= \|(I - T_s \bar{T}_m \tilde{T}_e)y_d\|,$$

$$= \|(I - T_s \tilde{T}_n)y_d\|.$$

But, this implies that $\|(I - T_s \tilde{T}_n)y_d\| \le \|(I - T_s T_n^*)y_d\|$, which is a contradiction. Hence the Theorem follows. **QED**

Theorem 3 says that to solve the causal, LTI learning control problem, OPT1, we simply solve the problem OPT2. This is a useful result because OPT2 is in a form that can often be solved. For instance if we choose an l_∞-optimality criterion then OPT2 can be solved by some recent results in l_∞ control [17].

Also note that we have formulated this problem for a fixed input $y_d(t)$. If we want to minimize the final error $e^*(t)$ for arbitrary inputs then our problem becomes

$$\min_{T_s,T_e \in \mathbf{X}} \|I - T_s(I - T_u + T_e T_s)^{-1} T_e\|_i$$

subject to $\|T_u - T_e T_s\|_i < 1$. Analogous to Theorem 3 we can show that this is equivalent to the problem $\min_{T_n \in \mathbf{X}} \|T_u - T_s T_n\|_i$. In this case we could use results from $\mathbf{H}_\infty$ or $\mathbf{L}_1$ [16] control theory to solve the optimization problem.

Now consider the implications of Theorem 3. Assume that we have a solution T_n^* to problem OPT2. To find T_u^* and T_e^* we must factor $T_n^* = T_m^* T_e^*$ with T_m^* invertible and $\|I - T_m^{*^{-1}}\|_i < 1$. This can always be done because we can choose $T_m^* = I$. Thus we may consider two cases.

CASE 1 : $T_m^* = I$ If we choose $T_m^* = I$ then we have $T_e^* = T_n^*$ and $T_u^* = T_n^* T_s$. This defines the optimal learning control law $u_{k+1} = T_u^* u_k + T_e^*(y_d - y_k) = T_u^* u_k + T_e^*(y_d - T_s u_k)$ or $u^*(t) = u_{k+1} = u_k = T_n^* y_d(t)$. Thus our input does not change from trial to trial. So *no iterative learning is required* and the input defined as $u^*(t) = T_n^* y_d(t)$ results in the minimum possible norm of the error without any learning iterations. This can be seen by substituting $T_u^* = T_n^* T_s$ and $T_e^* = T_n^*$ into the expression for e^* given in Theorem 1. This yields $e^*(t) = (I - T_n^* T_s)y_d(t)$. But, by definition of T_n^* this error will have minimum norm. All we have done is found the closest approximation to an inverse system for the input y_d.

CASE 2 : $T_m^* \ne I$ Now suppose we factor $T_n^* = T_m^* T_e^*$ with $T_m^* \ne I$ invertible and $\|I - T_m^{*^{-1}}\|_i < 1$. Let $T_u^* = I - T_m^{*^{-1}} + T_e^* T_s$. Then for the learning control iteration $u_{k+1} = T_u^* u_k + T_e^*(y_d - y_k)$ we can evaluate the fixed point u^* and the resultant error e^* from Theorem 1. This results in $u^*(t) = (I - T_u^* + Te^* T_s)^{-1} T_e^* y_d(t) = (I - (I - T_m^{*^{-1}} + T_e^* T_s) + T_e^* T_s)^{-1} T_e^* y_d(t) = T_n^* y_d(t)$, and $e^*(t) = y_d(t) - T_s u^*(t) = (I - T_n^* T_s)y_d(t)$. Comparing this result to Case 1 we see that the resulting optimal input $u^*(t)$ and the associated error $e^*(t)$ are the same for both cases. What this means is that the best input signal for the system (in terms of minimum norm of the error), when the plant is known, can be determined through a learning proccess or through an *a priori* design. In fact, to obtain the best learning control law it is first necessary to solve the *a priori* design problem to obtain T_n^*, from which T_u^* and T_e^* can be derived. Thus we conclude that the best possible learning controller of the class of causal, LTI learning controllers (given by $u_{k+1} = T_u u_k + T_e(y_d - y_k)$) can do no better than the open loop control law $u^*(t) = T_n^* y_d(t)$, where T_n^* is the best appoximation to the inverse system of T_s for the input $y_d(t)$, defined as the solution of the problem $\min_{T_n \in \mathbf{X}} \|(I - T_n T_s)y_d)\|$.

IV. LTI LEARNING CONTROL VIA PARAMETER ESTIMATION

Now let the plant have a known structure but unknown paramaters. In this case it is reasonable to assume that some type of learning iteration would be useful in improving the system response. Our assumptions on the plant will be that it is causal, LTI, stable, minimal, and its order is known. We will only consider the single-input, single-output case. Our approach is to use a parameter estimator to obtain estimates of the system and use these estimates in our learning controller. This is illustrated in Figure 2. During each trial we obtain an estimate for the plant. This estimate is then used to solve the problem OPT2 described above. This defines the optimal input for the next trial. With no loss of generality we give our result for a least-squares estimator and consider the case of an l_∞ optimality criterion (which implies a discrete-time system).

A. <u>Notation</u> Let $X(z)$ denoted the z-transform of the time signal $x(t)$, where t is an integer and $x(t) \in l_\infty$. l_∞ is the class of bounded sequences and the z-transform is defined by $X(z) = \sum_{n=0}^{\infty} x(n)z^n$. Let A denote the subring of stable, rational functions in z. A can be viewed as a space of bounded operators on l_∞. We assume that the plant $H(z) \in A$, has the form

$$H(z) = \frac{b_1 z^{n-1} + \cdots + b_{n-1}z + b_n}{z^n + a_1 z^{n-1} + \cdots + a_{n-1}z + a_n},$$

and that it is irreducible (i.e., the system is controllable and observable). If our original system is not stable then we first stabilize it with a conventional controller.

Now parametrize the system in the following way. At the kth trial we have $Y_k(z) = H(z)U_k(z)$. In the time domain this can be written as

$$y_k(t) = -\sum_{j=1}^{n} a_j y_k(t-j) + \sum_{j=1}^{n} b_j u_k(t-j).$$

If we define $\theta_0 = (-a_1, -a_2, \cdots, -a_n, b_1, b_2, \cdots, b_n)^t$ and

$$\phi_k(t-1) = (y_k(t-1), y_k(t-2), \cdots, y_k(t-n), u_k(t-1), u_k(t-2), \cdots, u_k(t-n))^t$$

then we can write $y_k(t)$ as $y_k(t) = \theta_0^t \phi_k(t-1)$. The estimation scheme will produce estimates $\hat{\theta}_k(t)$ of the parameter vector θ_0. These estimates can then be used to obtain an estimate of the plant transfer function.

B. <u>Estimation and learning control law</u> We utilize the standard least-square estimator [19] although similar results would hold for other estimators. The key to showing the convergence of the learning control scheme is the manner in which the estimator is initialized at the start of each trial. The essential feature is that at the start of each trial the estimator is initialized with the final covariance update from the previous trial. Specifically, we estimate the parameter vector by the following rule for each trial k. For $t = 1, 2, \cdots, N$ let

$$\hat{\theta}_k(t) = \hat{\theta}_k(t-1) + \frac{P_k(t-2)\phi_k(t-1)}{1 + \phi_k^t(t-1)P_k(t-2)\phi_k(t-1)}[y_k(t) - \phi_k^t(t-1)\hat{\theta}_k(t-1)]$$

$$P_k(t-1) = P_k(t-1) - \frac{P_k(t-2)\phi_k(t-1)\phi_k^t(t-1)P_k(t-2)}{1 + \phi_k^t(t-1)P_k(t-2)\phi_k(t-1)}$$

$$P_{k+1}(-1) = P_k(N-1)$$

$$\hat{\theta}_{k+1}(0) = \hat{\theta}_k(N),$$

where $P(-1)$ is an arbitrary positive definite matrix, $\hat{\theta}_1(0)$ is a given initial estimate, and each trial is conducted over the finite interval $(0, N)$. Now, using the results of section III we propose the learning control law $U_{k+1}(z) = N_k^*(z)Y_d(z)$, where $N_k^*(z)$ is the solution at the kth trial to the l_∞ minimization problem $\min_{N_k \in A} \|(I - N_k(z)\hat{H}_k(z))Y_d(z)\|_{l_\infty}$. In this problem $\hat{H}_k(z)$ is the estimate of the transfer function of the plant at the kth trial, constructed from $\hat{\theta}_k(N)$.

C. <u>Main result</u> In order to show the convergence of the learning control scheme we have described it will be necessary that the sequence of inputs $U_k(z)$ be persistently exciting. We also need a continuity condition on the solutions of the l_∞-minimization problem. The following two Lemmas provide the necessary results.

LEMMA 1 : For the learning control scheme described above the sequence of inputs $U_k(z)$ is persistently exciting of order $2n$ if the initial input $U_0(z)$ is persistently exciting of order $2n$.

LEMMA 2 : If $\lim_{k \to \infty} \hat{H}_k(z) = H(z)$ then $\lim_{k \to \infty} N_k^*(z) = N^*(z)$ where $N_k^*(z)$ is the solution of the l_∞ minimization problem $\min_{N_k(z) \in A} \|(I - N_k(z)\hat{H}_k(z))Y_d(z)\|_{l_\infty}$ and $N^*(z)$ is the solution to the problem $\min_{N(z) \in A} \|(I - N(z)H(z))Y_d(z)\|_{l_\infty}$.

Now we may state the following result.

THEOREM 4 : For the configuration of Figure 2, with the plant, estimator, and learning control law as described let (i) $H(z) \in A$ be irreducible and (ii) The initial input u_0 be persistently exciting of order $2n$ or more. Then $\lim_{k \to \infty} \hat{H}_k(z) = H(z)$, which implies that the input $U^*(z) = \lim_{k \to \infty} N_k^*(z)Y_d(z) = N^*(z)Y_d(z)$ is the best possible input to the system in terms of minimizing the l_∞-norm of the error $e^*(t) = y_d(t) - y^*(t)$.

PROOF : Here we sketch the main idea of the proof. Note that if we show that $\lim_{k \to \infty} \hat{H}_k(z) = H(z)$ then given Lemma 2 and Theorem 3 the result follows. To show $\lim_{k \to \infty} \hat{H}_k(z) = H(z)$ note that each trial has a finite duration of N samples. Thus consider a sequence of vectors defined by

$$\hat{\theta}(t) = \{\hat{\theta}_1(0), \cdots, \hat{\theta}_1(N), \hat{\theta}_2(0), \cdots, \hat{\theta}_2(N), \cdots, \hat{\theta}_k(0), \cdots, \hat{\theta}_k(N), \cdots\}.$$

That is,

$$\hat{\theta}(t) = \begin{cases} \hat{\theta}_k(t - (k-1)N), & \text{if } (k-1)N < t \leq kN; \\ \hat{\theta}_1(0), & \text{if } t = 0. \end{cases}$$

Similarly, for $1 \leq (k-1)N + 1 \leq t \leq kN$, define $\phi(t-1) = \phi_k(t - (k-1)N)$ and

$$P(t-1) = \begin{cases} P_k(t - (k-1)N - 1), & \text{if } (k-1)N + 1 \leq t \leq kN; \\ P_1(-1), & \text{if } t = 0. \end{cases}$$

Now just note that the sequence $\hat{\theta}(t)$ can be viewed as the output of a least squares etimator with a sequence of regression vectors $\phi(t-1)$ and associated covariance matrices $P(t-1)$. This is a result of the way we have defined the initial conditions $\hat{\theta}_k(0)$ and $P_k(-1)$ at each trial. Then by properties of the least-squares algorithm and our assumptions of stability, minimality, and persistence of excitation of the original input (which by Lemma 1 ensures persistence of excitation of the input sequence) we have $\lim_{t \to \infty} \hat{\theta}(t) = \theta_0$ [19]. But, if $\hat{\theta}(t)$ converges then so does any subsequence. In particular consider the subsequence defined by $\hat{\theta}(t_k) = \hat{\theta}(kN) = \hat{\theta}_k(N)$, for $k = 1, 2, \cdots$. Then $\lim_{k \to \infty} \hat{\theta}_k(N) = \theta_0$, and thus $\lim_{k \to \infty} \hat{H}_k(z) = H(z)$. **QED**

D. Comments Theorem 4 says that the learning control scheme based on parameter estimates will converge with a minimum normed error. Thus it is not necessary to have full knowlege of the plant to design the best possible LTI learning controller for an l_∞-optimality criterion. However, the same claim can be made for a conventional adaptive control scheme. This can be seen by considering again the configuration shown in Figure 2. Suppose now, however, we execute only one trial, but allow the duration of the trial to be very large. In this case we can use the least-square estimator to obtain an estimate of the plant that converges asymptotically to the true value under the same set of assumptions as above. In fact, the error in the estimate at time $t_1 = kN$ will be the same estimation error we have at the end of the kth-trial in the learning control scheme. So rather than executing k trials, each of duration N, we simply run a single trial of duration kN to obtain a convergent estimate of the plant. This estimate can then be used to solve the l_∞ problem, resulting in the best possible input as described in Theorem 3. Again, as in the case when we have full plant knowlege, we conclude that *no iterative learning is required.*

V. A TIME-VARYING LEARNING CONTROLLER FOR ROBOTICS Because a robotic manipulator typically performs a given task over and over it is natural to try to use learning control to exploit the repetition in the robot operation. Also, from the results of the previous sections it is clear that successful learning control schemes will likely be time-varying and/or nonlinear in nature. In this section we describe a time-varying learning control for a class of systems which includes the model of a typical robotic manipulator. The scheme is an improvement of the technique given by Bondi, Casalino, and Gambardella in [6] which uses a linear output feedback control law with a linear error feedback learning controller.

A. <u>System description</u> Consider the configuration of Figure 3. The system dynamics are given by

$$A(x_k)\ddot{x}_k + B(x_k,\dot{x}_k)\dot{x}_k + C(x_k) = u_k,$$

where u_k is an n-dimensional vector of input torques which drive the system at the kth trial and $x_k \in \mathbf{R}^n$ is a vector of the resulting output positions relative to some generalized coordinate space. The positive definite matrix A, the matrix B, and the vector C are continuous and locally Lipschitzian functions of their arguments. We assume that $C(x_d(0))$ is known, where x_d is the desired position trajectory defined on the interval $[0,T]$. Otherwise A, B, and C are unknown. For convenience, let $y_k = (x_k^T, \dot{x}_k^T, \ddot{x}_k^T)^T \in \mathbf{R}^{3n}$ denote the complete system output at the kth trial and $y_d = (x_d^T, \dot{x}_d^T, \ddot{x}_d^T)^T$ denote the desired trajectory. The trajectory error at the kth trial is given by $e_k = y_k - y_d$.

The control law for this system is a time-varying linear output feedback of the form $u_k = r_k - \alpha_k \Gamma y_k + C(x_d(0))$, where $\Gamma \in \mathbf{R}^{n \times 3n}$ is a matrix partitioned as $\Gamma = [P, L, K]$ with P, L, and K nonsingular and α_k a time-varying gain. The signal r_k can be viewed as a time-varying reference input. This signal is updated according to the learning control law $r_{k+1} = r_k - \alpha_k \Gamma e_k$.

We assume the system operates iteratively as follows. An initial reference input r_0 is chosen and the manipulator is commanded to execute its trajectory for several trials. During each trial the current reference r_k and the resulting error e_k are stored. This information is then used to compute the reference r_{k+1} for use on the next trial and to update the gain α_k. All initial conditions are reset to a common value at the start of each trial. For this type of learning control scheme Bondi, et $al.$ [6] give an existence result which has the following Corollary.

COROLLARY Let $\Gamma = [K, L, P]$ with P, L, and K nonsingular and $\det(Ks^2 + Ls + P)$ Hurwitz. Then there exists a gain $\bar{\alpha}$ such that if the sequence $\{\alpha_k\}$ is monotone increasing with $\alpha_0 \geq \bar{\alpha}$ then for all $t \in [0,T]$ we have $\lim_{k \to \infty} \|e_k\| = 0$.

The proof of this Corollary follows from an examination of the proof of Theorem 2 of [6], which considers the case of constant gains. Rather than reproduce the proof we simply describe the essential features and refer the interested reader to [6] for more details. The proof involves two parts. First, it is demonstrated that the uniform boundedness of the trajectory error at each trial can be guaranteed, given a gain α greater than some $\bar{\alpha}$. The argument of this part of the proof is easily extended to the case of an increasing sequence of α's as described in the Corollary. Second, using the uniform boundedness property and the other assumptions on the system it is shown that there exists a Lipschitz constant $\rho(\alpha) < 1$ which controls the convergence of the position error, thus leading to the result given in Theorem 5. This Lipschitz constant is monotonically decreasing toward zero for increasing α. This implies that for a monotonically increasing sequence $\{\alpha_k\}$ we have a monotonically decreasing sequence of Lipschitz constants satisfying $\rho(\alpha_{k+1}) \leq \rho(\alpha_k)$. From this the convergence result of the Corollary follows.

B. <u>Adaptive gain adjustment</u> To use the Corollary it is necessary to know a $priori$ the minimum gain $\bar{\alpha}$. We now describe an adaptive technique for obtaining a gain α that will ensure convergence without a $priori$ knowledge of the minimum gain. In the learning control law $r_{k+1} = r_k + \alpha_k \Gamma e_k$ the gain α_k is updated at each trial according to $\alpha_{k+1} = \alpha_k + \|e_k\|^m$, for $m > 0$. The idea is to make α_k larger at each trial so that eventually it will become bigger than $\bar{\alpha}$. When this happens the results of the Corollary can be applied to the successive iterations and we will have a convergent learning control scheme. This is formalized in the following Theorem.

THEOREM 5. Consider the configuration of Figure 4 with the previous assumptions on the system and its operation as described above. Let the time-varying learning control law be given by $r_{k+1} = r_k + \alpha_k \Gamma c_k$ and the associated time-varying feedback gain be given by $\alpha_k \Gamma$, where $\Gamma = [P, L, K]$ is arbitrary with P, L, and K nonsingular and $\det(Ks^2 + Ls + P)$ Hurwitz. Also let α_k be iteratively updated by $\alpha_{k+1} = \alpha_k + \|e_k\|^m$, for $m > 0$ and $\|\cdot\|$ a specified norm, possibly time-varying. Then for all $t \in [0, T]$ we have $\lim_{k \to \infty} \|\epsilon_k\| = 0$.

PROOF: With no loss of generality let $m = 1$. Suppose $\lim_{k \to \infty} \|e_k\| \neq 0$. Then we have two cases.

(1) $\lim \sup_{k \to \infty} \|e_k\| = \infty$. In this case there exists a sequence $\{k_i\}$ such that $\|e_{k_{i+1}}\| > \|e_{k_i}\| > 1$ for $i = 1, 2, \ldots$ Then $\alpha_{k_i+1} = \alpha_{k_i} + \|e_{k_i}\| > \alpha_{k_i} + 1$, so the subsequence $\{\alpha_{k_i}\}$ diverges as $i \to \infty$ and hence $\lim_{k \to \infty} \alpha_k = \infty$.

(2) $\lim \sup_{k \to \infty} \|c_k\| = C > 0$, where C is a constant. In this case there exists a sequence $\{k_i\}$ such that $\lim_{i \to \infty} \|e_{k_i}\| = C$. Thus there exists an integer N such that $\|e_{k_i}\| > C/2$ whenever $i > N$. But, when $i > N$ then $\alpha_{k_i+1} = \alpha_{k_i} + \|e_{k_i}\| > \alpha_{k_i} + C/2$, so the subsequence $\{\alpha_{k_i}\}$ diverges as $i \to \infty$ and hence $\lim_{k \to \infty} \alpha_k = \infty$.

Thus if $\lim_{k \to \infty} \|e_k\| \neq 0$ we conclude that the sequence α_k is increasing without bound. Then there exists an integer $\bar{k}$ such that $\alpha_{\bar{k}} \geq \bar{\alpha}$. But, $\alpha_{k+1} \geq \alpha_k$ for all k because $\|e_k\| \geq 0$. So we have $\alpha_{k+1} \geq \alpha_k \geq \bar{\alpha}$ for all $k > \bar{k}$ and by the Corollary above this implies that $\lim_{k \to \infty} \|e_k\| = 0$. However, this contradicts our assumption. **QED**

Notice that the scheme will work for any value of m. This variable can be chosen based on considerations of the rate of convergence. Ideally, one would like α_k to increase rapidly at first. However, after trial $\bar{k}$ such that $\alpha_k > \bar{\alpha}$ we may want α_k to slow down or stop its rate of increase so we will not drive the system or its actuators to saturation. Thus we may consider a time-varying rule for the value of m. Also note that the results here apply for any norm of the error, including those that are time-varying. For instance, use of the standard Euclidean norm defined at each instance of time in the interval $[0, T]$ would lead to a completely adaptive gain adjustment law which changes the gain at each instant of time during each trial. That is, $\alpha_{k+1}(t) = \alpha_k(t) + \|e_k(t)\|^m$. Alternately, we might use only a finite record of the error to update the gain. This could lead to a gain α that is constant throughout a given trial, but is updated at each iteration. For example, again using the Euclidean norm, let $\alpha_{k+1} = \alpha_k + \|e_k(T)\|^m$. Yet another approach would be to use a norm which gives a measure of the error over the entire interval $[0, T]$, such as the standard L_p-norm. Thus we might obtain an update law of the form $\alpha_{k+1} = \alpha_k + \|e_k\|^m_{L_p[0,T]}$. The choice of a rule for m and the use of different types of norms is an area for further study. Finally we note that an identical adaptive gain adjustment law can be given for a learning control method reported by Kawamura, Miyazaki, and Arimoto [4].

C. Simulation results To illustrate the adaptive gain adjustment technique a simulation was conducted using the two-joint manipulator shown in Figure 4. The vector x is taken as $x(t) = (\theta_1(t), \theta_2(t))^T$. The model for this system is well known and can be found in many references (see [15] for instance). For the masses and lengths shown in Figure 4 we obtain

$$A(x) = \begin{pmatrix} .54 + .27 \cos \theta_2 & .135 + .135 \cos \theta_2 \\ .135 + .135 \cos \theta_2 & .135 \end{pmatrix}$$

$$B(x, \dot{x}) = \begin{pmatrix} .135 \sin \theta_2 & 0 \\ -.27 \sin \theta_2 & -.135(\sin \theta_2)\dot{\theta}_2 \end{pmatrix}$$

$$C(x) = \begin{pmatrix} 13.1625 \sin \theta_1 + 4.3875 \sin(\theta_1 + \theta_2) \\ 4.3875 \sin(\theta_1 + \theta_2) \end{pmatrix}.$$

As noted in [6], it is not possible to directly feed back the acceleration terms without causing algebraic loops in the closed-loop system. Therefore we use a simple first-order filter to estimate $\ddot{\theta}_1$ and $\ddot{\theta}_2$ from $\dot{\theta}_1$

and $\dot{\theta}_2$, respectively. For the simulation the filters had the form $H(s) = 10s/(s+10)$. The gain matrices P, L, and K were chosen to be

$$P = \begin{pmatrix} 50.0 & 0 \\ 0 & 50.0 \end{pmatrix}, L = \begin{pmatrix} 65.0 & 0 \\ 0 & 65.0 \end{pmatrix}, K = \begin{pmatrix} 2.5 & 0 \\ 0 & 2.5 \end{pmatrix}.$$

To implement the learning controller with adaptive gain adjustment we update the gain by the rule $a_{k+1} = a_k + .1\|e_k(T)\|$, where we have used the standard Euclidean norm and we set $a_0 = 0.01$.

Figure 5 shows the resulting trajectory of θ_1 for a triangular input. On the first trial the observed error is very large. However, after only one iteration the improvement is quite impressive and by the eighth trial the system response cannot be distinguished from the desired response. This is also true of the other signals in the vector y. It should also be noted that without the adaptive gain adjustment the learning scheme does not converge for the initial value $a_0 = 0.01$. Thus the adaptive gain adjustment solves the problem of finding a suitable value of a *a priori*.

An important observation about the learning control algorithm is illustrated in Figure 6. This shows the reference input r_{θ_1} at the eighth trial. This is the signal that must drive the system in order to achieve the tracking property demonstrated in Figure 5. As can be seen the input has large "spikes" at the initial time and at the midpoint of the trajectory. Intuitively it is clear that such large magnitude inputs are necessary to overcome the system dynamics and obtain perfect tracking. However, in a practical implementation such signals may drive the system actuators to saturation. Thus we may have limiting constraints on the system inputs. More work must be done to see how such constraints will affect the convergence properties of the learning controller.

VI. **CONCLUSIONS** Learning control is a new approach to the problem of improving transient response behavior for processes that are repetitive in nature and operate over a fixed interval. Several learning control schemes have been presented in the literature. In this paper we have presented a general formulation of the learning control problem which includes the schemes reported in the literature as special cases. In particular we have examined the class of LTI, causal learning controllers. For this class we have noted that if an H_∞ optimality criterion is used then convergence of the learning control scheme with zero error is equivalent to a requirement of plant invertibility. We have also shown that if the plant is known the best possible LTI learning controller is no better than the best possible controller designed *a priori* from knowlege of the plant. In either case we obtain an open-loop control law which passes the desired input through the best approximation to an inverse system for the plant. We also showed that when the plant is unknown it is possible to devise a learning control scheme based on parameter estimation which converges with a minimum normed error. However, we noted that the same results can be obtain by using a conventional adaptive control scheme. A tentative conjecture at this point may be that for LTI systems with a known structure learning control can do no better than conventional or adaptive controller design techniques. An area for further study is to verify such a conjecture. We also described a time-varying learning control technique which can be applied to a class of nonlinear systems. The approach utilized an adaptive gain adjustment technique with the learning control scheme of [6] to iteratively improve the performance of a robotic manipulator. Simulation results have verified the effectiveness of the gain adjustment procedure. Future efforts will include an analysis of the rate of convergence properties of the learning control scheme with adaptive gains and a study of the effect of input magnitude constraints on the learning control scheme.

The authors are with the Dept. of Electrical Eng., Texas A&M University, College Station, TX 77843-3128.

REFERENCES

[1] S. Arimoto, S. Kawamura, F. Miyazaki, "Bettering Operation of Robots by Learning," *Journal of Robotic Systems*, Vol. 1, No. 2, pp. 123–140, 1984.

[2] ibid., "Bettering Operation of Dynamic Systems By Learning : A New Control Theory for Servomechanism or Mechatronic Systems," in *Proc. of 23rd Conference on Decision and Control*, Las Vegas, NV, pp. 1064–1069, December 1984.

[3] S. Arimoto, "Mathematical Theory of Learning with Applications to Robot Control," in *Proc. of 4th Yale Workshop on Applications of Adaptive Systems*, New Haven, CT, pp. 379-388, May 1985.

[4] S. Kawamura, F. Miyazaki, S. Arimoto, "Realization of Robot Motion Based on a Learning Method," *IEEE Trans. on Systems, Man, and Cybernetics*, Vol. 18, No. 1, pp. 126–134, Jan./Feb. 1988.

[5] C. Atkeson, J. McIntyre, "Robot Trajectory Learning Through Practice," in *Proc. of IEEE Confer. on Robotics and Automation*, San Francisco, Calif., April 1986.

[6] P. Bondi, G. Casalino, L. Gambardella, "On the Iterative Learning Control Theory for Robotic Manipulators," *IEEE Journal of Robotics and Automation*, Vol.4., No.1, pp. 14–22, February 1988.

[7] John J. Craig, "Adaptive Control of Manipulators Through Repeated Trials," in *Proc. of 1984 American Control Conference*, San Diego, Calif. pp. 1566–1572, June 1984.

[8] K. Furuta, M. Yamakita, "The Design of a Learning Control System for Multivariable Systems," in *Proc. of IEEE Int. Symposium on Intelligent Control*, Philadelphia, Penn., pp. 371–376, January 1987.

[9] Y.L. Gu, N.K. Loh, "Learning Control in Robotic Systems," in *Proc. of IEEE Int. Symposium on Intelligent Control*, Philadelphia, Penn., pp. 360–364, January 1987.

[10] S. Hara, Y. Yamamoto, T. Omata, M. Nakano, "Repetitive Control System : A New Type Servo System for Periodic Exogenous Signals," *IEEE Trans. Automatic Control*, Vol. AC-33, No. 7, pp. 659–668, July 1988.

[11] E. Harokopos, "Optimal Learning Control of Mechanical Manipulators in Repetitive Motions," in *Proc. of IEEE Int. Symposium on Intelligent Control*, Philadelphia, Penn., pp. 396–401, January 1987.

[12] T. Mita, E. Kato, "Iterative Control and its Application to Motion Control of Robot Arm – A Direct Approach to Servo-Problems," *Proc. of 24th Conference on Decision and Control*, Ft. Lauderdale, FL, pp.1393–1398, December 1985.

[13] M. Togai, O. Yamano, "Analysis and Design of an Optimal Learning Control Scheme for Industrial Robots : A Discrete System Approach," *Proc. of 24th Conference on Decision and Control*, Ft. Lauderdale, FL, pp. 1399-1404, December 1985.

[14] S. Wang "Inversion of Nonlinear Dynamical Systems," Technical Report, University of California, Davis, Spring, 1988.

[15] J. Craig, P Hsu, S. Sastry, "Adaptive Control of Mechanical Manipulators," in *Proc. of IEEE Confer. on Robotics and Automation*, San Francisco, Calif., pp. 190–195, April 1986.

[16] M.A. Dahleh, J.B.Pearson, Jr., "L^1-Optimal Compensators for Continuous-Time Systems," *IEEE Trans. Automatic Control*, Vol. AC-32, No. 10, pp. 889–895, October 1987.

[17] M.A. Dahleh, J.B.Pearson, Jr., "Minimization of a Regulated Response to a Fixed Input," *IEEE Trans. Automatic Control*, Vol AC-33, No.10, pp. 924–930, October 1988.

[18] B.A. Francis, *A Course in H_∞ Control Theory*, Lecture Notes in Control and Information Sciences, Springer-Verlag, 1987.

[19] G.C. Goodwin, K.S. Sin, *Adaptive Filtering, Prediction, and Control*, New Jersey, Prentice-Hall, 1984.

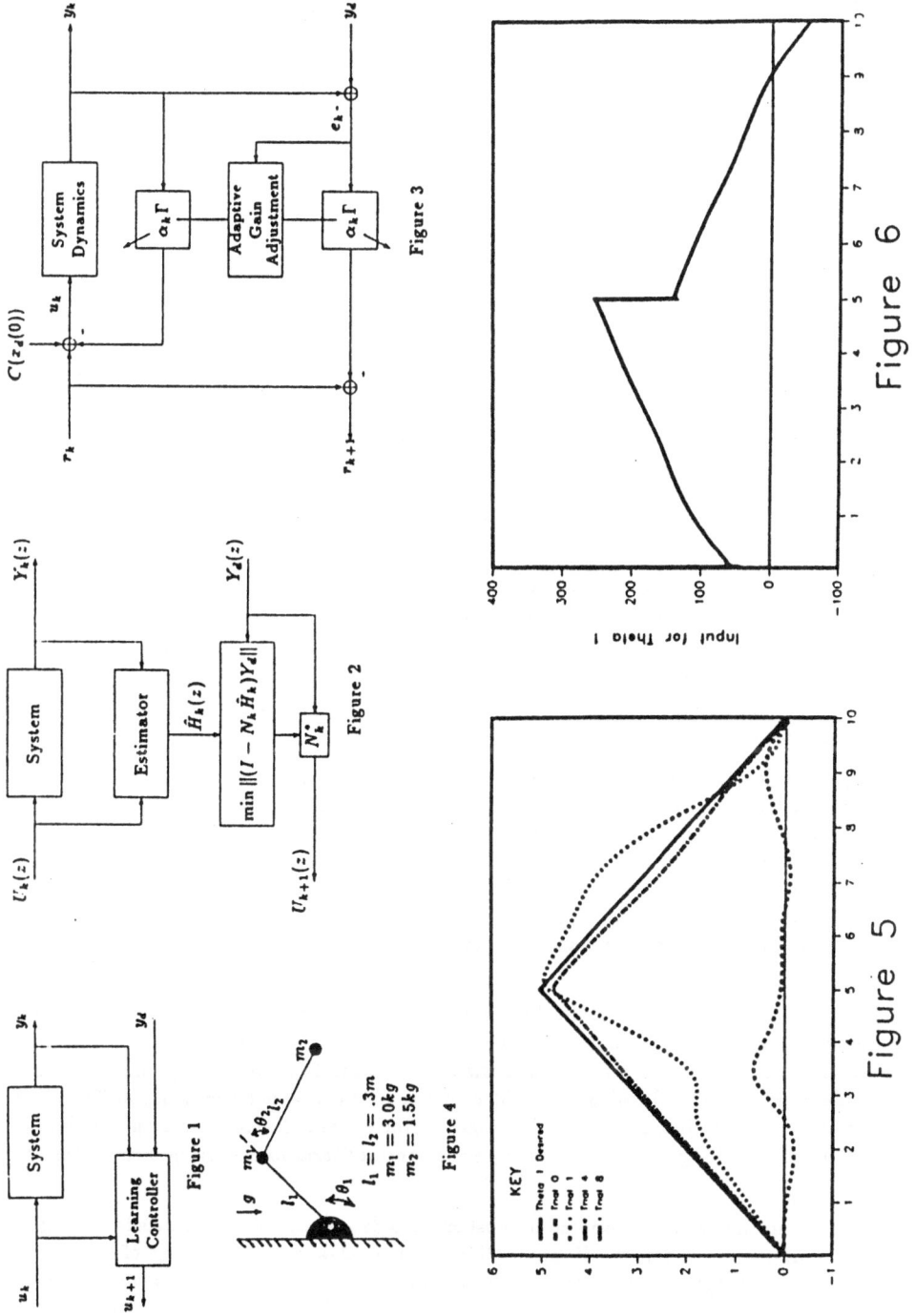

Figure 1

Figure 2

Figure 3

Figure 4

Figure 5

Figure 6

Instantaneous Trajectory Planning For Redundant Manipulator In The Presence of Obstacles

Chiung-Li Lee   Micheal W. Walker

Department of EECS , University of Michigan

1. Introduction. Manipulators presently used in the industry usually have a minimum number of joints required to accomplish their tasks. In general, six joints are used. Intuitively, manipulators with six joints are adequate to execute most given tasks. However, because of obstacles in the manipulator working environment and the existence of singular configurations of the manipulators, six joints may not be sufficient. As a consequence, manipulators with more than six joints, i.e. kinematically redundant manipulators, are proposed to solve these problems.

As a redundant manipulator has more joints than the degrees of freedom of the end-effector, the mapping from end-effector velocities to the joint velocities is not one-to-one. Hence, the solution for the joint velocities from end-effector velocities is not unique. In fact, an infinite number of solutions are possible. Therefore, additional constraints must be added to obtain an unique solution. Constraints which have been used in the past include minimum energy [12] [8], minimum torques [7], singularity avoidance [13], obstacle avoidance [10] and avoidance of the repeatability problem.[1][4]

Adding more joints to the manipulator has the advantage of flexibility in performing tasks, which means we have the freedom to choose a joint velocity among an infinite set of joint velocities. However, it does have the following drawbacks: the extra joints will increase the weight of the manipulator and the total energy consumption. The extra joints will introduce more round off errors and hence increase the position and orientation error. In addition, because of computational complexity, the controller for the redundant manipulator will be more costly. However, if the additional flexibility can effectively handle environmental constraints, these disadvantages can be offset.

The purpose of adding extra joints to the manipulator is to better handle those problems which the non-redundant manipulators usually have difficulty dealing with, such as singularities, joint position constraints and avoidance of obstacles. In order to solve these problems, Baillieul has expanded his extended Jacobian technique to the obstacle avoidance problem [2] [3]. An optimal joint velocity is obtained to avoid a single obstacle which is described as a cylinder, and the possibility of switching between different performance criteria in executing a single task has also been discussed.

Klein and Maciejewski [10] proposed an obstacle avoidance scheme which avoids moving objects in the working environment. However, this method relys on the accuracy of the sensors, which provide precise information regarding the minimum distance between the manipulator and the moving object, the nearest point on the manipulator to the moving obstacle and the near-

---

[1]Given a cyclic trajectory for the end-effector of a redundant manipulator, the joint trajectories may not be cyclic.

est point on the moving obstacle to the manipulator. The information processing time was not discussed. Nakamura [11] also proposed a similar scheme with additions of priorities associated with each type of constraint.

Our approach to the problem focuses on formulating the constraints and modeling the manipulator environment. The obstacles and the links of the manipulator are described as a union of convex polytopes, and the obstacles' trajectories are known. Also, three different types of constraint are unified into a single constraint, thus eliminating the priority problem.

In order to calculate the distance between the manipulator and objects in the working environment, some mathematical primitives are needed to describe the manipulator and obstacles. Gilbert and Johnson [6] provides a very general three dimensional model of objects by using unions of convex polytopes, and a fast *distance algorithm* for calculating the exact distance between two objects.

This paper is organized as follows. In section 2, the trajectory planning problem of the redundant manipulator is formulated and three constraints are imposed. In section 3, we unify the three different constraints into a single constraint. In section 4, an algorithm is proposed which ensures satisfaction of the constraint by utilizing the redundancy. A numerical example is presented in section 5 and conclusion and discussion are drawn in section 6.

2. Problem Formulation. In this section, we formulate the trajectory planning problem of the redundant manipulator which includes three different constraints: joint limit constraints, obstacle avoidance constraints and singularity avoidance constraints.

The manipulator forward kinematic equation can be expressed by the following discrete time function.

$$X(t_i) = f(q(t_i)), \quad t_i \in [t_0, t_1, \ldots, t_{K-1}, t_K] \tag{1}$$

where $X(t_i) \in R^m$ denotes the position and orientation vector of the manipulator's end-effector within the working space, and the $X(t_i)$ is given as part of the task information. t_i is time, f represents the known forward kinematic equation of the manipulator. $q(t_i) \in R^n$ denotes the vector of joint variables. For revolute joints, the joint variable is the joint angle, and for the prismatic joint, the variable is the joint distance. For the redundant manipulator, we have $n > m$.

The state equation of the manipulator can be obtained by linearizing equation (1):

$$\Delta X(t_{i+1}) = J(q(t_i))\Delta q(t_{i+1}) \tag{2}$$

where the $J(q(t_i)) \in R^{m \times n}$ is the Jacobian matrix, $\Delta X(t_{i+1}) = X(t_{i+1}) - X(t_i)$, and $\Delta q(t_{i+1}) = q(t_{i+1}) - q(t_i)$. The solution of the above equation is

$$\Delta q(t_{i+1}) = J^+ \Delta X(t_{i+1}) + (I - J^+ J)U(t_{i+1}) \tag{3}$$

where the $I \in R^{n \times n}$ is the identity matrix. $U(t_{i+1}) \in R^n$ is an unknown vector which is yet to be determined, and J^+ is the Moore-Penrose pseudoinverse of $J(q(t_i))$ [1].

The instantaneous redundant manipulator trajectory planning problem can be stated as follows. Given a trajectory of the end-effector, and a set of obstacles in the working space, find a joint space trajectory of the manipulator which minimizes a given performance index and satisfies all of the constraints. The performance index is to minimize the speed vector of the joint space.

We imposed three constraints in this paper: The joint position constraints, the singularity avoidance constraints and obstacle avoiding constraints. These constraints are respectively described by the following equations.

$$q^{min} \le q(t_i) \le q^{max} \tag{4}$$

$$| \triangle q(t_i) | \leq \triangle q^{max} \tag{5}$$

$$d_{ij}(q(t_i)) \geq d_{ij}^0 > 0, \quad t_i \in [t_0, t_1, \ldots, t_K], (i,j) \in I. \tag{6}$$

Equation (4) represents the joint limit constraints, where the q^{min} and q^{max} are respectively the minimum and maximum joint position vectors of the manipulator.

Because of the computational complexity of the singularity problem, we adopt an indirect approach to the problem for the real time computation. Since the manipulator will require a very large velocity at the neighborhood of singularity, we place constraints on the joint speed as in equation (5), where $\triangle q^{max}$ is the maximum change in joint position. However, since the maximum joint speed is dependent on the time, the configuration of the manipulator and the payload of the system, we take a conservative measure of the maximum speeds and set them as constants for simplicity.

The joint speed constraints are very restrictive in terms of possible joint speeds which the manipulator could generate. However, since equation (2) is a linearized equation of the nonlinear equation (1), the joint speed constraints will ensure the accuracy of the equation (2).

Obstacle avoidance constraints are imposed by equation (6) as described in [5]. Where $d_{ij}(q)$ is the minimum distance between a pair of potentially colliding parts of the manipulator and it's environment. d_{ij}^0 is the margin of error in avoiding the obstacles. $I \subset (1, \ldots, N_0)^2$ is the set of possible colliding pairs of parts, and N_0 is the number of parts. Throughout the context, the distance is defined as the Euclidean norm of two points in the space.

In the following section, we unify the above three different types of constraints into a single constraint equation. Hence there is no priority problem of selecting constraints. In section 4, we show how to utilize the redundancy to satisfy the constraint.

3. Unification Of Constraints. In this section, we transform the joint limit constraints, the obstacle avoidance constraints and the joint speed constraints into one unified constraint. By doing this, we eliminate the complicated equations of prioritizing different types of constraints.

The obstacle avoidance constraints of equation (6) can be expressed in more detail as follows. For each pair of possible colliding objects, once the distance between the pair is less than the given error margin d_{ij}^0, this constraint becomes active and the distance is expressed as the following.

$$d_{ij}(q(t_i)) = ((P_i - O_j)^T (P_i - O_j))^{1/2} \tag{7}$$

Where P_i is the nearest point of link i, and O_j is the nearest point of obstacle j, and (i,j) is a member of set I. In order to keep the link i away from obstacle j when the constraint is activated, we want the distance between the pair at next sampling period is greater than the threshold d_{ij}^0.

$$d_{ij}^0 \leq d_{ij}(q(t_{i+1})) = d_{ij}(q(t_i)) + \triangle d_{ij}(q(t_{i+1})) \tag{8}$$

In other words, we want

$$\triangle d_{ij}(q(t_{i+1})) \geq v_k \tag{9}$$

where

$$\triangle d_{ij}(q(t_{i+1})) \simeq \nabla_q d_{ij}(q(t_i)) \triangle q(t_{i+1}) \tag{10}$$

$$v_k = (d_{ij}^0 - d_{ij}(q(t_i))) \tag{11}$$

where ∇_q represents the gradient operator respect to vector q, and v_k is the distance that link i is moved away from obstacle j.

Rewrite equation (9), we have

$$g_k(t_i)^{min} \leq A_k(t_i) \triangle q(t_{i+1}) \leq g_k(t_i)^{max} \tag{12}$$

where

$$A_k(t_i) = \nabla_q d_{ij}(q(t_i))$$

$g_k(t_i)^{min} = v_k$, and $g_k(t_i)^{max}$ is set to ∞. The method for computing $\nabla_q d_{ij}(q(t_i))$ is given in appendix 2.

Following the same approach as the obstacle avoidance constraints, the joint limits and joint displacement limit constraints for each joint, equation (4), can be written in the same form as equation (12).

$$g_i(t_i)^{min1} \leq B_i(t_i)\Delta q(t_{i+1}) \leq g_i(t_i)^{max1} \tag{13}$$

where

$$g_i^{min1}(t_i) = q_i^{min} - q_i(t_i) \tag{14}$$

$$g_i^{max1}(t_i) = q_i^{max} - q_i(t_i) \tag{15}$$

and B_i is the i-th row of an identity matrix I, $1 \leq i \leq n$.

The inequality constraint of equation (5) can also be written as the same form as equation (12)

$$g_i(t_i)^{min2} \leq B_i(t_i)\Delta q(t_{i+1}) \leq g_i(t_i)^{max2} \tag{16}$$

where

$$g_i^{min2}(t_i) = -\Delta q_i^{max} \tag{17}$$

$$g_i^{max2}(t_i) = \Delta q_i^{max} \tag{18}$$

and B_i is the i-th row of an identity matrix I as defined before.

Note from above equations, we can combine equation (14) and (17), and equation (15) and (18) as :

$$g_i^{min}(t_i) = \max \left(g_i^{min1}(t_i) , g_i^{min2}(t_i) \right) \tag{19}$$

$$g_i^{max}(t_i) = \min \left(g_i^{max1}(t_i) , g_i^{max2}(t_i) \right) \tag{20}$$

Grouping the constraint equations (12), (19) and (20), we have

$$g^{min}(t_i) \leq C\Delta q(t_{i+1}) \leq g^{max}(t_i) \tag{21}$$

$$C = \begin{pmatrix} B \\ A \end{pmatrix}$$

where $g^{min}(t_i)$ and $g^{max}(t_i)$ are $n + k$ vectors, n is the number of joints and k is the number of active obstacle avoidance constraints, and $C \in R^{(n+k) \times n}$. From equation (3) and (21), all the constraints can be expressed by the following equation:

$$g^{min}(t_i) - CJ^+\Delta X(t_{i+1}) \leq C(I - J^+J)U(t_{i+1}) \leq g^{max}(t_i) - CJ^+\Delta X(t_{i+1}) \tag{22}$$

If any upper and lower limits of equation (22) intersects, the method fails and the manipulator is halted.

In this section, we unified three different types of constraints into one constraint equation. However, there are two problems unsolved. The first one is to find a suitable $U(t_{i+1})$ which satisfies the constraints defined in (22). The second one is to find an algorithm which is computational efficient in evaluating the active obstacle avoidance constraints defined in equation (6). In the following section, we will discuss our approaches to these two problems.

4. **Control Algorithm.** This section is divided into two parts. In the first part of this section, an approximate distance measure is proposed to reduce the computation time of calculating

the distance between each possible colliding pairs. In the second part, the extra joints of the manipulator are utilized to satisfy the constraint equation (22).

4.1 Approximated Distance Measure. The obstacle avoidance scheme is the most crucial part in our solution. The computation of the minimum distance (6) requires a considerable amount of time if there are several obstacles in the manipulator's working environment. For example, suppose there are three obstacles in the manipulators working environment and that the manipulator contains six links. At each sampling time, we must apply the *distance algorithm* eighteen times. This requires too much time for real time applications. For this reason, we use an approximate distance measure algorithm when the distances are large and the exact *distance algorithm* when the distances are small.

The method of utilizing the approximate distance measure is as follows. At each sampling time t_i, $t_0 \leq t_i \leq t_K$, we update the distance of each possible colliding pair of set I by using an approximate distance measure. If the result of the approximation for a possible colliding pair is less than the specified d_{ij}^0, the *distance algorithm* is invoked to calculate the exact distance d_{ij}. If d_{ij} is still less than d_{ij}^0, then the constraint associated with the pair becomes active and is included in equation (22). Otherwise, the constraint is not used in obtaining the next position of the manipulator.

The approximate distance measure can be stated as follows for determining the approximate distance between two convex objects, call them object A and object B. Object A will be part of an object in the manipulators working environment and object B will be part of one of the manipulator links. Recall that both objects and links are represented as the union of convex polytopes. At the beginning of motion, the nearest points between the objects and the vector between these points are determined using the exact *distance algorithm*. This vector and the nearest point on object A defines a plane called the separating plane, which is perpendicular to the vector and passing through the nearest point on the object. Since the objects are convex, object B will be entirely on one side of the plane and object A will be on the other side. It is assumed that object A is stationary so that if object B remains on the same side of this plan, then it is guaranteed not to collide with object A. The distance between the object B and the separating plane is easy to calculate, and we use this distance as the approximate distance measure.

At any step during the motion of the manipulator, if the approximate distance measure is less than d_{ij}^0, the *distance algorithm* is then again invoked to obtain a new separating plane between the objects, and a set of nearest points. These new information is then used to determine the successive approximate distance measure.

The following theorem shows that the approximate distance measure is always less than or equal to the actual distance.

Theorem 1 *The exact distance obtained from the distance algorithm will be greater or equal to the distance obtained by the approximate distance measure.*

The theorem is proved in the appendix 1.

4.2 Solution Algorithm. From the state equation (3) in section 2, the vector $(I - J^+ J)U(t_{i+1})$ is in the null space of the Jacobian matrix. Notice that this component of the joint displacement does not have any effect on the position of the manipulators end-effector. Hence, we have complete freedom in the choice of $U(t_{i+1})$ and can choose it such that the joint limit constraints given in equation (22) are satisfied.

In the constraint equation (22), there are k equations obtained from the obstacle avoidance constraints and n equations obtained from the joint and displacement (speed) limit constraints.

To simplify the solution processes, we separate the null space solution, $U(t_{i+1})$ into two parts,

$$U(t_{i+1}) = U1(t_{i+1}) + U2(t_{i+1}) \tag{23}$$

$U1(t_{i+1})$ is utilized to satisfy the absolute joint limits and the relative joint displacement (speed) limits and $U2(t_{i+1})$ is utilized to satisfy the most active obstacle avoidance constraint.

The joint displacement constraints are activated under the following conditions. We define a threshold g_k^{lim} for each joint displacement constraint. When joint k is within the range of g_k^{lim} of the displacement limit, the corresponding component of $U1(t_{i+1})$, $U1_k(t_{i+1})$, is activated to push the joint k out of g_k^{lim}. Otherwise, $U_k(t_{i+1})$ is set to zero.

Consider the first n constraint equations. The matrix C is reduced to a square identity matrix if the obstacle avoidance constraints are not present. We choose $U1$ such that if it were added to $J^+\Delta X(t_{i+1})$ the total joint displacement would satisfy the constraint equations. That is, we select $U1(t_{i+1})$ such that:

$$\hat{g}(t_i)^{min} \leq U1(t_{i+1}) \leq \hat{g}(t_i)^{max} \tag{24}$$

where

$$\hat{g}^{min}(t_i) = g^{min}(t_i) - J^+\Delta X(t_{i+1})$$

$$\hat{g}^{max}(t_i) = g^{max}(t_i) - J^+\Delta X(t_{i+1})$$

Define a function $h(\alpha, \beta)$, for $\beta \geq 0$ such that

$$h(\alpha, \beta) = \begin{cases} 0 & \text{if } |\alpha| \geq \beta \\ -(\alpha + \beta) & \text{if } |\alpha| < \beta \text{ and } \alpha < 0 \\ \beta - \alpha & \text{if } |\alpha| < \beta \text{ and } \alpha > 0 \end{cases} \tag{25}$$

Suppose $\hat{g}_i^{min}(t_i) < \hat{g}_i^{max}(t_i)$ for all i, $1 \leq i \leq n$. We let the $k - th$ component of $U1(t_i)$ be:

$$U1_k(t_{i+1}) = \begin{cases} h(\hat{g}_k^{max}(t_{i+1}), g_k^{lim}) & \text{if } h(\hat{g}_k^{max}(t_{i+1}), g_k^{lim}) > 0 \\ h(\hat{g}_k^{min}(t_{i+1}), g_k^{lim}) & \text{if } h(\hat{g}_k^{min}(t_{i+1}), g_k^{lim}) < 0 \\ 0 & \text{otherwise} \end{cases} \tag{26}$$

Since it is very time consuming to determine a $U2(t_{i+1})$ which satisfies all the k collision constraints, at each sample period we only use the constraint associated with two closest objects. That is, we choose the constraint with the largest current value of v_k, equation (11). We call this the most active obstacle avoidance constraint. For this constraint, v_k is the minimum distance the manipulator is required to move away from the object in order to satisfy the constraint equation. Thus, the approach here is to select $U2(t_{i+1})$ so the manipulator moves in v_k units along the direction vector, η, between the two nearest points between the manipulator and the object.

$$V_o = v_k\eta = S\Delta q(t_{i+1})$$

where

$$S = \frac{\partial P_i}{\partial q}$$

and P_i is the position of the nearest point on the manipulator. An algorithm for computing η and S is given in appendix 2. The value of $U2(t_{i+1})$ is obtained from:

$$V_o = S(J^+\Delta X(t_{i+1}) + (I - J^+J)U2(t_{i+1}))$$

Thus,

$$U2(t_{i+1}) = [S(I - J^+J)]^+(V_o - SJ^+\Delta X(t_{i+1})) \tag{27}$$

Adding this to $U1(t_{i+1})$ given above gives:

$$\Delta q(t_{i+1}) = J^+ \Delta X(t_{i+1}) + (I - J^+ J)(U1(t_{i+1}) + U2(t_{i+1})) \qquad (28)$$

In summary, the solution for the joint displacements are obtained as follows. At each time instance t_i, the controller will do the following.

1. Update the geometric information of the manipulator according to the calculated $q(t_i)$.

2. Check all the distances between links and obstacles using the approximate distance measure. If any distance is less than its given threshold, the *distance algorithm* is invoked to obtain the exact distance.

3. Formulate the constraint equation (22) and check the intersection of the upper and lower limits. If they intersect, then halt the program and report failure to obtain a solution.

4. Solve for $U1(t_{i+1})$ using equation (26).

5. Select the most active obstacle avoidance constraint and solve for $U2(t_{i+1})$ using equation (27).

6. Compute $\Delta q(t_{i+1})$ using equation (28).

7. if $t_i = t_K$, stop. Otherwise set $t_i = t_{i+1}$, and go to the step 1.

5. Numerical Example. This example is an illustration in the two dimensional space using a manipulator with six rotational joints. Geometrically, each link of the manipulator is described as a rectangle except the last link, which is a triangle. The kinematic parameters of the manipulator are given in table 1.

There are five obstacles in the working environment as shown in figure 1. The desired trajectory of the end-effector is shown as a dotted line.

Since the instantaneous obstacle avoiding scheme is a local scheme, the selection of the initial position of the manipulator is very important. The initial position and the joint limits of the manipulator are shown in table 2.

The solution for the joint space trajectories are shown in figures 2 through 7. These solutions utilize all of the joint constraints. For comparison, each figure also includes the minimum norm solution which does not consider any of the constraint equations.

In figure 8, the minimum distances obtained from the approximate distance algorithm are shown. In figure 9, the minimum distance obtained using the exact *distance algorithm* is shown for comparison. Notice the discontinuities in the figure. There are two reasons for them. First, when proximal links are closest to an object, a small displacement in the associated joints to avoid contact with the object produces a large displacements in the distal links. Note that the converse is not true. Second, in figure 8, the approximate distance measure is actually the distance between the link and the separating plane of the object. When this distance reaches d_{ij}^0, the exact *distance algorithm* is invoked to compute a new separating plane for use in computing the approximate distance. Hence, a dramatic change of the approximate distance occurs. For instance, around $i = 260$, the distance between link five and a separating plane of object zero is a continuous function of time. As the distance between link five and the separating plane reaches $d_{ij}^0 = 10$ centimeters, the exact distance is then calculated creating a discontinuity in the approximate distance measure.

The interaction between two possible colliding pairs also creates problems. For example, around $t = 200$ in Figures 10, link 3 and 4 are between objects 0 and 4. When the manipulator moves link 4 to avoid object 0, it moves link 3 in a direction closer to object 4.

The selection of d_{ij}^0 and g^{lim} is mainly determined by the following factors. First is the number of objects in the manipulator's working environment. If the manipulator is working in a congested space, the d_{ij}^0 and g^{lim} should be as small as possible. However, since the $U1$ and $U2$ are projected into the null space fo the Jacobian, the actual joint displacement may not satisfy the constraint equations. A small d_{ij}^0 and g^{lim} gives little Margin for error and increases the likelyhood for collision and violation of the joint position and displacement constraints.

As we mentioned in section 3, an approximate distance measure is employed if the distance between a possibly colliding pair is larger than a predefined threshold. Employing the approximate distance measure rather than always utilizing the exact *distance algorithm* reduces the trajectory planning time from 6 minutes and 4 seconds to 4 minutes and 23 seconds. For three dimensional problems, we expect the difference in trajectory planning time to be even greater. Both planning programs are run on a Silicon Graphics IRIS 2400 turbo workstation.

6. Conclusion. In this paper, a new and efficient obstacle avoidance algorithm was presented for local trajectory planning of redundant manipulators. Three types constraints were considered: joint limit constraints, singularity constraints and obstacle avoidance constraints. These constraints were then unified into a single constraint equation. The problem was reduced to finding the null space component of the joint displacement vector which ensures satisfaction of all motion constraints.

It is possible, in the initial state, for geometric information regarding the stationary objects and the manipulator to be retrieved from a data base. This would reduce the initialization time of the controller. Storing information in the data base will enable the manipulator to start at different initial configurations with the same set of stationary objects without a long set up time.

The bottle neck in the algorithm is the computational time required to update the distance of all the possible colliding pairs. If the number of the possible colliding pairs is too large, it is possible to use a more powerful computer such as a parallel processor. Since, each possible colliding pair can be updated independently, an N processor system could speed up the computation of distances by a factor of N.

For a manipulator which consist of the union of a large number of small convex polytopes, it is possible to further reduce the computation time by enclosing the link in a single imaginary polytope for use in the approximate distance algorithm. Future, work will be directed toward creating a hierarchical representation of the links. When the link is far away from the separating plane, a more "rough" representation will be used, when the link is closer, a more precise representation of the link will be used.

Acknowledgement This work was supported by the NASA sponsored Center for Autonomous and Man-Controlled Robotic and Sensing Systems at the Environmental Research Institute of Michigan, Ann Arbor, Michigan.

Appendix 1

Theorem 1 *The exact distance obtained from the distance algorithm will be greater or equal to the distance obtained by the approximate distance measure.*

The theorem can be proven as follows. From the Eidelheit separation theorem [9], let K_1 and K_2 be convex sets in a linear normed space X, such that K_1 has interior points and does not intersect K_2. Then there exist closed hyperplanes which separate K_1 and K_2.

Let us rewrite the distance function:

$$d(K_1, K_2) = \min |x_1 - x_2| \; ; x_1 \in K_1, x_2 \in K_2 \qquad (29)$$

and construct a hyperplane which separates K_1 and K_2 as follows.

$$H_1 = \{x : p^T(x - x_1) = 0\} \qquad (30)$$

where $p = x_2 - x_1$, and x_1 and x_2 are the points which minimize the above distance between K_1 and K_2. Clearly H_1 is a support hyperplane of K_1.

Our objective is to prove that:

$$d(H_1, K_2) \leq d(K_1, K_2) \qquad (31)$$

From the triangle inequality of a linear normed space

$$d(H_1, K_2) \leq d(K_1, K_2) + d(H_1, K_1) \qquad (32)$$

Since the supporting hyperplane intersects K_1, $d(H_1, K_1) = 0$, Hence,

$$d(H_1, K_2) \leq d(K_1, K_2) \qquad (33)$$

$\square$

Appendix 2. From [5], once the minimum distance is determined, the generalized gradient of $d_{ij}(q(t))$ does exist and can be expressed as:

$$\nabla_q d_{ij}(q(t)) = \eta \cdot S \qquad (34)$$

where

$$\eta = \frac{(P_i - O_j)^T}{d_{ij}(q(t))} \qquad (35)$$

P_i is the nearest point on the manipulator link and O_j is the nearest point on the object, and the $k - th$ column of S is $\acute{s}_k$, where

$$\acute{s}_k = \begin{cases} 0 & \text{if } k > i \\ Z_k \times (P_i - P_k^0) & \text{if } k \leq i \text{ and joint } k \text{ is rotational} \\ Z_k & \text{if } k \leq i \text{ and joint } k \text{ is translational} \end{cases} \qquad (36)$$

Z_k is the axis of rotation (or translation) of joint k, and P_k^0 is the position of any point along that axis.

References

[1] Ben-Israel A. and Greville T.N.E. 'Generalized Inverses: Theory and Application' John Wiley & Son, New York, 1980.

[2] Baillieul J. "Design of Kinematically Redundant Mechanisms" Proc. of 24th Conference on Decision and Control, Ft. Launderdale, Fl. December 1985, P18-21.

[3] Baillieul J. "Avoiding Obstacles and Resolving Kinematic Redundancy", Proc. of IEEE Int. Conf. on Robotics and Automation, San Francisco, CA. April 7-11 1986. P1698-1704.

[4] Dubey R. and Walker M.W. "Control Scheme for Redundant Manipulator" IEEE Proc. 17th Southeastern Symposium on System Theory. April 1985. P36-38.

[5] Gilbert E.G. and Johnson D.W. "Minimum Time Robot Path Planning in the Presence of Obstacles" Proc. of 24th Conference on Decision and Control, Ft. Launderdale, Fl. December, 1985. P1748-1753.

[6] Gilbert E.G., Johnson J.W. and Keerthi S.S. "A Fast Procedure for Computing the Distance Between Complex Objects in Three Space" Report of Center of Research on Integrated Manufacturing. RSD-TR-26-86. 1986.

[7] Hollerbach J.M. and Suh K.C. "Redundancy Resolution of Manipulator through Torque Optimization" Proc. IEEE Int. Conf. Robotics and Automation, St. Louis. March 25-28, 1985. P1016-1021.

[8] Klein C.A. and Huang C.H. "Review of Pseudoinverse Control for Use with Kinematically Redundant Manipulators" IEEE Trans. on System, Man and Cybernetics. Vol smc-13, No. 3, March/April 1983. P245-250.

[9] Luenberger David G. 'Optimization By Vector Space Method', John Wiley & Sons, Inc. New York. 1969.

[10] Maciejewski A.A. and Klein C.A. "Obstacle Avoidance for Kinematically Redundant Manipulators in Dynamically Varying Environments" The Int. Journal of Robotic Research, Vol 4, No. 3, Fall 1985. MIT. P109-117.

[11] Nakamura Y., Hanafusa H. and Yoshikawa T. "Task-Priority Based Redundancy Control of Robot Manipulators" The Int. Journal of Robotics Research, Vol. 6, No.2, Summer 1987. P3-15.

[12] Whitney D.E. "Resolved Motion Rate Control of Manipulators and Human Prostheses" IEEE Trans. Man-Machine System, Vol. MMS-10 1969 P47-53.

[13] Yoshikawa T. "Analysis and Control of Robotic Manipulator with Redundancy" Robotics Research, The First International Symposium, MIT press, Cambridge, MA. 1984.

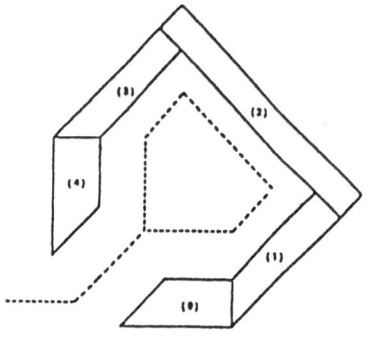

Figure 1: The Given path and obstacle

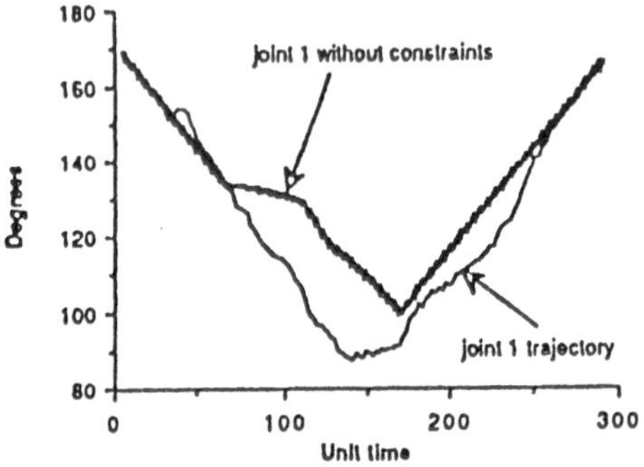

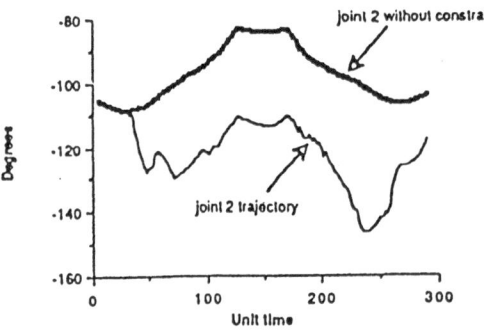

Figure 3: Joint two

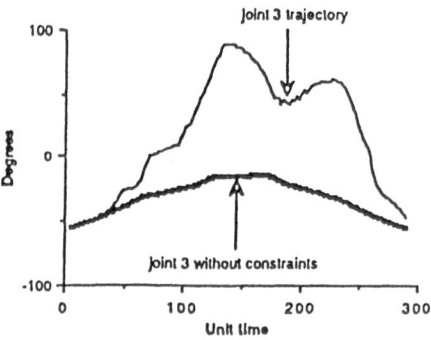

Figure 4: Joint three

Figure 2: Joint one

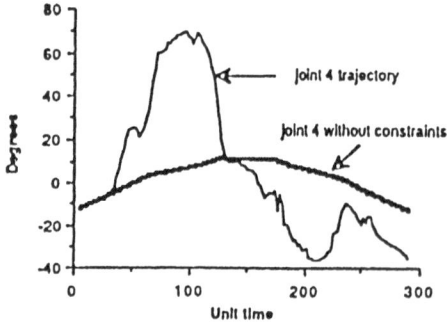

Figure 5: Joint four

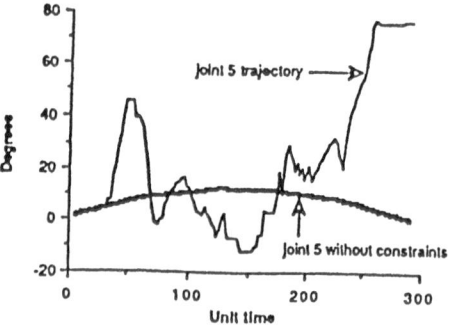

Figure 6: Joint five

	initial	min.	maz.
joint 0	166.16	-120.0	180.0
joint 1	-107.30	-150.0	120.0
joint 2	-60.0	-120.0	120.0
joint 3	15.0	-120.0	120.0
joint 4	1.15	-120.0	120.0
joint 5	0.57	-120.0	120.0

Table 2: Initial joint position and joint limits in degrees

link 1	1000 mm
link 2	750 mm
link 3	400 mm
link 4	300 mm
link 5	200 mm
link 6	100 mm

Table 1: Kinematic data

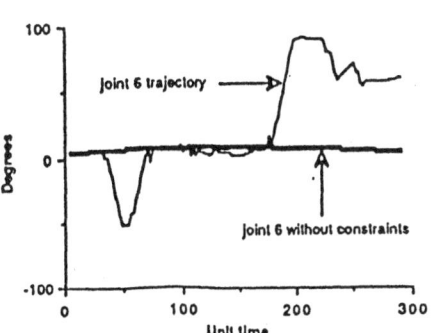

Figure 7: Joint six

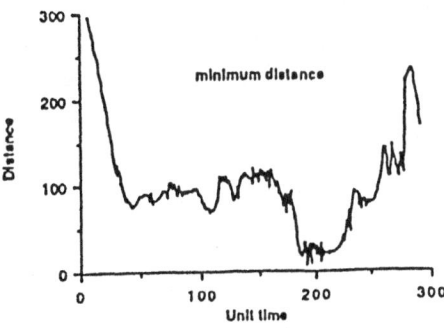

Figure 8: The approximate minimum distance

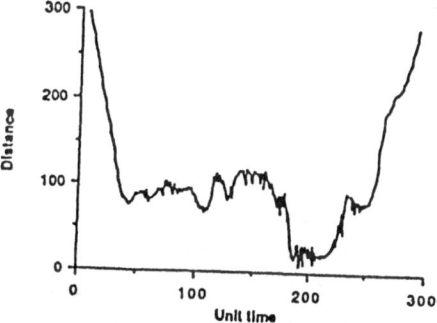

Figure 9: The exact minimum distance

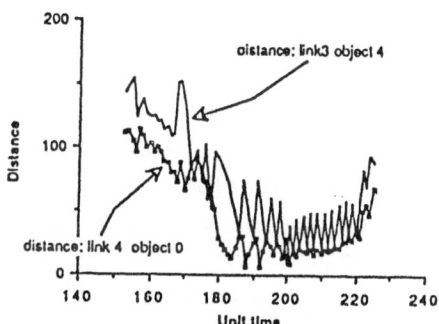

Figure 10: Distance between link 3 and object 4,
link 4 and object 0

COORDINATED CONTROL IN A MULTI-MANIPULATOR WORKCELL

Peng-Yung Woo   C. Nelson Dorny
Northern Illinois University   University of Pennsylvania

The Workcell and Its Function.   Efficient small quantity production requires a more flexible automatic workcell than is currently
available. We define a workcell as two or more manipulators which co-
operatively perform tasks. Each manipulator in the workcell specializes
in activities which are consistent with its capabilities, as in human
cooperation or in specialized operations by different machines in a
production line. Workcell functions which require cooperation include
the following: (1)   Handling of an object which is too large for a
single manipulator; (2)   Mating of complicated workpieces; (3)
Simultaneous operations, such as grinding, drilling, and polishing on
a single workpiece in order to increase production speed; (4)   Operations which require significantly different capabilities such as heavy
load lifting together with precise manipulation for assembly; (5)
Changing of the grasping position of an object by a succession of ex-
changes between two manipulators; and (6)   Cooperation of robotic fin-
gers in grasping and manipulation of an object. Although some of the
above tasks can be carried out by a single manipulator in cooperation
with passive fixtures, dexterous cooperation of two manipulators is more
effective than the use of a single manipulator.

Cooperative Control Structure. We propose a general control
structure for manipulator cooperation. The structure is based on the
following observations concerning practical operations in a workcell:
1. The purpose of the workcell is to provide fast, accurate operations
on a workpiece. Thus the focus of our attention is on the workpiece -
- the kinematic path of the workpiece, the operations on the workpiece,
and the limitations on that path and those operations owing to the
workcell environment. Therefore, job specification and path planning
for a particular operation include: specification, in world coordinates,
of the trajectory of the end-effector of the manipulator which holds the
primary workpiece; specification, in world coordinates, of the trajectories of the end-effectors of the manipulators holding the tools (or
mating workpieces) prior to their arrivals in the vicinity of the primary
workpiece; and specification in a workpiece coordinate frame of the
relative position trajectory and the interacting force trajectory between

each tool (or mating workpiece) and the primary workpiece after they approach mutual contact.

2. In carrying out a task, the end-effector of each manipulator must be carefully controlled during four distinct episodes of operation: Absolute position control in world coordinates to bring the end-effector of each manipulator to the vicinity of the primary workpiece; relative position control to produce acceptable mutual contact between the workpiece and the end-effector of each cooperating manipulator; "collision" management during the process of contact; and hybrid control of relative position and interacting force, in workpiece coordinates, after their mutual contact. The performance requirements necessary for a satisfactory job are different for each of these episodes.

3. The most important quantities to be controlled are usually the relative position (referred to as d) and the interacting force (referred to as F) between the workpiece and the end-effector of each cooperating manipulator during the physical operations on that workpiece.

4. The task-oriented input specifications mentioned in observation 1 must be converted to individual manipulator commands which take into account the differences in the capability limits of the various manipulators. The subsystem that carries out this task-assignment function can be viewed as a workcell coordinator. Then the set of manipulators should be feedback-controlled as a unit by a workcell controller. This coordination and control of the workcell as a single unit is a different concept than the conventional leader-follower approach.

Based on the four observations mentioned above, we propose the general control structure which is illustrated for a single pair of manipulators in Fig. 1. Manipulator A holds the workpiece and manipulator B the tool. The job-specification signal vectors X_{Ad} and d_d are the desired trajectory of the workpiece (in world coordinates) and the relative tool-to-workpiece position difference (in workpiece coordinates), respectively, as discussed in observations 1 and 3. The coordinator mentioned in observation 4 converts the job-specification signal vectors to separate command-signal vectors, labeled W_A and W_B (in units of acceleration), for the two manipulators. The nature of these command signals is described further below. The controller acts on the pair of command signal vectors W_A and W_B and on measurements of the absolute position and velocity of the workpiece, of the relative position of the tool, and of the absolute velocity of the tool, denoted X_A, $\dot{X}_A$, d, and $\dot{X}_B$, seeking to achieve $X_A = X_{Ad}$ and $d = d_d$. The feedback-controlled signals V_A and V_B (in units of acceleration) generated by the controller direct the actions of the two manipulators as described further below. Note that attention is focused on control of d in episode 2 and on hybrid control of d and F in episode 4. The hybrid control requires that a hybrid job-specification of d and F be provided and that both d and F be measured. We focus on episode 2 throughout the rest of the article.

Workcell Control System Design Procedure. The cooperative workcell control system which we propose applies both feedback and feedforward control in a systematic way which achieves user-specified workcell performance requirements. We first present a functional conceptualization of the components of the workcell, then summarize the design procedure in terms of those functional components, and finally illustrate the procedure via an example. The various components of the workcell can be viewed from inner loop to outer loop in the following way:

1. Each manipulator constitutes a dynamic nonlinear coupled relationship between the world-coordinate trajectory of the workpiece (X_A) or tool (X_B) and the vector U_A or U_B of electrical signals which are applied to the link actuators of the manipulator.

2. A nonlinear feedback system (proposed in [1]) is used to convert that relationship to a linear decoupled relationship between a new set of control signals V_A and V_B (in units of acceleration) and the same world-coordinate trajectories X_A and X_B. The ith component of the control signal V is the acceleration of the ith world-coordinate component of the corresponding trajectory X. Since $V = \ddot{X}$, the poles of the linearized and decoupled manipulator dynamics are all at the origin of the complex plane. At this level the manipulators are still controlled independently.

3. The controller treats the pair of manipulators as a single unit. The controller uses state-variable feedback from the sensors to convert the above linear output-decoupled relationship to a new linear relationship between the same world-coordinate trajectories X_A and X_B and a new set of command signals W_A and W_B (in units of acceleration). This new relationship matches a prototype pole-zero pattern which is chosen to meet certain of the performance requirements mentioned in observation 2. The new relationship couples the dynamics of the two manipulators, but maintains the decoupling of the world-coordinate dimensions.

4. The coordinator takes into account the prototype relationship generated by the controller and uses the remaining performance requirements of observation 2 as criteria for selection of feedforward command signals W_A and W_B for the individual manipulators so as to cause the system to approximate $X_A = X_{Ad}$ and $X_B - X_A = d_d$ satisfactorily.

The procedure for design of the cooperative control system can be summarized as follows:

1. Choose world-coordinate variables appropriate to the workcell and

establish static and dynamic requirements on the workcell behavior in terms of those variables.

2. Determine the nonlinear feedback (Tarn, et.al. [1]) necessary to convert the nonlinear coupled dynamics of each manipulator to dynamics in which the relations between the control inputs (V's) and the world-coordinate variables are linear and decoupled. Regroup the dynamic equations for the workcell into subsets each of which involve only variables from a single world-coordinate dimension rather than from a single manipulator. Each dimension of the workcell will be controlled independently.

3. Choose appropriate world-coordinate workcell state variables; for each world-coordinate dimension assume a linear state-variable feedback structure and determine the transform relationship (in terms of the undetermined feedback coefficients) between the components of the command inputs (W's) for that dimension and the world-coordinate outputs which are important to the workcell function for that dimension.

4. Use the workcell behavior requirements as criteria to choose a prototype form of world-coordinate response for each world-coordinate dimension of the workcell. That is, select an appropriate pole-zero pattern for the transform relationship of step (3). Different prototypes can be used for each dimension. Choose the values of the coefficients in the transform relationship to match the coefficients of the prototype response. This step amounts to design of the feedback control behavior of each dimension of the workcell.

5. Determine values for the state-variable feedback coefficients which correspond to the values of the coefficients of the transform relationship obtained in step 4 for each dimension.

6. For each world-coordinate dimension use the workcell behavior requirements to determine the components of the command inputs (W's) for that dimension which will produce the best behavior of the workcell in that dimension. This step constitutes design of feedforward control signals for each dimension.

We emphasize that the workcell performance requirements and prototype behavior which we suggest below are not unique. The design procedure can be used with other performance requirements and prototype behaviors. Furthermore, the prototype behavior and performance requirements can be different for different world-coordinate dimensions. For example, if we were to control relative position in one dimension and interacting force in another, we would probably not use the same performance requirements in both dimensions.

Example: Cooperating ℓ_1, ℓ_2 Manipulators. We demonstrate the procedure for design of a cooperative workcell control system by means of the pair of two-dimensional manipulators shown in Fig. 2. We proceed through the six design steps in the order specified above.

Step 1: We choose to control the workcell in the world-coordinate dimensions R and θ. We set the following requirements on the behavior of the workcell and apply the requirements in each of those two world-coordinate dimensions:

a. The response time delay should be the same for each manipulator and for each world coordinate.

b. The relative tool-to-workpiece error d should be essentially non-negative at all times, and the magnitude of the approach velocity d should be smaller than a specified value in order to avoid damaging collision on contact.

c. The shape of the desired trajectory X_A of the end of manipulator A (the workpiece trajectory) should have no effect on the shape of the trajectory d (= X_B - X_A) of the task performed on the workpiece.

d. The steady-state value of the error in d should be zero.

e. The error in the task d may be influenced by the initial value of the tool-to-workpiece distance d, but not by the initial value of the position X_A of the workpiece.

Step 2: The equations for the dynamics of a single θ_1, θ_2 manipulator are given in detail in [1]. We represent them together with the equation for the dynamics of the pair of actuators for the manipulator in the abstract notation of [2],

$$r = D\ddot{q} + p \tag{1}$$

$$\ddot{q} = a\dot{q} + fr + bU \tag{2}$$

where r is a 2x1 vector of joint torques, q is a 2x1 vector of joint angles (θ_1 and θ_2), D is a 2x2 inertia matrix, p is a 2x1 vector of centripetal and coriolis torques, U is a 2x1 vector of electrical actuator voltages, and a, b, and f are 2x2 diagonal matrices. The combination of (1) and (2) yields

$$(I - fD)\ddot{q} = a\dot{q} + fp + bU \tag{3}$$

The world-coordinate representation X of the tip of the manipulator is related to the joint-angle vector by the kinematic equation

$$X = h(q) \tag{4}$$

where h is a 2x1 vector function. In this example, X has as its components the world coordinates R and θ mentioned earlier. Each of the manipulators of Fig. 2 behave in the nonlinear coupled manner described by (3) and (4). We follow the approach of [1] and apply nonlinear feedback from measurements of X so as to make each manipulator behave like

$$\ddot{X} = V \tag{5}$$

where V is a 2x1 input vector under our control. We choose the nonlinear feedback in the form

$$U = \alpha(q) + \beta(q) V \tag{6}$$

where U is the 2x1 vector of electrical control voltages applied to the link actuators. We must determine appropiate 2x1 nonlinear functions $\alpha(q)$ and $\beta(q)$. Insertion of (5) into (6) and use of the derivatives $\dot{X} = J \dot{q}$ and $\ddot{X} = J \ddot{q} + \dot{J} \dot{q}$ of (4) produces

$$u = \alpha(q) + \beta(q)(J \ddot{q} + \dot{J} \dot{q}) \tag{7}$$

where J is the Jacobian matrix of h. We insert (7) into (3) to obtain

$$[fD - I + b \beta(q) J]\ddot{q} + [a\dot{q} + fp + b \alpha(q) + b \beta(q)\dot{J}\dot{q}] = 0 \tag{8}$$

From (8) we see that the desired behavior (5) is obtained if we choose

$$\alpha(q) = -b^{-1}[a\dot{q} + fp + (I - fD) J^{-1} \dot{J} \dot{q}] \tag{9}$$

$$\beta(q) = b^{-1}(I - f D) J^{-1} \tag{10}$$

After the nonlinear feedback is applied to each manipulator, the form of the dynamics of manipulator A is of the simple open-loop form

$$\begin{bmatrix} \ddot{R}_A \\ \ddot{\theta}_A \end{bmatrix} = \begin{bmatrix} V_{AR} \\ V_{A\theta} \end{bmatrix} \tag{11}$$

Manipulator B has a similar form. We regroup the variables according to world-coordinate dimension rather than according to manipulator. For the radial dimension the grouping is

$$\begin{bmatrix} \ddot{R}_A \\ \ddot{R}_B \end{bmatrix} = \begin{bmatrix} V_{AR} \\ V_{BR} \end{bmatrix} \tag{12}$$

These equations determine the state trajectories of the two manipulators in the R world-coordinate dimension. A similar grouping and determination of state trajectories exists for the θ dimension.

Step 3: For the radial dimension, state-variable feedback is applied in the form

$$\begin{bmatrix} V_{AR} \\ V_{BR} \end{bmatrix} = \begin{bmatrix} W_{AR} \\ W_{BR} \end{bmatrix} + K \begin{bmatrix} R_A \\ \dot{R}_A \\ d_R \\ \dot{R}_B \end{bmatrix} \tag{13}$$

where K is a 2x4 feedback coefficient matrix which must be determined so as to provide prototype behavior as discussed below. We have chosen d_R, rather than R_B, as one of the state variables because it is the primary quantity of concern in performance of the workcell function; it should be measured directly in order that it be as accurate as possible. Measurements of the other state variables, R_A, $\dot{R}_A$, and $\dot{R}_B$, are easily computed with sufficient accuracy from measurements of the joint angles. The variables W_{AR} and W_{BR} are the radial-coordinate feedforward signals which are also designed below. Similar state-variable feedback is applied in the θ dimension.

For the state equations (12) and the state-variable feedback specified in (13) the transform relationship between the radial components of the command signal vector W and the output state vector $(R_A, d_R)^t$ is

$$
\begin{bmatrix} R_A \\ \\ d_R \end{bmatrix} = \begin{bmatrix} s^2 - K_{24}s + K_{23} & K_{14}s - K_{13} \\ \\ s^2 - (K_{24} + K_{22})s - K_{21} & -s^2 + (K_{12} + K_{14})s + K_{11} \end{bmatrix} \begin{bmatrix} W_{AR} \\ \\ W_{BR} \end{bmatrix}
$$

$$
+ \begin{bmatrix} R_A(0)\,[\,s^3 - (K_{12} + K_{24})s^2 + (K_{12}K_{24} - K_{14}K_{22} - K_{13} + K_{23})s \\ \quad\quad + (K_{22}K_{13} - K_{23}K_{14} + K_{13}K_{24} - K_{23}K_{12})\,] \\ \\ R_A(0)\,[\,(K_{11} - K_{21})s + (K_{21}K_{14} + K_{21}K_{12} - K_{24}K_{11} - K_{11}K_{22})\,] \\ \quad + d_R(0)\,[\,s^3 - (K_{12} + K_{24})s^2 + (K_{12}K_{24} - K_{11} - K_{22}K_{14})s \\ \quad\quad\quad + (K_{11}K_{24} - K_{21}K_{14})\,] \end{bmatrix} \quad (14)
$$

where $\Delta = s^4 - (K_{12} + K_{24})s^3 + (K_{12}K_{24} - K_{11} - K_{14}K_{22} - K_{13} + K_{23})s^2$
$\quad\quad + (K_{22}K_{13} - K_{14}K_{23} + K_{13}K_{24} - K_{23}K_{12} + K_{11}K_{24} - K_{21}K_{14})s$
$\quad\quad + (K_{21}K_{13} - K_{11}K_{23})$

and we have assumed that the initial velocities of the manipulators are zero.

Step 4: For successful coordination, according to requirement (a), it is the relative timing of the various actions rather than the amount of the time delay, which is important to the function of the workcell. Pure time delay of the whole workcell would not hamper the workcell functions. Control of the two manipulators as a single unit in each decoupled world-coordinate dimension will result in the delays of all variables associated with that dimension having the same time delay as specified in requirement (a). The transfer function of a system with pure time delay of magnitude t_0 is of the form

$\quad H(j\omega) = e^{-j\omega t}$ \hfill (15)

The magnitude of (14) is independent of frequency and its phase lags in direct proportion to frequency. According to the theory of analog filters, for a fixed number of poles the transfer function of a Bessel filter provides the best approximation to linear phase lag ([3], [4]). Furthermore, a Bessel filter of order higher than two will respond more quickly than a critically-damped second-order system and and yet exhibit essentially no overshoot. The coefficients of the characteristic equation for a Bessel filter of given order are the known coefficients of the Bessel polynomial of the same order. We choose the coefficients of the characteristic polynomial Δ of (14) to match the coefficients of the Bessel polynomial in order to achieve maximally linear phase. For this example we choose as the values of the four coefficients of Δ the Bessel coefficients -15.62, 109.75, 400, and 625. These Bessel coefficients are based on selection of a time delay of $t_0 = 0.64$ second.

Matching of the four coefficients of the characteristic polynomial Δ of (14) to the four Bessel coefficients selected in step (4) produces four equations among the eight unknown feedback coefficients. The use of the Bessel polynomial as the prototype characteristic polynomial would also assure that the overshoot in d_R would be essentially nonnegative as specified in requirement (b) if the numerator of the transient (or initial condition) term for d_R in (14) were a constant. Fortunately, the feedback coefficients which produce the Bessel polynomial also cause the coefficients of the non-constant terms of the numerator to be small relative to the constant term, and that portion of requirement (b) is satisfied.

According to requirement (e), we want the transient portion of d_R to be independent of the initial value $R_A(0)$. According to (14), the $R_A(0)$ term of d_R will be zero, and this requirement will be satisfied if we choose to make the following four additional relations among the feedback coefficients,

$$K_{11} = K_{21}, \qquad K_{12} = K_{24}, \qquad K_{14} = K_{22}, \qquad K_{13} = 0 \qquad (16)$$

Step 5: A solution to the eight equations in the eight feedback coefficients is $K_{11} = K_{21} = -22.23$, $K_{12} = K_{24} = -7.81$, $K_{14} = K_{22} = -1.26$, $K_{13} = 0$, and $K_{23} = 28.12$. For this choice of feedback coefficients the transform relation (14) becomes

$$\begin{bmatrix} R_A \\ d_R \end{bmatrix} = (1/\Delta) \begin{bmatrix} s^2 + 7.81 \, s + 28.12 & -1.26 \, s \\ s^2 + 9.07 \, s + 22.23 & -s^2 -9.07 \, s -22.23 \end{bmatrix} \begin{bmatrix} W_{AR} \\ W_{BR} \end{bmatrix}$$

$$+ (1/\Delta) \begin{bmatrix} R_A(0) \, [s^3 + 15.62 \, s^2 + 87.53 \, s + 255.05] \\ d_R(0) \, [s^3 + 15.62 \, s^2 + 84.82 \, s + 145.6] \end{bmatrix} \qquad (17)$$

where $\Delta = s^4 + 15.62s^3 + 109.75s^2 + 400s + 625$. The same choice of feedback coefficients is used for control in the θ dimension.

Step 6: We now choose coordinator output signals W_{AR} and W_{BR} which will cause the system to satisfy requirements (c) and (d) in the radial world-coordinate dimension. The steady-state requirement (d) is satisfied automatically by (17) if the coordinator outputs themselves approach a static condition. According to requirement (c), the task behavior d_R must not depend on the trajectory R_{Ad} selected for the workpiece. We satisfy this requirement by selecting W_{AR} and W_{BR} so that their linear combination as shown in (17) is a function only of the desired trajectory d_{Rd}. Suppose that we pick

$$W_{AR} = \ddot{R}_{Ad} + p_1\dot{R}_{Ad} + p_0 R_{Ad} \tag{18}$$

$$W_{BR} = W_{AR} + (\ddot{d}_{Rd} + r_1\dot{d}_{Rd} + r_0 d_{Rd})$$

$$= R_{Bd} + r_1\ddot{R}_{Bd} + r_0\dot{R}_{Bd} + (p_1 - r_1)\dot{R}_{Ad} + (p_0 - r_0)R_{Ad} \tag{19}$$

Then by appropriate choice of the coefficients in (18) and (19) we can
adjust the zeros in (17) so as to cancel the poles. Specifically, if
$p_1 = 9.07$, $p_0 = 22.23$, $r_1 = 6.55$, and $r_0 = 28.11$, we precisely cancel the
characteristic polynomial in the driven parts of both trajectories of
(17), and obtain the response

$$\begin{bmatrix} R_A \\ d_R \end{bmatrix} = \begin{bmatrix} R_{Ad} \\ d_{Rd} \end{bmatrix} + (1/\Delta) \begin{bmatrix} -1.26s(s^2+6.55s+28.11)d_{Rd} \\ +R_A(0)(s^3+15.6s^2+87.5s+255) \\ d_R(0)(s^3+15.6s^2+84.8s+146) \end{bmatrix} \tag{20}$$

Hence, if the desired trajectories are sufficiently smooth that the
feedforward signals chosen in (18) and (19) do not cause saturation of
the actuators, the only errors incurred by the control will be those
associated with the initial errors, measurement errors, and imperfections
in the linearization and decomposition. A similar set of coordinator
signals is selected for the θ dimension.

The above choice of coordinator signals does not address the need to
limit the collision velocity at contact as specified in requirement (c).
According to (20), the collision can be limited by limiting the initial
tool-to-workpiece distance $d_R(0)$ by a more slow and careful episode 1.
Alternatively, the approach of the tool to the workpiece in episode 2
can be slowed by use of a longer delay t_0 in the Bessel prototype. It
is intuitively clear that such a slowing will lead to lower velocities
throughout the episode.

<u>Simulation Of Manipulator Cooperation</u>.

A computer simulation of the specific control scheme described above
for the θ_1, θ_2 robot of Fig. 2 is shown in Figs. 3 and 4. The manipulator
links are uniform rods of length 24 in and mass 0.25 oz s^2/in. The
actuators each have inertia 0.025 oz in s^2 and torque constant 16 oz
in/v. The curves in the figures verify satisfaction of the specified
requirements for perfect decoupling. They also demonstrate the degrad-
ed performance which results from a 20% error in the torque constants
used in the model. This error arises primarily from imperfect output
decoupling of the manipulators. A similar control system design and
simulation has been carried out for the θ dimension [5]. The controls
are essentially independent even if the decoupling is not perfect.

The sensitivity of the decoupling process to parameter errors can be
compensated for by on-line calibration of the parameters of the model.
This calibration can be accomplished in either of two ways, by
measurement of critical parameters (actuator constant and load) or by

adjusting the parameters in the model in order to improve observed performance. Efficiency of the latter method requires knowledge of the sensitivity of the actuator torques to errors in the model parameters. We have developed an analytical procedure for estimating the sensitivity of such output-decoupled systems to errors in the parameters and errors in the measurements. Simulation of hybrid position-force control has also been carried out using this control scheme [5].

Summary: We have established an intuitively sound description of a workcell consisting of manipulators cooperating on tasks. The focus of attention is on the interaction of the tool with the workpiece. The critical quantities to be controlled are the differential distance and differential approach velocity. After contact, interacting force is also important. We present a unified control structure appropriate for manipulator cooperation, and present a systematic design procedure consistent that control structure. The design procedure applies unchanged to force control. We suggest use of the Bessel prototype as a generalization of critical damping for pole placement in such high-order systems.

References.

[1] H. Asada and J.-J. E. Slotine, *Robot Analysis and Control*, New York: John Wiley, 1986.

[2] T. J. Tarn, A. K. Bejczy, and X. Yun, "Coordinated Control of Two Robot Arms," Int. Conf. Robotics Auto., San Francisco, CA, pp. 1375-1380, April 1986.

[3] Harry Y-F. Lam, *Analog and Digital Filters*, Englewood Cliffs, NJ: Prentice-Hall, 1979.

[4] L. P. Huelsman, and P. E. Allen, *Introduction to the Theory and Design of Active Filters*, New York: McGraw-Hill, 1980.

[5] Woo, Peng-Yung, "Coordinated Control in a Multi-Manipulator Workcell," Ph.D. Dissertation, University of Pennsylvania, Philadelphia, August 1988.

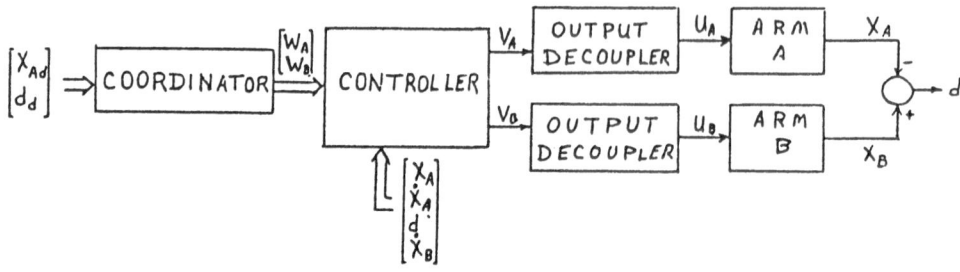

Fig. 1   Cooperative Position Control Structure

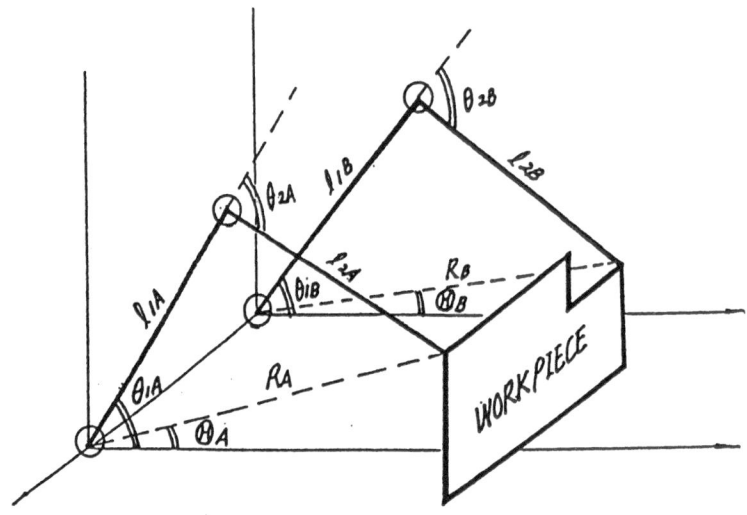

Fig. 2   Cooperating Manipulators

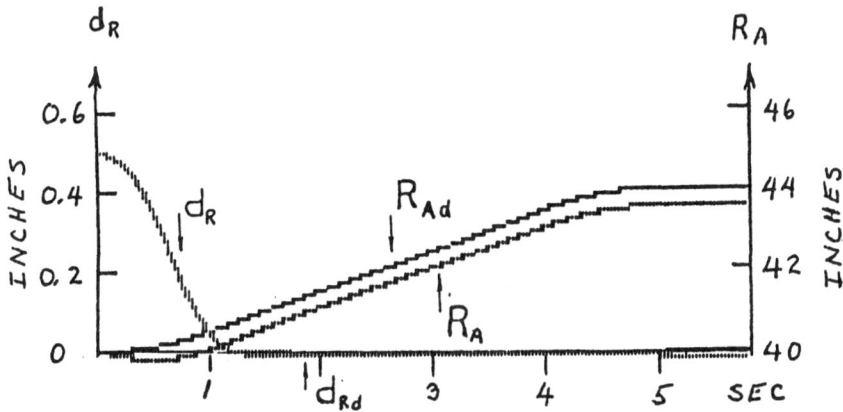

Fig. 3   Position Control With Perfect Decoupling

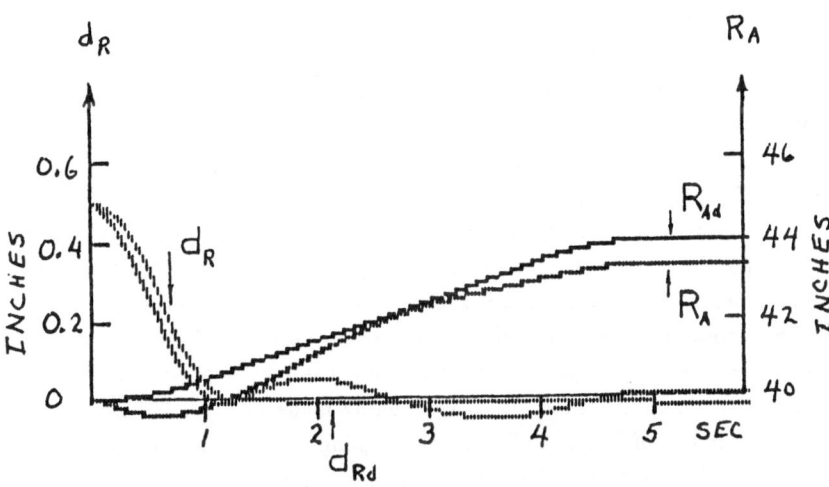

Fig. 4   Position Control With 20% Error In Torques

DIGITAL CONTROL OF LINEARIZABLE SYSTEMS

Hong-Gi Lee,
Louisiana State University

Aristotle Arapostathis and Steven I. Marcus
University of Texas at Austin

1. Introduction. The increasing use of robotics and other high technology systems (for example, aircraft) makes necessary a sustained research effort in the application of modern control principles. Advanced strategies are necessary to satisfy the strict requirements imposed by the operation of such complex nonlinear systems. Due to this, nonlinear system theory has been rapidly evolving over the past twenty years. The problem of feedback linearization of control systems has attracted a great deal of attention in the literature and consists of determining a state-feedback law as well as new coordinates in the state space such that the resulting closed-loop system becomes a linear one. The motivation for this approach is that, if a nonlinear system is equivalent to a linear system, then linear control techniques can be used to control it. Many authors have studied this linearization problem (Su [11], Jakubczyk and Respondek [6], Hunt et.al. [4], Grizzle [2], Lee and Marcus [10], Lee et.al. [8], and Jakubczyk [5], etc.).

Feedback linearization is a very effective control technique, when applicable. For example, T.J. Tarn et.al. have applied this technique to control a single-arm robot [12] and two cooperating robot arms [13] and have shown its effectiveness. Their technique is very similar to that which stems from Freund [1]. However, a digital feedback law is inevitable in practice, and discretization can destroy linearizability [2,3]. This is because the control input should be constant between the sampling times. The method which is usually employed is that of neglecting the error due to sampling by increasing the sampling frequency. However, increasing the sampling frequency is not always possible or desirable. Furthermore, it is shown in the next section that this method is equivalent to the classical Euler method which is known to be lacking in accuracy. Some preliminary work on the effect of sampling on linearizability has been reported in [2,7,9] and much work remains to be done. An effort has been recently made by J.W. Grizzle and P.V. Kokotovic [3] to avoid this loss of linearizability under sampling, by using multi-rate sampling. They show that sampling the input signal faster than the state can preserve linearizability.

2. Main Results. Consider a nonlinear continuous time control system of the form

$$\dot{x}(t) = F(x(t)) + G(x(t))u(t), \tag{1}$$

where $x \in \Re^n$, $u \in \Re$, and $F(x)$ and $G(x)$ are smooth vector fields which are analytic at $x = 0$ with $F(0) = 0$.

It is well known that if the system (1) is feedback linearizable, then there exist new coordinates $z = T(x)$ such that

$$\dot{z}_k(t) = z_{k+1}(t), \qquad 1 \le k \le n-1$$
$$\dot{z}_n(t) = c(z(t)) + d(z(t))u(t). \tag{2}$$

If we sample the feedback which linearizes (2), we obtain

$$u(t) = \frac{1}{d(T(x^k))}(-c(T(x^k)) + v^k), \qquad kh \le t < (k+1)h,$$

where $x^k = x(kh)$ and v^k is the new digital input. Thus, the resulting closed-loop system is

$$\dot{z}_k(t) = z_{k+1}(t), \qquad 1 \le k \le n-1$$
$$\dot{z}_n(t) = c(z(t)) - \frac{d(z(t))}{d(z^k)}c(z^k) + \frac{d(z(t))}{d(z^k)}v^k, \tag{3}$$

where $z^k = z(kh)$. The resulting system is only approximately linear; it contains higher order terms with respect to the sampling interval h in its Taylor series expansion, i.e.,

$$z^{k+1} = A_h z^k + b_h v^k + 0(h^{\rho+1}), \tag{4}$$

where

$$A_h = \begin{bmatrix} 1 & h & \frac{h^2}{2!} & \cdots & \frac{h^{n-2}}{(n-2)!} & \frac{h^{n-1}}{(n-1)!} \\ 0 & 1 & h & \cdots & \frac{h^{n-3}}{(n-3)!} & \frac{h^{n-2}}{(n-2)!} \\ \vdots & \vdots & \vdots & \ddots & \vdots & \vdots \\ 0 & 0 & 0 & \cdots & 1 & h \\ 0 & 0 & 0 & \cdots & 0 & 1 \end{bmatrix}, \quad b_h = \begin{bmatrix} \frac{h^n}{n!} \\ \frac{h^{n-1}}{(n-1)!} \\ \vdots \\ \frac{h^2}{2} \\ h \end{bmatrix}$$

and $\rho = 1$. This technique is in essence the classical Euler integration method. In many cases, the Euler method is not satisfactorily accurate to integrate a differential equation. Most software packages use the 4th order Runge-Kutta method (i.e., $\rho = 4$) or more accurate ones. Hence we propose to develop a new digital control technique which approximately linearizes system (1) (by using state coordinate change and digital feedback) up to a higher order term in the sampling interval h. That is, we consider the following question:

Question: Do there exist a state coordinate change $z^k = T_h(x^k)$ and feedback $u(t) = \gamma_h(x^k, v^k)$ for $kh \le t < (k+1)h$ such that $T_0(x)$ is invertible and, for some $\delta > 0$ and positive integer ρ, $z^{k+1} = A_h z^k + b_h v^k + 0(h^{\rho+1})$ for $h \in (0, \delta)$.

A preliminary result on this technique is the following:

Theorem: The answer to the above question is "yes" with $\rho = 4$ if and only if:

(i) the system (1) is feedback linearizable.

(ii) $ad_G ad_F G \in sp\{G\}$.

(iii) $L_{ad_F^j G} L_G L_F^n(\varphi) + L_{ad_F^j G} L_F L_G L_F^{n-1}(\varphi) = 0$, $0 \le j \le n-3$ for any $\varphi(x)$ which satisfies $L_{ad_F^j G}(\varphi) = 0$, $0 \le j \le n-2$, and $L_{ad_F^{n-1} G}(\varphi) \ne 0$.

Proof: (Necessity) Let z_i, $i = 1, \ldots, n$, denote the new coordinates. Expanding their Taylor series in h we represent them by $z_i(x) = \sum_{q=0}^{\infty} h^q \varphi_i^q(x)$, where φ_i^q are smooth functions. For a smooth function α we define $\langle L_G^0, L_F^i \rangle(\alpha) = L_F^i(\alpha)$, $i = 0, 1, \ldots$, and by induction, for $k \ge 0$,

$$\langle L_G^{k+1}, L_F^i \rangle = \sum_{j=0}^{i} L_F^j L_G \langle L_G^k, L_F^{i-j} \rangle(\alpha).$$

Consider the identity

$$z_i \circ \Phi_h^{F+uG} = \sum_{q=0}^{\infty} \sum_{r=0}^{q} \sum_{j=0}^{q} \frac{h^q}{(q-j)!} u^r \left\langle L_G^r, L_F^{q-j-r} \right\rangle (\varphi_i^j).$$

Since $z_n \circ \Phi_h^{F+uG} = z_n + hv + O(h^5)$, we obtain

$$v = \sum_{q=1}^{4} \sum_{r=0}^{q} \sum_{j=0}^{q-1} \frac{h^{q-1}}{(q-j)!} u^r \left\langle L_G^r, L_F^{q-j-r} \right\rangle (\varphi_n^j) + O(h^4) \tag{5}$$

and equating the coefficients of powers of h yields $L_G(\varphi_n^0) \neq 0$. Since $z_{n-1} \circ \Phi_h^{F+uG} = z_{n-1} + hz_n + \frac{h^2}{2!} v + O(h^5)$,

$$O(h^5) = \sum_{q=1}^{4} \sum_{r=0}^{q} \sum_{j=0}^{q-1} \frac{h^q}{(q-j)!} u^r \left\langle L_G^r, L_F^{q-j-r} \right\rangle (\varphi_{n-i}^j)$$

$$- \sum_{q=0}^{3} h^{q+1} \varphi_n^q - \sum_{q=1}^{3} \sum_{r=0}^{q} \sum_{j=0}^{q-1} \frac{h^{q+1}}{(q-j)!2!} u^r \left\langle L_G^r, L_F^{q-j-r} \right\rangle (\varphi_n^j)$$

which implies that

$$\varphi_n^0 = L_F(\varphi_{n-1}^0), \qquad L_G(\varphi_{n-1}^0) = 0 \tag{6a}$$

$$\varphi_n^1 = L_F(\varphi_{n-1}^1), \qquad L_G(\varphi_{n-1}^1) = 0 \tag{6b}$$

$$\varphi_n^2 = L_F(\varphi_{n-1}^2) - \tfrac{1}{12} L_F^2(\varphi_n^0) \tag{6c}$$

$$L_G(\varphi_{n-1}^2) = \tfrac{1}{12} \langle L_G, L_F \rangle (\varphi_n^0) \tag{6d}$$

$$L_G^2(\varphi_n^0) = 0 \tag{6e}$$

$$\varphi_n^3 = L_F(\varphi_{n-1}^3) - \tfrac{1}{12} L_F^2(\varphi_n^1) \tag{6f}$$

$$L_G(\varphi_{n-1}^3) = \tfrac{1}{12} \langle L_G, L_F \rangle (\varphi_n^1) \tag{6g}$$

Using the same procedure, we obtain, for $1 \leq i \leq n-2$,

$$\varphi_{i+1}^j = L_F(\varphi_i^j), \qquad L_G(\varphi_i^j) = 0, \qquad 0 \leq j \leq 3 \tag{7}$$

From (6a) and (7)

$$L_G L_F^j(\varphi_1^0) = 0, \qquad 0 \leq j \leq n-2$$
$$L_G L_F^{n-1}(\varphi_1^0) \neq 0. \tag{8}$$

Since $T_0(x)$ is invertible, it is easy to see that system (1) is feedback linearizable [11]. Also, equation (6e) implies that

$$ad_G ad_F^{k-1} G \in sp\{G, ad_F G, \ldots, ad_F^{k-2} G\}, \qquad 2 \leq k \leq n. \tag{9}$$

From (6d) and (7),

$$L_{ad_F^j G}(\varphi_1^2) = 0, \qquad 0 \leq j \leq n-3$$
$$L_{ad_F^{n-2} G}(\varphi_1^2) = \tfrac{1}{12} \{ L_G L_F^n(\varphi_1^0) + L_F L_G L_F^{n-1}(\varphi_1^0) \}, \tag{10}$$

where φ_1^0 satisfies (8). For the existence of φ_1^2 which satisfies (10) it is necessary and sufficient that for $0 \leq j \leq n-3$

$$L_{ad_F^j G} L_G L_F^n(\varphi_1^0) + L_{ad_F^j G} L_F L_G L_F^{n-1}(\varphi_1^0) = 0. \tag{11}$$

(Sufficiency): Suppose that conditions (i)-(iii) hold. Then we can find φ_i^0, φ_i^1 and φ_i^2 for $1 \leq i \leq n$ which satisfy (6) and (7). Let $z_i(x) = \sum_{q=0}^{2} h^q \varphi_i^q(x)$. Hence equation (4) holds if we apply any feedback law which satisfies (5). (Q.E.D.)

Remark: When $n = 2$, condition (iii) is redundant.

Corollary 1: The answer to the above question is "yes" with $\rho = 3$ if and only if the answer to the question is "yes" with $\rho = 4$.

Corollary 2: The answer to the above question is "yes" with $\rho = 1$ and $\rho = 2$ if and only if system (1) is feedback linearizable.

Example: Consider

$$\dot{x} = \begin{pmatrix} x_2 \\ x_1 + x_2^2 \end{pmatrix} + \begin{pmatrix} 0 \\ 1 \end{pmatrix} u = F(x) + G(x)u. \tag{12}$$

Then $ad_F G = \begin{pmatrix} -1 \\ -2x_2 \end{pmatrix}$ and $ad_G ad_F G = \begin{pmatrix} 0 \\ -2 \end{pmatrix} \in sp\{G\}$. Therefore, the hypotheses of the Theorem are satisfied. The corresponding state coordinates and feedback law can be found as follows: Since $L_G(\varphi_2^0) \neq 0$, then from (6a) $\varphi_1^0 = x_1$ and $\varphi_2^0 = L_F(\varphi_1^0) = x_2$. Utilizing (6d) we obtain $\varphi_1^2 = \frac{1}{12}x_2^2$, while from (6c) $\varphi_2^2 = -\frac{1}{12}x_2$. Note that $\varphi_1^1 = \varphi_2^1 = \varphi_1^3 = \varphi_2^3 = 0$ satisfy (6b), (6f), and (6g). Hence $z_1 = x_1 + \frac{h^2}{12}x_2^2$ and $z_2 = x_2 - \frac{h^2}{12}x_2$. Finally, a feedback law can be obtained from (5) as a solution of

$$v = x_1 + x_2^2 + u + \frac{h}{2!}\{x_2(1 + 2x_1 + 2x_2^2) + 2x_2 u\}$$

$$+\frac{h^2}{3!}\{2x_2^2 + (x_1 + x_2^2)(\tfrac{1}{2} + 2x_1 + 6x_2^2) + (\tfrac{1}{2} + 4x_1 + 8x_2^2)u + 2u^2\}.$$

Example: Consider

$$\dot{x} = \begin{pmatrix} x_2 \\ 0 \end{pmatrix} + \begin{pmatrix} 0 \\ 1 + x_2^2 \end{pmatrix} u = F(x) + G(x)u. \tag{13}$$

Then $ad_F G = \begin{pmatrix} -(1 + x_2^2) \\ 0 \end{pmatrix}$ and $ad_G ad_F G = \begin{pmatrix} -2x_2(1 + x_2^2) \\ 0 \end{pmatrix} \notin sp\{G\}$. Thus, the hypotheses of the Theorem are not satisfied. However, by Corollary 2, system (13) can be controlled with $\rho = 2$ and the corresponding state coordinates and the feedback law are $z_1 = x_1$, $z_2 = x_2$, and

$$v = (1 + x_2^2)u + hx_2(1 + x_2^2)u^2.$$

Acknowledgement. This research was supported in part by the Air Force Office of Scientific Research under grant AFOSR-86-0029, in part by the National Science Foundation under grant ECS-8617860, in part by the DoD Joint Services Electronics Program through the Air Force Office of Scientific Research (AFSC) Contract F49620-86-C-0045, and in part by the Texas Higher

Education Advanced Technology Program.

Authors' addresses are: Hong-Gi Lee, Department of Electrical and Computer Engineering, Louisiana State University, Baton Rouge, Louisiana 70803-5901; Aristotle Arapostathis and Steven I. Marcus, Department of Electrical and Computer Engineering, University of Texas at Austin, Austin, Texas 78712-1084.

References

[1] E. Freund, "Fast Nonlinear Control with Arbitrary Pole-Placement for Industrial Robots and Manipulators," in *Robot Motion*, Brady, Hollerback, Johnson, Lozano-Perez and Mason, eds., MIT Press, Cambridge, MA, 1982.

[2] J.W. Grizzle, "Feedback Linearization of Discrete-Time Systems," in *Lecture Notes in Control and Information Sciences*, 83, 1986, 273-281.

[3] J.W. Grizzle and P.V. Kokotovic, "Feedback Linearization of Sampled-Data Systems," *IEEE Trans. Automatic Control*, AC-33, 1988, 857-859.

· [4] L.R. Hunt, R. Su and G. Meyer, "Design for Multi-Input Nonlinear System," in *Differential Geometric Control Theory*, R.W. Brockett, et.al., eds., Birkhauser, Boston, 1983, 268-293.

[5] B. Jakubczyk, "Feedback Linearization of Discrete-Time Systems," *Systems and Control Letters*, 9, 1987, 411-416.

[6] B. Jakubczyk and W. Respondek, "On the Linearization of Control Systems," *Bull. Acad. Polon. Sci. Ser. Math. Astron. Physics*, 28, 1980, 517-522.

[7] B. Jakubczyk and E.D. Sontag, "The Effect of Sampling on Feedback Linearization," *Proc. 26th IEEE Conf. on Decision and Control*, Los Angeles, CA, December 1987, 1374-1379.

[8] H.G. Lee, A. Arapostathis and S.I. Marcus, "On the Linearization of Discrete Time Systems," *Int'l. J. Control*, 45, 1987, 1803-1822.

[9] H.G. Lee, A. Arapostathis and S.I. Marcus, "Remarks on Discretization and Linear Equivalence of Continuous Time Nonlinear Systems," *Proc. 26th IEEE Conf. on Decision and Control*, Los Angeles, CA, December 1987, 1783-1785.

[10] H.G. Lee and S.I. Marcus, "Approximate and Local Linearization of Nonlinear Discrete-Time Systems," *Int'l. J. Control*, 44, 1986, 1103-1124.

[11] R. Su, "On the Linear Equivalents of Nonlinear Systems," *Systems and Control Letters*, 2, 1982, 48-52.

[12] T.J. Tarn, A.K. Bejczy, A. Isidori and V. Chen, "Nonlinear Feedback in Robot Arm Control," *Proc. 23rd IEEE Conf. on Decision and Control*, Las Vegas, NV, 1984, 736-751.

[13] T.J. Tarn, A.K. Bejczy and X. Yun, "Coordinated Control of Two Robot Arms," *Proc. 25th IEEE Conf. on Decision and Control*, Athens, Greece, 1986, 1193-1202.

REPRESENTATIONS OF SYSTEMS WITH FINITE OBSERVATIONS

Clyde F. Martin

Texas Tech University

Introduction The problem of determining the state of a dynamical system from a set of discrete valued observations arise in many contexts. In [4], Taylor examines the relation of this problem to the general methods of symbolic dynamics and relates the problem to the problem of observability of systems defined on the topological group $2^{\aleph_0}$. In [5], Drager and Martin looked at the problem of whether or not all sequences of zeros and ones could be generated by a system evolving on the unit interval and using ergodicity conditions were able to show that there is a large class of observable systems that generate large sets of sequences of zeros and ones and are observable.

Our object in this paper is to examine in some detail systems evolving on the space, $2^{\aleph_0}$. We consider the space, $2^{\aleph_0}$, as the countable product of two element topological spaces with the product topology–the Cantor set of general topology. Let a system be defined on this space by a function generating a discrete dynamical system and a scalar valued observation:

$$x_{n+1} = f(x_n)$$

$$y_n = h(x_n).$$

Where f is a map from $2^{\aleph_0}$ to $2^{\aleph_0}$ and h is a map form $2^{\aleph_0}$ to 2. Now we define a map from $2^{\aleph_0}$ to $2^{\aleph_0}$ by the following:

$$T(x) = (h(x), h \circ f(x), h \circ f^2(x), \cdots, h \circ f^n(x), \cdots).$$

It is clear that the system is observable if and only if the map T is one to one. The novelty of this paper is to show that this map can be analyzed in some detail. We give for the space, $2^{\aleph_0}$, two different proofs in most cases. We first prove results using the combinatorial and topological structure of the space, $2^{\aleph_0}$ and then show that the same result follows from very elementary but noninciteful techniques. However the construction of this map can be generalized to more general spaces and systems.

Let M be any space and let the system as defined above be defined with the dynamics acting on M. The map T defines a mapping into the Cantor space, $2^{\aleph_0}$. We will show that there is a great deal of analysis that can be done in this context. In particular we give necessary and sufficient conditions for T to be realized by an observable system.

We also show that there exists a system on the space, $2^{\aleph_0}$, that is unobservable and does not posses any nontrivial symmetry. This result is unexpected, at least in this context. There has been a great deal of recent work that was beginning to suggest that the presence of symmetries was equivalent to unobservability. We show this by the construction of a particular example but the proof would suggest that it is not an isolated case.

Observability Let $2^{\aleph_0}$ denote the countable product of two element discrete topological spaces with the product topology. Let f be a mapping from $2^{\aleph_0}$ to $2^{\aleph_0}$ continuous in the product topology and let g be a mapping form 2^{N_0} to 2 continuous in the product topology. Define a system as the following

$$x_{n+1} = f(x_n) \tag{1}$$
$$y_n = g(x_n). \tag{2}$$

Recall that a system is observable if and only if distinct initial data yields distinct sequences $\{y_n\}$.

Theorem 1 *If the system (1) and (2) has the property that every output sequence is eventually periodic then it is not observable.*

Proof: Let $a \in 2^{N_0}$. Construct the set, V_a, that is the minimal set that contains a and is closed under both the action of f and of f^{-1}. Note that if the system is observable then the set V_a is at most countable. This follows from the fact that $f^{-1}(a)$ has cardinality at most two. Thus there must exist an uncountable number of such sets. Each of the sets V_a contains a unique periodic orbit of period N_a. Now note that there are uncountably many orbits of some period N. We restrict our attention to this set of orbits. Now since 2^{N_0} is compact the union of this set of periodic orbits contains a limit point, say v. Now let $a_{i,1}$ be a countable sequence of points that converge to v with $a_{i,1}$ being in what we will now denote by orbit P_i. Now note that since g is continuous then the sign of $g(a_{i,1})$ is constant on a terminal set of points $a_{i,1}$, $i > n_1$. Now consider the set of points $\{f(a_{i,1})\}$ where $i > n_1$. This set is again infinite and thus has a limit point and again g must be constant on a cofinal set of this sequence. We continue in this manner and because the number N is finite we terminate in a finite number of steps. Thus there exist a countable number of periodic orbits which are not distinguished by the function g and hence the system is not observable.

A corollary to the proof of Theorem 1 is the following.

Corollary 1 *If the system (1) and (2) has more than finitely many periodic orbits of any given period then the system is not observable.*

In fact, more can be said. The real problem is to determine how many periodic orbits of a given period exist. This is just the problem of labeling the nodes of an k-cycle uniquely. In [2] this problem is solved and the number, $N(k)$, is given by:

Corollary 2 ([2] Corollary 8.16.1) *With $\mu(n)$ the Möbius μ-function we have*

$$N(k) = \frac{1}{k} \sum_{d/k} \mu(k/d) q^d.$$

Using the above corollary we have

Corollary 3 *If the system (1) and (2) has more than $N(k)$ periodic orbits then it is not observable.*

The question now arises; Are there observable systems of the form (1) and (2)? We will show that there is one and in the next section we sill show that in fact there is essentially only one.

Let S be a function from $2^{\aleph_0}$ to $2^{\aleph_0}$ defined by

$$S(x)(n) = x(n+1) \qquad n \in \mathbb{Z}^+$$

where we are considering points of $2^{\aleph_0}$ to be functions of the form

$$(x(0), x(1), x(2), \ldots).$$

Define a function p from $2^{\aleph_0}$ to 2 by

$$p(x) = x(0).$$

Clearly the system

$$x_{n+1} = x(x_n) \tag{3}$$

$$y_n = p(x_n) \tag{4}$$

is observable since

$$(y_0, y_1, \ldots) = x.$$

Let P be a labeled k-cycle and choose some initial point x_0 in P. Then we have the following graph

$$(x_0, n(x_0)), (x_1, n(x_1)), \ldots, (x_{k-1}, n(x_{k-1}))$$

where n is a function with range in 2. Let x be the point in $2^{\aleph_0}$

$$x = (n(x_0), n(x_1), \ldots, n(x_{k-1}), n(x_0), \ldots)$$

Now note that the sequence

$$a, s(a), s^2(a), \ldots$$

is periodic of period k and has the given labeling. Since (3) and (4) are observable there exists exactly one such orbit. Thus the shift produces exactly the maximal number of periodic orbits.

Existence of Observable Systems In this section we prove the existence of a large class of observable systems by proving a representation theorem for one to one onto maps of $2^{\aleph_0}$. Consider a system of the form (1) and (2). For an arbitrary $x \in 2^{\aleph_0}$ we have the output sequence

$$(g(x), g(f(x)), g(f(f(x))), \cdots, g \circ f^n(x), \cdots).$$

Now note that this defines a map from $2^{\aleph_0}$ to $2^{\aleph_0}$ and note that this map is one to one if and only if the system is observable. The question arises of whether or not every one to one function from $2^{\aleph_0}$ to $2^{\aleph_0}$ arises in this manner. If there is a representation theorem of this type then it will assure us that there is indeed a large class of observable systems. We will show a somewhat weaker result that every one to one and onto function has such a representation.

Let T be such a map and let the coordinate functions be denoted by T_i, $i = 1, 2, \cdots$. Define the sets

$$V_{n,0} = \{x \in 2^{\aleph_0} : T_n(x) = 0\}$$

and

$$V_{n,1} = \{x \in 2^{\aleph_0} : T_n(x) = 1\}.$$

Note that $V_{n,0} \cup V_{n,1} = 2^{\aleph_0}$ and $V_{n,0} \cap V_{n,1} = \emptyset$ and note that they are both compact open sets in the topology of $2^{\aleph_0}$. We first prove the following lemma.

Lemma 1 *For any function $k(\ )$ from the integers to 2 we have that $\bigcap_n V_{n,k(n)}$ has exactly one element if and only if T is one to one and onto.*

Proof: Suppose that there are at least two elements, x and y, in the intersection. Thus we have that $T_n(x) = T_n(y)$ for all n and hence that $T(x) = T(y)$ so that T is not one to one. Suppose that T is onto. Then there is an x such that $T_n(x) = k(n)$. By definition x is in $V_{n,k(n)}$ for all n and hence in the intersection $\bigcap_n V_{n,k(n)}$. It follows immediately that if the intersection has cardinality exactly one then T is one to one and onto.

We first observe that if T has the desired representation that we must have $T_0(x) = g(x)$ for all x and hence that g is immediately determined. The more difficult task is to construct f. We begin by observing that if such an f exists then it must have the property

$$f : V_{n+1,\delta} \to V_{n,\delta}$$

for all n and $\delta = 0$ or 1. Let $x \in 2^{\aleph_0}$. For each n we have that $x \in V_{n,0}$ or $x \in V_{n,1}$ and hence that $f(x) \in V_{n-1,0}$ or $f(x) \in V_{n-1,1}$ respectively. Thus there exists a unique element $f(x)$ that satisfies these inclusion relations by the lemma. We have thus proved the following theorem.

Theorem 2 *If T is a one to one onto function on $2^{\aleph_0}$ there exists a unique observable system (f, g) such that $T_n(x) = g \circ f^n(x)$.*

The natural question arise as to whether or not the function f is continuous. Clearly the function g is continuous however there is no a prior reason to believe that the function f need be continuous. So we assume that the function T is one to one, onto and continuous and hence that the component functions, T_n, are also continuous. We observe the following simple fact: the family $\{T_n : n \in Z\}$ separates points. This follows, of course, from the fact that T is one to one. Recall that $2^{\aleph_0}$ has a basis of open and closed sets. Let $\mathcal{O}$ be such a set. Let x be an arbitrary point in $\mathcal{O}$ and let y be an arbitrary point in $2^{\aleph_0} \setminus \mathcal{O}$. Choose T_{n_y} such that $T_{n_y}(x) \neq T_{n_y}(y)$. Now the set $\mathcal{O} \setminus T_{n_y}^{-1}(T_{n_y}(y))$ is open and hence the set

$$\{\mathcal{O}\} \cup \{\mathcal{O} \setminus T_{n_y}^{-1}(T_{n_y}(y)) : y \in 2^{\aleph_0} \setminus \mathcal{O}\}$$

is an open cover for $2^{\aleph_0}$. Since $2^{\aleph_0}$ is compact there is a finite subcover generated by the functions which we will now denote by T_{α_i}. Note that

$$\bigcap T_{\alpha_i}^{-1}(T_{\alpha_i}(x))$$

is open, since the intersection is finite, and is contained in the set $\mathcal{O}$. Thus the set $\mathcal{O}$ can be formed from the union of finite intersections of the sets $V_{n,\delta}$ and in fact the sets $V_{n,\delta}$ form a subbasis for the topology. The only fact we have used is that the function T is one to one and continuous. Now from the definition of f it follows immediately that f is continuous and we state this in the form of a theorem.

Theorem 3 *If the function T from $2^{\aleph_0}$ to $2^{\aleph_0}$ is continuous, one to one and onto then there exists a unique system (f, g) such that for all x $T_n(x) = g \circ f^n(x)$ and f and g are continuous.*

The proofs of Theorems 1 and 2 are basic but we will now show that there is a simpler way to achieve the same facts when T is one to one. We first note that the identity function id has a representation in terms of the shift (3) and (4).

$$id = (p, ps, ps^2, \ldots, ps^n, \ldots)$$

Now if T is one to one and onto then T^{-1} exists and is continuous if and only if T is continuous. Now we note that

$$id(T(x)) = (p(T(x)), ps(T(x)), ps^2(T(x)), \ldots)$$

and hence that

$$T = (pT, psT, ps^2T, ps^3T, \ldots).$$

Since T^{-1} exists we have

$$T = (pT, pT(T^{-1}sT), pTT^{-1}sTT^{-1}sT, \ldots)$$

and hence that

$$T = (pT, pT \cdot T^{-1}sT, pT \cdot (T^{-1}sT)^2, \ldots, pT \circ (T^{-1}sT)^n, \ldots)$$

Thus if T is one to one and onto then it is represented by the system

$$x_{n+1} = T^{-1}sT(x_n) \tag{5}$$

$$y_n = pT(x_n). \tag{6}$$

Now if T is only one to one, the problem is a bit harder, for not every T has such a representation. Consider for example the function R defined by

$$R(x)(n) = \begin{cases} 0 & n = 0 \\ x(n-1) & n > 0 \end{cases}$$

This function, the right shift can clearly have no such representation since it would require $g \equiv 0$. We note the following Lemma.

Lemma 2 *If T is represented by an observable system then the range of T is invariant under the shift.*

Proof: Let $T = (g, gf, gf^2, \ldots)$ and calculate.

$$\begin{aligned} S \circ T(x) &= (gf(x), fg^2(x), \ldots) \\ &= T(f(x)) \end{aligned}$$

We now prove the following theorem.

Theorem 4 *If T is one to one, then the range of T is invariant under S if and only if T is represented by an observable system.*

Proof: The proof of this is dependent on the construction in theorem 2. Suppose that the range of T is invariant under the shift operator S. Let $x \in 2^w$ and we have that $x \in V_{n,T_n(x)}$. Since T is shift invariant there exists a $y \in RngT$ such that $y \in V_{n-1}, T_n(x)$. Thus we have that $\cap_{n=1}^{\infty} V_{n,k(x)} \neq \phi$ implies that $\cap_{n=1}^{\infty} V_{n-1,k(n)} \neq \phi$. Since T is one to one we then have that $\cap_{n=1}^{\infty} V_{n-1,T(n)}$ has a unique element and hence we can define f as in Theorem 2.

We can now prove the following theorem.

Theorem 5 *Let T be a one to one mapping of $2^{\aleph_0}$ to $2^{\aleph_0}$. There exists a unique observable system (1) and (2) that represents T if and only if the range of T is invariant under the left shift operator. Furthermore, we then have that the system is given by*

$$x_{n+1} = T^{-1} \cdot s \cdot T(x_{n+1})$$
$$y_n = p \cdot T(x_n).$$

The proof of the above theorem is clear once we have shown that the operator $T^{-1} \cdot s \cdot T$ is defined but that is the essence of Theorem 4.

We have shown an interesting and surprising fact about observable systems on $2^{\aleph_0}$. Up to a change of bases under homeomorphism, there is exactly one system, namely the system defined by the projection operator and the shift. Thus we have that every observable system has a graph that is the complete graph of the intersection with a subset of $2^{\aleph_0}$. This subset can be reasonably complicated as we will see in the next section.

Unobservable Systems In this section we sill demonstrate a representation theorem for unobservable continuous systems. Sussmann, [1], showed that in the case of real smooth systems defined on a manifold M there is a quotient manifold for which the quotient system is observable. We will show a similar result for systems of the type (1) and (2). We begin by recalling the lemma of the previous section which stated that if a function T is represented by a system then the range is shift invariant. This lemma was independent of T.

We begin by assuming T is shift invariant but not necessarily one to one or onto. Define an equivalence relation on $2^{\aleph_0}$, "$\sim$", by $a \sim b$ if and only if $T(a) = T(b)$. First it is clear that "$\sim$" is an equivalence relation and furthermore if T is continuous then "$\sim$" is closed. Thus there exists a quotient M of $2^{\aleph_0}$ defined by $2^{\aleph_0} / \sim$. It is known that [3], every compact metric space is a quotient of $2^{\aleph_0}$ but M is not that general. Since T maps into $2^{\aleph_0}$ and $\tilde{T} : M \rightarrow 2^{\aleph_0}$ we have that M is homeomorphic to a subspace of $2^{\aleph_0}$. We will come back to the question of characterizing M later. However, now we note that M is homeomorphic to a closed subspace of $2^{\aleph_0}$.

We define $\tilde{T}$ as a map from M to $2^{\aleph_0}$ by $\tilde{T}(a) = T(x)$ for $x \in a \in M$. This we'll defined by the definition of M. We define $\tilde{T}_n(a) = T_n(x)$, $x \in a$, and again note that the map is well defined. We now define, as in the previous section, the sets $\tilde{V}_{n,\delta}$ to be

$$\tilde{V}_{n,\delta} = \{a \in M : \tilde{T}_n(a) = \delta\}.$$

These sets have the same properties as the sets $V_{n,\delta}$. We proceed with construction of f.

Let $a \in M$ then $a \in \tilde{V}_{n,\tilde{T}_n(a)}$ for all n. As before, since the range of T is shift invariant we have that there exists a unique b such that

$$b \in \cap_{n=1}^{\infty} \tilde{V}_{n-1,T-n(a)}.$$

We define $f(a) = b$. Now note that f maps M to M and if we define

$$g(a) = \tilde{T}_0(a)$$

we have that

$$\tilde{T}(a) = (g(a), gf(a), gf^2(a), \ldots).$$

Thus there exists an observable system on M that generates $\tilde{T}$.

We note the following. If $a \in M$ and $a \in \tilde{V}_{n, \tilde{t}_n(a)}$ then $\tilde{T}^{-1}s\tilde{T}(a) \in \tilde{V}_{n-1, \tilde{t}_n(a)}$ and hence f is represented as

$$f = \tilde{T}^{-1}s\tilde{T}.$$

Likewise note that

$$g(a) = p\tilde{T}$$

where p is the projection onto the first component. Thus we have the result that we state in the form of a proposition.

Proposition 1 *Let T be a mapping from $2^{\aleph_0}$ to $2^{\aleph_0}$. The map T has a factor that is represented by a conjugate of the shift operator if and only if the range of T is shift invariant.*

A Generalization Let M be any metric space and let a system be defined on M by a function $f : M \to M$ and an observation function $g : M \to 2$. Now we construct a map form M into $2^{\aleph_0}$ exactly in the same manner as before, that is, $T(x) = (g(x), g(f(x)), \cdots)$. Now the proof of Theorem 5 generalizes to give a representation for the system as

$$f = T^{-1} \cdot s \cdot T$$

and

$$g = p \cdot T.$$

Likewise we can show by the above methods that any map from M to $2^{\aleph_0}$ is represented by the shift if and only if the range is shift invariant.

Now it is also clear that the key to all of the arguments has been that the space two is finite. It is easy to generalize the results of this paper to any system whose output is in a finite set.

An Example A recurring theme in the theory of observability is that a system is observable if and only if there is no non-trivial symmetry. We will show by example in the case of $2^{\aleph_0}$ that this is not true. Let the system defined by the shift operator s and the observation given by

$$h(x) = x_1 + x_2 + x_1 x_2.$$

The function h is zero if and only if the first two entries of x are zero and hence the system will be identically one for any initial point that has sequences of zeros of at most length one. For example the two points given by

$$x_1 = 0, \quad x_n = 1, \; n > 1$$

and

$$y_2 = 0 \quad y_n = 1 \ n \neq 2$$

give identical output sequences. We will show that if there exists a function T that commutes with s and at the same time has the property that $h(T(x)) = h(x)$ then $T = id$. We begin by characterizing those functions which commute with s.

Lemma 3 T commutes with s if and only if there exists a scalar valued function g such that

$$T(x) = (g(x), g(s(x)), g(s^2(x)), \cdots).$$

Proof: First it is clear that if T has the above form then T does indeed commute with s. So suppose that T commutes with s. Writing T in coordinate form we have

$$T(s(x)) = (T_1(s(x)), T_2(s(x)), T_3(s(x)), \cdots)$$

and

$$s(T(x)) = (T_2(x), T_3(x), T_4(x), \cdots).$$

We see that there is a recursion given by

$$T_{n+1}(x) = T_n(s(x)).$$

Solving this recursion we have

$$T_n(x) = T_1(s^{n-1}(x)).$$

The function T is completely determined by the arbitrary function T_1 and this finishes the proof of the lemma.

We will now show that if there is a function of this form that fixes h i.e., $h(T(x)) = h(x)$ then g is the projection onto the first coordinate and consequently $T = id$.

Lemma 4 If $h(x) = x_1 + x_2 + x_1 x_2$, $T(x) = (g(x), g(s(x)), g(s^2(x)), \cdots)$ and $h(T(x)) = h(x)$ then $g(x) = x_1$.

Proof: The proof is a series of calculations based on the following identity which is just $h(T(x)) = h(x)$.

$$g(x) + g(s(x)) + g(x)g(s(x)) = x_1 + x_2 + x_1 x_2.$$

We evaluate this identity for special values of x. First let

$$x = (0, 0, x_3, x_4, \cdots).$$

We have then

$$g(x) + g(s(x)) + g(x)g(s(x)) = 0.$$

Now if $g(x) = 1$ we would have that $1 = 0$ and hence $g(x) = 0$. We then have that $g(s(x)) = 0$ and so

$$g(0, x_2, x_3, \cdots) = 0.$$

Now evaluating at the point

$$y = (1, 0, y_3, y_4, \cdots)$$

we have

$$g(y) + g(s(y)) + g(y)g(s(y)) = 1.$$

If $g(y) = 0$ then we would have that $g(0, y_3, y_4, \cdots) = 1$ which is a contradiction. Thus $g(y) = 1$. Now evaluating at

$$z = (1, 1, z_3, \cdots)$$

we again have that

$$g(z) + g(s(z)) + g(z)g(s(z)) = 1.$$

If $g(z) = 0$ then for all values of z_3 we would have that $g(1, z_3, \cdots) = 1$. Using the identity a second time we have that $g(z_3, \cdots) = 1$. But, this contradicts the above conditions if $z_3 = 0$. Thus we have shown that g is the projection onto the first coordinate.

Supported in part by NSA Grant #MDA904-85-H0009

References

[1] H. Sussman, "A generalization of the closed subgroup theorem to quotients of arbitrary manifolds," J. of Differential Geometry, vol. 10, pp. 151-166, 1975.

[2] L.L. Dornhoff and F.E. Hohn, "Applied Modern Algebra," Macmillain Publishing Co., Inc.

[3] J. Hocking and G. Young, "Topology", Addison-Wesley Publishing Co. Inc., Reading, 1961.

[4] T. Taylor, "An example of global observability of a chaotic system," Proceedings of the 26th conference on Decision and Control, Los Angeles, Ca. 1987, pp. 972-974.

[5] L. Drager and C. Martin, "Global observability of a class of nonlinear discrete time systems," Systems and Control Letters, Vol. 6, pp. 65-68, 1985.

ON-LINE IDENTIFICATION OF ELECTRIC LOAD MODELS
FOR LOAD MANAGEMENT

Roland Malhamé   Sofiène Kamoun
Ecole Polytechnique de Montréal   Ecole Polytechnique de Montréal

Denis Dochain
Universite Catholique de Louvain

1. INTRODUCTION .

In order to provide electric power reliably, economically, while satisfying the various voltage and frequency constraints in a power system, it is essential to have a reasonable understanding of load dynamics.

While "black box" model identification methods provide acceptable results for normal power system operating conditions [2], they fail to provide adequate models in situations - such as load management - where the ordinary load behavior is profoundly modified. In this case, it is necessary to resort to more complex , physically-based,load modeling methodologies.

In a previous work [1], we have shown that it is possible to use the theory of stochastic processes to generate, much like in statistical mechanics, dynamic physically-based models of the average aggregate load behavior, for particular applications in load management by direct device control.

The particular loads aimed at here, are electric space heating or cooling loads, which because they are associated with energy storage, are often cycled, during peak demand hours, in order to achieve peak load shaving. The models we have developed [1], can predict post-interruption load dynamics and house hold temperature distributions, and thus can be a tool in generating power cycling strategies.

In order to make the approach practical, however, it is necessary to be able to estimate models parameters, while minimizing data requirements. Also, since the model parameters are essentially time-varing, it would be highly desirable to generate parameter estimates, on-line.

In this paper, we address the above issues. In order to have a "controlled" situation, measurements are simulated at the household level, using a noise source and a dynamic household thermal model, both with completely known characteristics. The "measurements" consist of a series of thermostat "on" and "off" durations, together with values of ambient temperatures (considered constant for all the numerical simulations reported here). We show that, on the basis of the above simulated measurements, various algorithms can yield the values of the unknown parameters.

The paper is organized as follows. In section 2, we discuss the structure of the measurements simulator. In section 3, we give a recursive least-squares parameter estimation algorithm, where it is assumed that noise characteristics are known. In section 4, we drop this last requirement and obtain instead a maximum likelihood estimator of all parameters, which is computed recursively. Then , the log likelihood function is maximized using a recursive numerical estimation algorithm with accelerated convergence. In section 5,some simulation results are reported. Conclusions and future directions are summarized in section 6.

2. SIMULATION OF ON-LINE MEASUREMENTS.

In this section we first review a simple stochastic model of elementary space heating loads which was used to derive the aggregate dynamic load model in [1]. Subsequently we give details of our measurements simulator.

2.1 The model.

It is assumed that the operating state of an individual electric space heater is controlled by a thermostat whose state it self depends on a temperature function.

The model proposed by Chong and Debs in [3] is used. It comprises a continuous state characterizing the energy storage portion of the device and a discrete state characterizing the switching mechanism. The model is as follows :

- *continuous state* : Let $x(t)$ (a temperature) characterize the state of energy storage of the house. $x(t)$ evolves according to the stochastic differential equation :

$$C dx(t) = -a'(x(t) - X_a(t))dt + R'm(t)b(t)dt + dv'(t) \tag{1}$$

where we have :

C the average thermal capacity of the house
a' the average heat loss rate through the floors, walls ceilings
     etc. , .
$x(t)$ an average temperature of the house
$Xa(t)$ the ambient temperature
R' the rate of heat gain supplied by the the heater
$m(t)$ the operating state of the device (1 for "ON" and 0 for "OFF")
$v'(t)$ a noise process
$b(t)$ the state of the power supply (1 for " ON" and 0 for "OFF")

$b(t)$ represents the control applied by the utility to the particular device , $v'(t)$ represents modeling errors and all processes of heat loss or heat gain gain which have not been accounted for explicit. We assume $v'(t)$ to be a Wiener process of zero mean and some variance parameter.

Division of (1) by C yields :

$$dx(t) = -a(x(t) - X_a(t))dt + Rm(t)b(t)dt + dv(t) \tag{2}$$

where the definitions of a and R are obvious, $v(t)$ is now assumed to have a variance parameter σ^2.
- *Discrete state :* The evolution of the discrete state $m(t)$ is governed by a thermostat with temperature

setting x_- and deadband $(x_+ - x_-)$. $m(t)$ switches from 1 to 0 when $m(t)$ reaches x_+ and from 0 to 1 when $m(t)$ reaches x_- .

Mathematically , for an arbitrary small time increment δt

$$m(t + \delta t) = m(t) + \pi(x(t),x_+,x_-) \tag{3}$$

where :

$$\pi(x,m,x_+,x_-) = \begin{cases} 0 & x_- < x < x_+ \\ -m & x > x_+ \\ 1-m & x < x_+ \end{cases}$$

2.2 The Measurements simulator .

It consists first of a module that integrates the linear stochastic differential equation that describes the continuous state $x(t)$, using a Runge Kutta method.[9]

The power supply is assumed to be always "ON" so that $b(t)$ takes the constant value 1.

The noise process is generated using the Box-Muller method [9]. This method provides random deviates with a normal (Gaussian) distribution of zero mean and unit variance, using a non linear transformation between two uniform deviates on [0,1] x_1 , x_2 given by :

$$y_1 = (-2\ln(x_1))^{1/2} \cos(2\pi x_2) \tag{4}$$

$$y_2 = (-2\ln(x_1))^{1/2} \sin(2\pi x_2) \tag{5}$$

The Wiener process is then obtained by taking y_1 or y_2 multiplied by $\sigma^2 \delta t$ as increments of the process $v(t)$ where δt is a time increment that is sufficiently small.

Thermostat switching behavior is generated simply by comparing the temperature $x(t)$ with x_- and x_+, while memorizing whether the system was "ON" or "OFF" before setting $m(t)$ to 0 or 1.

At this stage, a sample path of the $[x(t),m(t)]$ process is generated. The next step is to record the duration of the "ON" and "OFF" states which together with ambient temperature, represent the set of measurements, used as a basis for parameter estimation.

3. LEAST SQUARES RECURSIVE IDENTIFICATION .

3.1 Formulation of the parameter estimation problem.

The hybrid-state stochastic model presented above can be viewed as an alternated concatenation of two models : one corresponding to the "ON" state of the device, the other to the OFF state, and described respectively by the following linear stochastic differential equations:

$$dx(t) = -a(x(t) - X_a(t))dt + Rdt + dv(t) \tag{6}$$

$$dx(t) = -a(x(t) - X_a(t))dt + dv(t) \tag{7}$$

These two switching structures characterize two systems having T_{on} and T_{off} as outputs, where T_{on} and T_{off} are respectively the "ON" and "OFF" durations.

We now establish some approximate analytic expressions of T_{on} and T_{off}.

Let us consider the "ON" state; The differential equation (2) can be approximated using the Euler's method as follows :

$$x_{k+1} - x_k = -a(x_k - X_a)\delta t + R\delta t + dv \tag{8}$$

Thus we can estimate T_{on} by taking x_{k+1} and x_k respectively as x_+, and x_-, that is:

$$T_{on} = \frac{x_+ - x_-}{R - a(x_- - X_a)} + \frac{\displaystyle\int_0^{T_{on}} dv}{R - a(x_- - X_a)} \tag{9}$$

We will denote by $\hat{T}_{on}$ the estimate of T_{on} :

$$\hat{T}_{on} = \frac{x_+ - x_-}{R - a(x_- - X_a)} \tag{10}$$

Let δ be the deadband and r the temperature upwards drift rate :

$$\delta = x_+ - x_- \tag{11}$$

$$r = R - a(x_- - X_a) \tag{12}$$

it yields :

$$\hat{T}_{on} = \frac{\delta}{r} \tag{13}$$

Similarly,during the "OFF" period,

$$T_{off} = \frac{x_+ - x_-}{a(x_+ - X_a)} + \frac{\displaystyle\int_0^{T_{off}} dv}{a(x_+ - X_a)} \tag{14}$$

$$\hat{T}_{off} = \frac{\delta}{c} \tag{15}$$

where c is defined as the temperature downwards drift rate:

$$c = a(x_+ - X_a) \tag{16}$$

We assume x_+ and x_- known and the ambient temperature X_a measured. Also, for the time being, the noise variance parameter is assumed to be known. Unknowns R and a in model (8) are now estimated by first generating a least-squares estimate of r and c.

For convenience, we rewrite (13) and (15) as follows,

$$\hat{T}_{on} = r_o \delta \tag{17}$$

$$\hat{T}_{off} = c_o \delta \tag{18}$$

where

$$r_o = \frac{1}{r} \tag{19}$$

$$c_o = \frac{1}{c} \tag{20}$$

It is clear, from above the above assumptions, that we are dealing here with two independent parameter estimation problems.

Let us denote by θ the unknown parameter for either one of the two systems, and $T(t,\theta)$ the corresponding estimated output at time t.

The "quality" of the approximation will be the following scalar criterion

$$V_N(\theta) = \frac{1}{N} \sum_{t=1}^{N} L(\hat{T}(t,\theta),T(t)) \tag{21}$$

where we choose

$$L(\hat{T}(t,\theta),T(t,\theta)) = \frac{1}{2}(\hat{T}(t,\theta) - T(t,\theta))^2 \tag{22}$$

In [4], it is shown that one possible approach to the minimization of $V_N(\theta)$ above, is the following sequential least-squares algorithm:

$$\hat{\theta}_N = \hat{\theta}_{N-1} + P_{N-1}\psi_{N-1}(T(N) - \hat{T}(N,\hat{\theta}_{N-1})) \tag{23}$$

where

$$P_{N-1} = P_{N-2} - \frac{p_{N-2}^2 \psi_{N-1}^2}{1 + \psi_{N-1}^2 P_{N-2}} \tag{24}$$

$$\psi_{N-1} = \frac{d\hat{T}(N,\theta)}{d\theta}\bigg|_{\theta=\theta_{N-1}} \tag{25}$$

Using this approach, the complete algorithm can now be formulated as follows:

$$\hat{c}_o(N) = \hat{c}_o(N-1) + P_{off}(N-1)\psi_{off}(N-1)(T_{off}(N) - \hat{T}(N,\hat{c}_o(N-1))) \tag{26}$$

where

$$P_{off}(N-1) = P_{off}(N-2) - \frac{P_{off}^2(N-1)\psi_{off}^2}{1 + \psi_{off}^2(N-1)P_{off}(N-2)} \tag{27}$$

$$\hat{T}_{off}(N,\hat{c}_o(N-1)) = c_o\delta \tag{28}$$

$$\psi_{off}(N-1) = \frac{d\hat{T}_{off}(N,c_o)}{dc_o}\bigg|_{c_o=\hat{c}_o(N-1)} = \delta \tag{29}$$

$$\hat{a}(N) = \frac{1}{\hat{c}_o(N)(x_+ - X_a)} \tag{30}$$

$$\hat{r}_o(N) = \hat{r}_o(N-1) + P_{on}(N-1)\psi_{on}(N-1)(T_{on}(N) - \hat{T}_{on}(N,\hat{r}_o(N-1))) \tag{31}$$

where

$$P_{on}(N-1) = P_{on}(N-2) - \frac{P_{on}^2(N-1)\psi_{on}^2}{1 + \psi_{on}^2(N-1)P_{on}(N-2)} \tag{32}$$

$$\hat{T}_{on}(N,\hat{r}_o(N-1)) = r_o\delta \tag{33}$$

$$\psi_{on}(N-1) = \frac{d\hat{T}_{on}(N,r_o)}{dr_o}\bigg|_{r_o=\hat{r}_o(N-1)} = \delta \tag{34}$$

$$\hat{R}(N) = \frac{1}{\hat{r}_o(N)} + \hat{a}(N)(x_- - X_a) \tag{35}$$

Some simulation results are reported at the end of the paper.

4. MAXIMUM LIKELIHOOD IDENTIFICATION.

In this section, we first construct the maximum likelihood functions. Subsequently, we use this function to obtain an "exact" optimal estimator which is computed recursively.

The basic principle of the maximum likelihood identification, is to estimate θ from an observation y by choosing the value of θ which maximizes the probability density of y given θ, called the likelihood function.

4.1 Probability densities of "ON" and "OFF" durations.

"ON" and "OFF" duration times correspond to first-passage times from x to x_* and from x_* to x for a brownian motion, with drift parameters, r and c respectively. The probability density function of such first-passage times is well known [5] and given in this case, by:

$$p_{off}(T_{off},\theta) = \frac{\delta}{\sigma(2\pi t^3)^{1/2}}\exp-\frac{(\delta-cT_{off})^2}{2\sigma^2 T_{off}} \tag{36}$$

$$p_{on}(T_{on},\theta) = \frac{\delta}{\sigma(2\pi t^3)^{1/2}}\exp-\frac{(\delta-rT_{on})^2}{2\sigma^2 t_{on}} \tag{37}$$

were θ is the scalar or vector parameter to be estimated, ((r, c, θ) in this case)

4.2 Construction of the likelihood functions.

Given a series of durations T_{on} and T_{off}, let us express the corresponding likelihood function. Assuming that all "ON" or "OFF" durations measurements are independent, the joint probability density function of all measurements viewed as random variables, is then the product of the individual probability density functions.

Two approaches are now considered : the first one consists of splitting up the system into two decoupled ("ON" and "OFF") structures,thus defining , two likelihood functions, while the second one unifies the two likelihood functions into a single global one.

4.2.1 The Separate likelihood functions approach:

The probability densities of the { T_i} (i = 1,2 ....N) is :

$$p_s = \prod_{i=1}^{N} p(T_i,\theta) \tag{38}$$

The product form of (38) suggests the use of the log-likelihood function,

$$L = \ln \prod_{i=1}^{N} p(T_i,\theta)$$

$$= \sum_{i=1}^{N} \ln(p(T_i,\theta))$$

$$= \sum_{i=1}^{N} \ln(\frac{\delta}{\sigma(2\pi T_i^3)^{1/2}} - \frac{(\delta - pT_i)^2}{2\sigma^2 T_i} \qquad (39)$$

We now obtain a log-likelihood function for each type of measurements, "ON" or "OFF", respectively:

$$L_{off} = \sum_{i=1}^{N} \ln\left(\frac{\delta}{\sigma(2\pi T_{off}(i)^3)^{1/2}}\right) - \frac{(\delta - cT_{off}(i))^2}{2\sigma^2 T_{off}(i)} \qquad (40)$$

$$L_{on} = \sum_{i=1}^{N} \ln\left(\frac{\delta}{\sigma(2\pi T_{on}(i)^3)^{1/2}}\right) - \frac{(\delta - rT_{on}(i))^2}{2\sigma^2 T_{on}(i)} \qquad (41)$$

where data are arranged as follows,

$$i = \quad 1 \quad 2 \quad 3 \quad 4 \ldots$$
$$T_{on1} \quad T_{on2} \quad T_{on3} \quad T_{on4} \cdot \cdot \cdot$$
$$T_{off1} \quad T_{off2} \quad T_{off3} \quad T_{off4} \cdot \cdot \cdot$$

4.2.2 Global likelihood function approach .

By splitting the two sets of measurements as in **-4.2.1-**, one obtains a simple parameter estimation algorithm. However, in view of the fact that the variance parameter is common to both "ON" and "OFF" structures, a more efficient estimation could be achieved using the global likelihood function.

The global log-likelihood function is the product of all the probability densities of the system outputs. However, the observations must be an alternation of T_{on} and T_{off}. Consequently , we must introduce an alternated eliminating factor into the likelihood function. This leads to:

$$L_G = \ln \prod_{i=1}^{N} p(T_{on}(i),\theta) \times p(T_{off}(i),\theta)$$

$$= \frac{(-1)^i + 1}{2} \sum_{i=1}^{N} \ln\left(\frac{\delta}{\sigma(2\pi T_{on}(i)^3)^{1/2}}\right) - \frac{(\delta - cT_{on}(i))^2}{2\sigma^2 T_{on}(i)}$$
$$+ \frac{(-1)^{i+1} + 1}{2} \sum_{i=1}^{N} \ln\left(\frac{\delta}{\sigma(2\pi T_{off}(i)^3)^{1/2}}\right) - \frac{(\delta - rT_{off}(i))^2}{2\sigma^2 T_{off}(i)} \qquad (42)$$

$$= \sum_{i=1}^{N} \left(\frac{(-1)^i + 1}{2}\right)\left(\ln\left(\frac{\delta}{\sigma(2\pi T_{on}^3)^{1/2}}\right) - \frac{(\delta - rT_{on})^2}{2\sigma^2 T_{on}}\right)$$
$$+ \left(\frac{(-1)^{i+1} + 1}{2}\right)\left(\ln\left(\frac{\delta}{\sigma(2\pi T_{off}^3)^{1/2}}\right) - \frac{(\delta - cT_{off})^2}{2\sigma^2 T_{off}}\right) \qquad (43)$$

where data are arranged as follows :

$$i = 1 \quad 2 \quad 3 \quad 4 \quad \ldots$$

$$T_{\text{off}1} \quad T_{\text{on}2} \quad T_{\text{off}3} \quad T_{\text{on}4} \quad \ldots$$

Here $T_{\text{off}}(2k+1)$ and $T_{\text{on}}(2k)$ take any arbitrary value .

4.3 Optimal estimators .

The analytic expressions of the optimal parameter estimators are obtained simply by setting the first partial derivates of the log-likelihood function with respect to these parameters, to zero .

4.3.1 σ^2 Known .

(41) and (42) yield :

$$\frac{\partial L_{\text{off}}}{\partial c} = \frac{\partial L_G}{\partial c} = \sum_{i=1}^{N} \frac{c T_{\text{off}}(i) - \delta}{\sigma^2} \tag{44}$$

$$\frac{\partial L_{\text{on}}}{\partial r} = \frac{\partial L_G}{\partial r} = \sum_{i=1}^{N} \frac{r T_{\text{on}}(i) - \delta}{\sigma^2} \tag{45}$$

It is clear here that ,whether or not we use the global log likelihood function (43), the results will be the same, because - for σ^2 known- the estimations of r and c are essentially uncoupled.

(44),(45) yield :

$$c_{\text{opt}} = \frac{N\delta}{\sum\limits_{i=1}^{N} T_{\text{off}}(i)} \tag{46}$$

$$r_{\text{opt}} = \frac{N\delta}{\sum\limits_{i=1}^{N} T_{\text{on}}(i)} \tag{47}$$

4.3.2 σ^2 Unknown.

Using the global log likelihood function (43), one observes that while the optimal estimators of r and c remain strictly dependent only on the "ON" and "OFF" structures respectively, the interaction between the two structures is obvious in the variance estimate. In fact, the optimal variances given by the separate likelihood functions, do not maximize the global one.

from (43),

$$\frac{\partial L_G(N)}{\partial \sigma} = 0$$

$$\Rightarrow \hat{\sigma}^2 = \frac{1}{N} \sum_{i=1}^{N} \frac{(-1)^i + 1}{2} \frac{(\delta - rT_{on}(i))^2}{T_{on}(i)} + \frac{(-1)^{i+1} + 1}{2} \frac{(\delta - cT_{off}(i))^2}{T_{off}(i)} \tag{48}$$

where,

$$i = 1 \quad 2 \quad 3 \quad 4 \ldots$$

$$T_{off1} \quad T_{on2} \quad T_{off3} \quad T_{on4} \cdot \cdot \cdot$$

Here $T_{off}(2k+1)$ and $T_{off}(2k)$ take any arbitrary value .

4.4. Recursive Computation of the Maximum Likelihood Estimator .

Given the optimal estimators expressions derived from the global likelihood function, we now build an algorithm which computes the parameters recursively.

4.4.1 The algorithm .

By rewriting adequately (44) , (45) we reach the following convenient recursive form of the computations,

$$\frac{1}{\hat{c}(N+1)} = \frac{\frac{1}{N+1}\sum\limits_{i=1}^{N+1} T_i}{\delta} \tag{49}$$

$$= \frac{N}{N+1}\left(\frac{1}{N\delta}\sum_{i=1}^{N} T_{off}(i)\right) + \frac{1}{(N+1)\delta}T_{off}(N+1)$$

and,

$$\frac{1}{\hat{c}(N+1)} = \frac{N}{(N+1)\hat{c}(N)} \frac{1}{} + \frac{T_{off}(N+1)}{(N+1)\delta} \tag{50}$$

An equivalent recursion holds for r :

$$\frac{1}{\hat{r}(N+1)} = \frac{N}{N+1}\frac{1}{\hat{r}(N)} + \frac{T_{on}(N+1)}{(N+1)\delta} \tag{51}$$

and

$$\hat{a}(N+1) = \frac{\hat{c}(N)}{(x_+ - X_a)} \tag{52}$$

$$\hat{R}(N+1) = \hat{r}(N) + \hat{a}(N)(x_- - X_a) \tag{53}$$

However , for the variance, the same type of recursion cannot be used since (48) requires the true values of r and c . Therefore, the variance is evaluated by updating partial sums of the optimal expression, and subsequently, taking the average value.

4.5. Accelerated Recursive Maximum Likelihood Identification .

In this section a completely recursive algorithm, based on a Newton search, with acceleration factors is developed.

4.5.1 Reformulation of the log-likelihood function .

As formulated, the loglikelihood function is not smooth enough for a derivative-based non-linear programming optimization method. The following change of variable appears to yield an objective function, which is better conditioned for Newton search.

$$\alpha = \frac{1}{\sigma^2} \tag{54}$$

(43) becomes:

$$
\begin{aligned}
L_G(N,\theta) = \sum_{i=1}^{N} &\frac{(-1)^i + 1}{2} \ln\left(\frac{0.5\alpha\delta^2}{(2\pi T_{off}(i)^3)}\right) - \frac{\alpha(\delta - cT_{off}(i))^2}{2T_{off}(i)} \\
+ &\frac{(-1)^{i+1} + 1}{2} \ln\left(\frac{0.5\alpha\delta^2}{(2\pi T_{on}(i)^3)}\right) - \frac{\alpha(\delta - rT_{on}(i))^2}{2T_{on}(i)}
\end{aligned}
\tag{55}
$$

4.5.2 A recursive estimation algorithm .

a.The algorithm :
Let,

$$V_N(\theta_N, Z^N) = L_G(N, \Theta) \tag{56}$$

$$V_N(\theta_N, Z^N) = \sum_{i=1}^{N} F(\theta_N, Z^N) \tag{57}$$

We can write ,

$$V_{N+1}(\theta_{N+1},Z^{N+1}) = V_N(\theta_N,Z^N) + F(\theta_N,Z^{N+1}) \tag{58}$$

Expanding in Taylor series we get :

$$V_{N+1}(\theta_{N+1},Z^{N+1}) = V_N(\theta_N,Z^N) + \dot{V}_N(\theta_N,Z^N)(\theta_{N+1}-\theta_N) + \frac{1}{2}\ddot{V}_N(\theta_N,Z^N)$$

$$\times(\theta_{N+1}-\theta_N)^2 + F(\theta_N,Z^{N+1}) + \dot{F}(\theta_N,Z^{N+1})(\theta_{N+1}-\theta_N) + \frac{1}{2}\ddot{F}(\theta_N,Z^N)$$

$$\times(\theta_{N+1}-\theta_N)^2 \tag{59}$$

At the optimum, we have :

$$\frac{\partial V_{N+1}}{\partial \theta_{N+1}}(\theta_{N+1},Z^{N+1}) = 0 \tag{60}$$

(60)It yields ,

$$\dot{V}_N(\theta_N,Z^N) + \dot{F}(\theta_N,Z^{N+1}) + (\ddot{V}_N(\theta_N,Z^N) + \ddot{F}(\theta_N,Z^{N+1}))(\theta_{N+1}-\theta_N) = 0 \tag{61}$$

i.e.,

$$\theta_{N+1} = \theta_N - \frac{\dot{V}_N(\theta_N,Z^N) + \dot{F}(\theta_N,Z^{N+1})}{\ddot{V}_N(\theta_N,Z^N) + \ddot{F}(\theta_N,Z^{N+1})} \tag{62}$$

where

$$\dot{V}_N(\theta_N,Z^N) = \dot{V}_{N-1}(\theta_{N-1},Z^{N-1}) + \dot{F}(\theta_{N-1},Z^N) + (\ddot{V}_{N-1}(\theta_{N-1},Z^{N-1})$$

$$+ \ddot{F}(\theta_{N-1},Z^N))(\theta_N - \theta_{N-1}) \tag{63}$$

and

$$\ddot{V}_N(\theta_N,Z^N) = \ddot{V}_{N-1}(\theta_{N-1},Z^{N-1}) + \ddot{F}(\theta_{N-1},Z^N) \tag{64}$$

In the above equations, $\dot{V}_N$ and $\ddot{V}_N$ are respectively the gradient and Jacobian matrix of L_G. The parameter vector is :

$$\theta = [r,c,\alpha]^T \tag{65}$$

b. Initialization :

Good starting values for the numerical search, are important to ensure convergence to the absolute maximum. The solution proposed here is to use during the first few iterations , the recursive theoretical estimation algorithm ,since it gives the optimal value of the parameters for the available outputs. Furthermore, acceptable initial values of the Hessian Matrix are searched recursively in order to make the transition sufficiently smooth .

5 . SIMULATION RESULTS :

We now report a number of numerical results illustrating the efficiency of the above identification methods .

The following fictitious parameters were used for the base case:

$x_+ = 21$ deg $x_- = 20$ deg $X_\bullet = 12$ deg $\sigma^2_\bullet = 0.5$

$a = 0.02$ min^{-1} $R = 0.7$ deg min $^{-1}$

The first and immediate result is that "ON" and "OFF" systems have similar dynamics , and so do the estimates $\hat{a}$ and $\hat{R}$. The following comparative estimation charts, have been plotted by taking the initial values as one tenth of the true values.

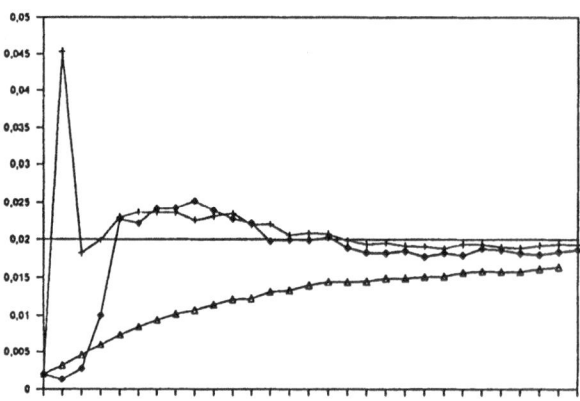

Figure 1. a Estimators
- True value
+ Least squares estimate (L.S)
△ Theoretical Maximum likelihood estimate (T.M.L)
□ Accelerated Recursive Maximum Likelihood Estimate (A.R.M.L)

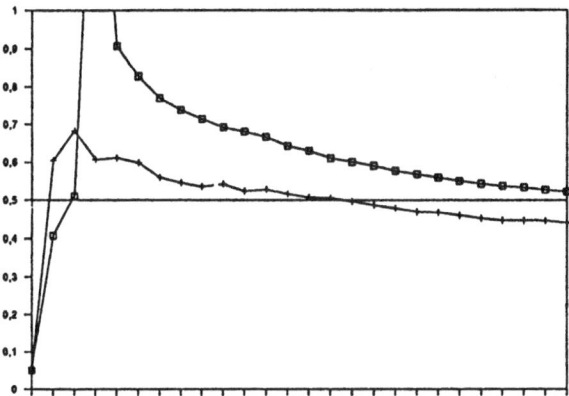

Figure 2. σ Estimators
+ Theoretical Maximum Likelihood Estimate
□ Accelerated Recursive Maximum Likelihood Estimate

For intermediate noise levels, both L .S and A.R.M.L estimators of a and R converge. However, the A.R.M.L exhibits the smoother behavior and does not seem to introduce the bias encountered with the L.S algorithm at weak noise levels. The bias is clearly related to an "insufficiency of excitation" condition.

The Theoretical Maximum likelihood estimator has a guaranteed convergence as the sample size increases, however experimental results seem to suggest that this convergence is cautious for a and R estimation. For σ^2 however, it seems to move fairly rapidly within close range of the exact value. In conclusion , by allowing a fast unbiased estimation of all unknown parameters ,the global accelerated maximum likelihood estimator appears to be a superior algorithm under the assumptions that the parameters are invariant or very slowly time-varying.

6.CONCLUSION

We have proposed various on-line identification algorithms for electric space heating or cooling, physically-based, electric load models.

In particular, we have developed. an accelerated recursive maximum likelihood, identification algorithm, which appears to exhibit excellent convergence properties. The data requirements are quite minimal (ambient temperature, and thermostat sequences of ON/OFF durations), and this is essential in order to keep implementation costs as low as possible. However , we have not as yet tested the algorithms over a long period of time, where ambient temperature can change. Also, non-linearities inherent in the thermal model have not been accounted for they may have an important effect. Finally, one should test the algorithms on real data. All these issues will be addressed in future work.

REFERENCES :

[1] R.MALHAME and C.Y.CHONG "Electric Load Model Synthesis by Diffusion Approximation of a High-Order Hybrid-State Stochastic System " IEEE Transactions on   Automatic
   Control Vol AC-30 N0 9 September 85 pp.854-860
[2] F.D GALIANA,et AL,"Identification of Stochastic Electric Load Models from Physical
   Data"
   IEEE Trans,Automat.Control,AC-19 (1974),pp.887-893.
[3] C.Y CHONG and A.S Debs "Statistical Synthesis of Power system Load Model" in
   Proc.IEEE Conf. Decision Contr. Fort Lauderdale FL (1979),pp. 264-269.
[4] G.C GOODWIN "Adaptive Filtering Prediction And Control" New Jersey,Prentice Hall 1984
[5] D.R COX and H.D Miller "The Theory of Stochastic Processes". New York: Willey,1965
[6] L.LJUNG "System Identification : Theory For The User" New York, Prentice-hall 1987
[7] K,J ASTROM " Maximum Likelihood and Prediction error Methods" Automatica, Vol 16,(1980)
   pp.551-574.
[8] N.K GUPTA and R.K MEHRA " Computational Aspects of Maximum Likelihood
   Estimation And Reduction In Sensitivity Function Calculations" IEEE Trans on Autom.
   Control Vol AC-19 N 6 Dec 74,pp. 774-783.
[9] W.PRESS, B.P. FLANNERY, S.A TEUKOLSKY, WT VETTERLING " Numerical Recipes:
   The art of scientific Computing" Cambridge University press ,1986

ACKNOWLEDGMENT

 This research was funded in part by Canada NSERC grant OGP000628220, and Quebec FCAR grant E002627. This support is gratefully acknowledged.

UNIFORM STABILIZATION OF A THIN ELASTIC
PLATE BY NONLINEAR BOUNDARY FEEDBACK

J. E. Lagnese

Georgetown University

1.Setting the Problem. Let Ω by a bounded, open connected set in $\mathbb{R}^2$ having a smooth boundary Γ. We assume that $\Gamma = \Gamma_0 \cup \Gamma_1$, where Γ_0 and Γ_1 are nonempty, relatively open subsets of Γ with $\Gamma_0 \cap \Gamma_1 = \emptyset$. We denote by $\nu = \{\nu_1, \nu_2\}$ the unit normal to Γ pointing out of Ω and set $\tau = \{-\nu_2, \nu_1\}$, a positively directed unit vector tangent to Γ. We further introduce the notation $\Delta = \partial^2/\partial x^2 + \partial^2/\partial y^2$ for the ordinary Laplacian, $\Delta^2 = \Delta\Delta$ and $' = \partial/\partial t$. Consider the following boundary value problem, which describes the small vibrations of a thin, homogeneous, isotropic elastic plate (see, e.g., [5, Chapter I]):

$$w'' + \Delta^2 w = 0, \quad \{x,y\} \in \Omega \times (0,\infty), \tag{1.1}$$

$$w = \frac{\partial w}{\partial \nu} = 0, \quad \{x,y\} \in \Gamma_0 \times (0,\infty), \tag{1.2}$$

$$\begin{cases} \Delta w + (1-\mu)B_1 w = 0, \\[2mm] \dfrac{\partial \Delta w}{\partial \nu} + (1-\mu)\dfrac{\partial B_2 w}{\partial \nu} = v, \quad \{x,y\} \in \Gamma_1 \times (0,\infty), \end{cases} \tag{1.3}$$

where

$$B_1 w = 2\nu_1\nu_2 \frac{\partial^2 w}{\partial x\, \partial y} - \nu_1^2 \frac{\partial^2 w}{\partial y^2} - \nu_2^2 \frac{\partial^2 w}{\partial x^2},$$

$$B_2 w = (\nu_1^2 - \nu_2^2)\frac{\partial^2 w}{\partial x\, \partial y} + \nu_1\nu_2\left(\frac{\partial^2 w}{\partial y^2} - \frac{\partial^2 w}{\partial x^2}\right).$$

μ is a constant called *Poisson's ratio* and satisfies $0 < \mu < 1/2$ in physical situations.

Conditions (1.2) mean that the plate is clamped along Γ_0. The first equation in (1.3) means that the bending moment about the vector τ vanishes, while in the second equation v represents shear force in the vertical direction on the edge of the plate. We consider v to be the *control variable*, and the point is to choose v as a *feedback* in $\{w, w'\}$ in such a way that (1.1)–(1.3) is *uniformly asymptotically stable* in a sense which will be made precise below.

To be more specific, we introduce a bilinear form a(u;v) defined by

$$a(u;v) = \int_\Omega [\frac{\partial^2 u}{\partial x^2}\frac{\partial^2 v}{\partial x^2} + \frac{\partial^2 u}{\partial y^2}\frac{\partial^2 v}{\partial y^2} + \mu(\frac{\partial^2 u}{\partial x^2}\frac{\partial^2 v}{\partial y^2} + \frac{\partial^2 u}{\partial y^2}\frac{\partial^2 v}{\partial x^2}) + 2(1-\mu)\frac{\partial^2 u}{\partial x\, \partial y}\frac{\partial^2 v}{\partial x\, \partial y}]dxdy \tag{1.4}$$

and we set

$$a(u) = a(u;u). \tag{1.5}$$

If w satisfies (1.1)–(1.3), then $\frac{1}{2}a(w)$ represents the *strain energy in bending*. The *total energy in bending* is

$$E(t) = \frac{1}{2} \int_\Omega w'^2 dxdy + \frac{1}{2} a(w), \tag{1.6}$$

the first term representing *kinetic energy in bending.*

Let $\{w^0, w^1\}$ be the *initial data* for (1.1)–(1.3):

$$w(0) = w^0, \quad w'(0) = w^1, \quad x \in \Omega. \tag{1.7}$$

Uniform Stabilization Problem. Determine functions f: $\mathbb{R}^2 \to \mathbb{R}$ such that, if

$$v(t) = f(w(t), w'(t)) \tag{1.8}$$

the closed loop system (1.1)–(1.3), (1.8) satisfies

$$\lim_{t \to \infty} E(t) = 0 \tag{1.9}$$

uniformly with respect to $\{w^0, w^1\}$ in each bounded set

$$E(0) = \frac{1}{2} \int_\Omega (w^1)^2 dxdy + \frac{1}{2} a(w^0) \le M$$

Remark 1.1. A strictly weaker concept of asymptotic stability for (1.1)–(1.3), (1.8) is *strong stability*, wherein (1.9) is only required to hold for each *fixed* pair $\{w^0, w^1\}$.

In order to determine *candidates* for uniformly stabilizing feedbacks (1.8), we calculate

$$E'(t) = \int_\Omega w'w''dxdy + a(w'; w). \tag{1.10}$$

The last term in (1.10) is transformed according to the following *Green's formula.*

LEMMA 1. *For all u,v in* $H^4(\Omega)$ *and* $\mu \in \mathbb{R}$,

$$a(u; v) = \int_\Omega u\Delta^2 v dxdy + \int_\Gamma \{[\Delta v + (1-\mu)B_1 v]\frac{\partial u}{\partial \nu} - [\frac{\partial \Delta v}{\partial \nu} + (1-\mu)\frac{\partial B_2 v}{\partial \tau}]u\}d\Gamma. \tag{1.11}$$

$H^4(\Omega)$ is the standard Sobolev space of order 4. For a proof of Lemma 1 see, e.g., [3, p. 206].

Use of (1.11) in (1.10) shows that

$$E'(t) = -\int_\Gamma vw'd\Gamma. \tag{1.12}$$

We therefore set

$$v = f(w') \tag{1.13}$$

where

f *is a single valued, continuous, nondecreasing function on* $\mathbb{R}$ *and* $f(0) = 0$. $\tag{1.14}$

Then it follows from (1.12) that $E'(t) \le 0$. In what follows, it will be shown that the system (1.1)–(1.3), (1.13) is uniformly asymptotically stable provided $f(\xi)$ satisfies (1.14) and *growth conditions* at $\xi = 0$ and at $\xi = \infty$, and provided the triple $\{\Omega, \Gamma_0, \Gamma_1\}$ satisfies certain *geometric conditions*.

Remark 1.2. When the function f is *linear*, i.e., $v = kw'$ ($k > 0$), the uniform asymptotic energy estimate $E(t) \le Ce^{-\omega t}E(0)$ ($\omega > 0$) was established in [4] under the assumption that the configuration $\{\Omega, \Gamma_0, \Gamma_1\}$ satisfies the following geometric conditions: there is a point $\{x_0, y_0\} \in \mathbb{R}^2$ such that

$$(\{x,y\} - \{x_0,y_0\}) \cdot \nu \le 0, \quad \{x,y\} \in \Gamma_0, \tag{1.15}$$

$$(\{x,y\} - \{x_0,y_0\}) \cdot \nu \ge 0, \quad \{x,y\} \in \Gamma_1. \tag{1.16}$$

2. Well–posedness of the System.

We introduce the spaces

$$H = L^2(\Omega), \quad W = H^2_{\Gamma_0}(\Omega) = \{v \mid v \in H^2(\Omega), \; v = \frac{\partial v}{\partial \nu} = 0 \text{ on } \Gamma_0\}.$$

(f,g) will denote the usual scalar product of f and g in H, and W has the topology induced by $H^2(\Omega)$. H and W are Hilbert spaces and W is densely and continuously embedded in H. We identify H with its dual space. We then have the dense and continuous embeddings $W \subset H \subset W'$ where W' denotes the dual of W.

A key point is the following: $a(u;v)$ is a continuous and *strictly coercive* form on W (here $\Gamma_0 \ne \emptyset$ is essential). This fact is known as *Korn's Lemma* (see, e.g., [3, Chapter III]). As a consequence, there is a linear operator A mapping W *onto* W' such that

$$\langle Au,v \rangle = a(u;v), \quad \forall u,v \in W, \tag{2.1}$$

where $\langle \cdot, \cdot \rangle$ denotes the pairing in the $W' - W$ duality.

Let B be the nonlinear operator from W into W' defined by

$$\langle Bu,v \rangle = \int_{\Gamma_1} f(u)v d\Gamma, \quad \forall u,v \in W.$$

Note that since dimension $= 2$, $H^2_{\Gamma_0}(\Omega) \subset C(\bar{\Omega})$ with continuous injection (Sobolev Embedding Theorem) and therefore $f(w(X))$ is continuous on Γ_1, so that Bw is a well defined element of W' for each $w \in W$. Moreover, the map $w \to Bw$ is *continuous* from W into W'. In addition, the operator B is *monotone* if and only if f is nondecreasing, since

$$\langle Bw - Bv, w - v \rangle = \int_{\Gamma_1} (f(w) - f(v))(w - v) d\Gamma. \tag{2.2}$$

If f satisfies (1.14), the problem (1.1)–(1.3), (1.13) may be solved by nonlinear semigroup methods. To this end, we first write the problem in the *variational form*

$$\langle w'',v \rangle + \langle Bw',v \rangle + \langle Aw,v \rangle = 0, \quad \forall v \in W. \tag{2.3}$$

(It may be verified with the aid of Green's formula (1.11) that (2.3) and (1.1)–(1.3), (1.13) are *equivalent* for classical solutions.) Equation (2.3) may in turn be formally written as a system

$$U' + \mathcal{A}U = 0, \quad U(0) = \begin{bmatrix} w^0 \\ w^1 \end{bmatrix}, \tag{2.4}$$

where

$$\mathcal{A} = \begin{bmatrix} 0 & -I \\ A & B \end{bmatrix}, \quad U = \begin{bmatrix} w \\ w' \end{bmatrix}.$$

(2.3) is an equation in W'. However, we are interested in *strong solutions* of (2.3), i.e., we wish to consider (2.3) in the space H. We therefore define the domain of the operator $\mathcal{A}$ by

$$D(\mathcal{A}) = \{\{u,v\} \mid u \in W, \; v \in W, \; Au + Bv \in H\}. \tag{2.5}$$

That (2.4) is a well–set problem follows from

THEOREM 1. *$\mathcal{A}$ is maximal monotone on* W×H.

Proof. If $U_1 = \{u_1, v_1\}$ and $U_2 = \{u_2, v_2\}$ we have

$$(\mathcal{A}U_1 - \mathcal{A}U_2, U_1 - U_2)_{W \times H}$$

$$= (\{v_2 - v_1, A(u_1 - u_2) + (Bv_1 - Bv_2)\}, \{u_1 - u_2, v_1 - v_2\})_{W \times H}$$

$$= -a(v_1 - v_2; u_1 - u_2) + \langle A(u_1 - u_2), v_1 - v_2 \rangle + \langle Bv_1 - Bv_2, v_1 - v_2 \rangle$$

$$= \langle Bv_1 - Bv_2, v_1 - v_2 \rangle \geq 0,$$

so that $\mathcal{A}$ is monotone.

To see that $Range(I + \mathcal{A}) = W \times H$, let $\{f,g\} \in W \times H$ and consider the equation

$$\{u,v\} + \mathcal{A}\{u,v\} = \{f,g\},$$

that is,

$$u - v = f \in W, \quad v + Au + Bv = g \in H. \tag{2.6}$$

Inserting $u = f + v$ into the second equation in (2.6) gives

$$Av + Bv + v = g - Af \in W'. \tag{2.7}$$

Since A is strictly coercive and $\langle Bv, v \rangle \geq 0$ on W, it is easily seen that $(A + B + I)^{-1}$ (defined on $Range(A + B + I)$) maps bounded sets in W' to bounded sets in W. Since, moreover, $A + B$ is monotone and continuous from W into W', it follows that $Range(A + B + I) = W'$ (see, e.g., [2, Chapter I]). If v solves (2.7) and $u = f + v$, then from (2.6) we have $Au + Bv = g - v \in H$. Thus $\{u,v\} \in D(\mathcal{A})$.

Since $\mathcal{A}$ is maximal monotone, it follows that $-\mathcal{A}$ generates a continuous semigroup of (*nonlinear*) *contractions* $\{S(t) \mid t \geq 0\}$ on $\overline{D(\mathcal{A})}$. Therefore, if $\{w^0, w^1\} \in D(\mathcal{A})$, there is a unique function $U = \{w, w'\} = S(\cdot)\{w^0, w^1\}$ which satisfies

$$t \to \{w(t), w'(t)\} \text{ is Lipschitz continuous from } [0, \infty) \text{ into } W \times H;$$

$w'(t) \in W, \quad Aw(t) + Bw'(t) \in H, \quad t > 0;$

$t \to \{w(t), w'(t)\}$ is strongly right differentiable and weakly differentiable from $(0,\infty)$ into $W \times H$;

$t \to Aw(t) + Bw'(t)$ is weakly continuous and strongly right continuous from $(0,\infty)$ into H;

$w''(t) + (Bw'(t) + Aw(t)) = 0, \quad t > 0.$

For proofs of the above properties see, e.g., [1, Chapter III].

Remark 2.1. If $\{w^0, w^1\} \in \overline{D(\mathscr{A})}$, then $S(t)\{w^0, w^1\}$ defines what we shall call *the mild solution* of (1.1)–(1.3), (1.13).

Remark 2.2. Let us assume that $f(0) = 0$. Then $D(\mathscr{A})$ is dense in $W \times H$. In fact, suppose $\{u,v\} \in D(\mathscr{A})$. Then $u \in W$, $v \in W$ and

$$a(u;\hat{w}) + \langle Bv, \hat{w} \rangle = (F, \hat{w}), \quad \forall \hat{w} \in W, \tag{2.8}$$

for some $F \in H$. Since $f(0) = 0$, (2.8) signifies that u is a weak solution of the boundary value problem

$$u \in H^2_{\Gamma_0}(\Omega), \quad \Delta^2 u = F \in L^2(\Omega), \tag{2.9}$$

$$\Delta u + (1-\mu)B_1 u = 0, \quad \frac{\partial \Delta u}{\partial \nu} + (1-\mu)\frac{\partial B_2 u}{\partial \nu} = f(v) \quad \text{on } \Gamma_1. \tag{2.10}$$

Since $f(0) = 0$ it follows from (2.9), (2.10) that, in particular, $D(\mathscr{A}) \supset W_0 \times H^2_0(\Omega)$, where $u \in W_0$ means that $u \in H^4(\Omega) \cap H^2_{\Gamma_0}(\Omega)$ and satisfies *homogeneous* boundary conditions

$$\Delta u + (1-\mu)B_1 u = 0, \quad \frac{\partial \Delta u}{\partial \nu} + (1-\mu)\frac{\partial B_2 u}{\partial \nu} = 0 \quad \text{on } \Gamma_1.$$

But $W_0 \times H^2_0(\Omega)$ is obviously dense in $H^2_{\Gamma_0}(\Omega) \times L^2(\Omega)$.

Remark 2.3. Suppose that $\Gamma_0 \cap \Gamma_1 = \emptyset$, $f(0) = 0$ and f is differentiable with $f' \in L^2_{loc}(\mathbb{R})$. Then $\{u,v\} \in D(\mathscr{A})$ implies $u \in H^4(\Omega)$. Indeed, $f(v)|_{\Gamma_1} \in H^1(\Gamma_1)$ and so, by elliptic regularity theory, solutions of (2.9), (2.10) satisfy $u \in H^4(\Omega)$.

3. **Uniform Asymptotic Stability of Solutions.** In this section we shall assume that Γ_0 and Γ_1 satisfy the geometric assumptions (1.15), (1.16). It is also assumed that the function f in (1.13) has the form

$$f = (m \cdot \nu)g, \quad m(x,y) = \{x,y\} - \{x_0, y_0\}, \tag{3.1}$$

where g is a differentiable, nondecreasing function with $g'(\xi)$ bounded on bounded sets of ξ and with $g(0) = 0$. In addition, g is required to satisfy the following growth conditions:

$$\begin{cases} \xi g(\xi) \geq c_0 |\xi|^{p+1}, & |\xi| \leq 1, \\ \xi g(\xi) \geq c_0 |\xi|^2, & |\xi| > 1, \end{cases} \tag{3.2}$$

for some $c_0 > 0$, $p \geq 1$; and

$$|g(\xi)| \leq C_0 |\xi|^q, \quad |\xi| \geq 1, \tag{3.3}$$

for some $C_0 > 0$ and $q \geq 1$.

Under these conditions we shall prove

THEOREM 2. *Let* $\{w^0, w^1\} \in W \times H$ *and* w *be the solution of* (1.1)–(1.3), (1.13). *There is a constant* $\omega > 0$ *such that*

$$\int_t^\infty [E(s)]^{(p+1)/2} ds \leq \frac{1}{\omega} E(t), \quad t \geq 0. \tag{3.4}$$

COROLLARY. *(a) If* $p = 1$,

$$E(t) \leq e \cdot e^{-\omega t} E(0). \tag{3.5}$$

(b) If $p > 1$, *there is a constant* λ *depending on* $E(0)$ *such that*

$$E(t) \leq 3E(0)(1 + \lambda t)^{-2/(p-1)}. \tag{3.6}$$

Remark 3.1. The constant ω in (3.4) will in general depend on $E(0)$, unless $p = q = 1$ in (3.2), (3.3). Note that even if $p = q = 1$, g may still be nonlinear.

Remark 3.2. Estimates analogous to (3.5) and (3.6) have been established by E. Zuazua [8] for solutions of the wave equation on a bounded region $\Omega \subset \mathbb{R}^n$ with energy dissipation effected through the nonlinear Neumann boundary condition $\partial w / \partial \nu = -f(w')$ on Γ_1.

The proof of Theorem 2 requires the following lemma.

LEMMA 2. *For all* $w \in H^4(\Omega)$ *and for all* $\mu \in \mathbb{R}$,

$$\int_\Omega (m \cdot \nabla w) \Delta^2 w \, dX = a(w) + \frac{1}{2} \int_\Gamma m \cdot \nu [w_{xx}^2 + w_{yy}^2 + 2\mu w_{xx} w_{yy} + 2(1-\mu) w_{xy}^2] d\Gamma$$

$$+ \int_\Gamma \{ [\frac{\partial \Delta w}{\partial \nu} + (1-\mu) \frac{\partial B_2 w}{\partial \tau}] m \cdot \nabla w - [\Delta w + (1-\mu) B_1 w] \frac{\partial (m \cdot \nabla w)}{\partial \nu} \} d\Gamma. \tag{3.7}$$

In (3.7) we have used the notation w_{xx} for $\partial^2 w / \partial x^2$, etc.

Proof. Suppose $w \in H^4(\Omega)$ and apply Green's formula (1.11) with $u = m \cdot \nabla w$ and $v = w$. We obtain

$$\int_\Omega (m \cdot \nabla w) \Delta^2 w \, dX = a(w; m \cdot \nabla w)$$

$$+ \int_\Gamma \{ [\frac{\partial \Delta w}{\partial \nu} + (1-\mu) \frac{\partial B_2 w}{\partial \tau}] m \cdot \nabla w - [\Delta w + (1-\mu) B_1 w] \frac{\partial (m \cdot \nabla w)}{\partial \nu} \} d\Gamma. \tag{3.8}$$

Let us calculate $a(w; m \cdot \nabla w)$. We have

$$a(w; m \cdot \nabla w) = \int_\Omega \{w_{xx}(m \cdot \nabla w)_{xx} + w_{yy}(m \cdot \nabla w)_{yy}$$
$$+ \mu[w_{xx}(m \cdot \nabla w)_{yy} + w_{yy}(m \cdot \nabla w)_{xx}] + 2(1-\mu)w_{xy}(m \cdot \nabla w)_{xy}\}dX.$$

When the differentiations of $(m \cdot \nabla w)$ beneath the integral are carried out, one finds that $a(w; m \cdot \nabla w)$ may be written

$$a(w; m \cdot \nabla w) = 2a(w) + \frac{1}{2}\int_\Omega m \cdot \nabla[w_{xx}^2 + w_{yy}^2 + 2\mu w_{xx}w_{yy} + 2(1-\mu)w_{xy}^2]dX$$

$$= a(w) + \frac{1}{2}\int_\Omega \operatorname{div}\{m[w_{xx}^2 + w_{yy}^2 + 2\mu w_{xx}w_{yy} + 2(1-\mu)w_{xy}^2]\}dX$$

$$= a(w) + \frac{1}{2}\int_\Gamma m \cdot \nu[w_{xx}^2 + w_{yy}^2 + 2\mu w_{xx}w_{yy} + 2(1-\mu)w_{xy}^2]d\Gamma. \tag{3.9}$$

Inserting (3.9) into (3.8) yields (3.7).

Proof of Theorem 2. We start with initial data $\{w^0, w^1\} \in D(\mathscr{A})$, and apply Lemma 2 with w the solution of (1.1)–(1.3), (1.13). After an integration in t from 0 to T we obtain

$$-\int_0^T \int_\Omega (m \cdot \nabla w)w'' dXdt = \int_0^T a(w)dt + \frac{1}{2}\int_0^T a_\Gamma(w)dt$$

$$-\int_0^T \int_{\Gamma_0} (\Delta w + (1-\mu)B_1 w)\frac{\partial}{\partial \nu}(m \cdot \nabla w)d\Gamma dt + \int_0^T \int_{\Gamma_1} (m \cdot \nu)g(w')m \cdot \nabla w d\Gamma dt \tag{3.10}$$

where

$$a_\Gamma(w) = \int_\Gamma m \cdot \nu[w_{xx}^2 + w_{yy}^2 + 2\mu w_{xx}w_{yy} + 2(1-\mu)w_{xy}^2]d\Gamma.$$

The left side of (3.10) may be written

$$\int_0^T \int_\Omega (m \cdot \nabla w')w' dXdt - Y_1 = \frac{1}{2}\int_0^T \int_\Omega \operatorname{div}(mw'^2)dXdt - \int_0^T \int_\Omega w'^2 dXdt - Y_1$$

$$= \frac{1}{2}\int_0^T \int_{\Gamma_1} (m \cdot \nu)w'^2 d\Gamma dt - \int_0^T \int_\Omega w'^2 dXdt - Y_1 \tag{3.11}$$

where

$$Y_1 = (w', m \cdot \nabla w)|_0^T. \tag{3.12}$$

Use of (3.11) in (3.10) allows (3.10) to be rewritten

$$Y_1 + \int_0^T \int_\Omega w'^2 dX dt + \int_0^T a(w)dt = \frac{1}{2} \int_0^T \int_{\Gamma_1} (m \cdot \nu)w'^2 d\Gamma dt$$

$$- \int_0^T \int_{\Gamma_1} (m \cdot \nu)g(w')m \cdot \nabla w d\Gamma dt + \int_0^T \int_{\Gamma_0} (\Delta w + (1-\mu)B_1 w)\frac{\partial}{\partial \nu}(m \cdot \nabla w)d\Gamma dt$$

$$- \frac{1}{2} \int_0^T a_{\Gamma_0}(w)dt - \frac{1}{2} \int_0^T a_{\Gamma_1}(w)dt. \tag{3.13}$$

Since $w = \partial w/\partial \nu = 0$ on $\Gamma_0 \times (0,T)$ we have $B_1 w = B_2 w = 0$ there. In addition,

$$\left\{\begin{array}{l} \frac{\partial}{\partial \nu}(m \cdot \nabla w) = (m \cdot \nu)\frac{\partial^2 w}{\partial \nu^2} = (m \cdot \nu)\Delta w, \\[2mm] w_{xx}^2 + w_{yy}^2 + 2\mu w_{xx}w_{yy} + 2(1-\mu)w_{xy}^2 = (\Delta w)^2 \quad \text{on } \Gamma_0 \times (0,T) \end{array}\right.$$

since $w_{xx}w_{yy} - w_{xy}^2 = 0$ on Γ_0 (as a consequence of the fact that $\nabla w_x|_{\Gamma_0}$ and $\nabla w_y|_{\Gamma_0}$ are linearly dependent). The integrals over $\Gamma_0 \times (0,T)$ in (3.13) thus combine to give

$$\frac{1}{2} \int_0^T \int_{\Gamma_0} (m \cdot \nu)(\Delta w)^2 d\Gamma dt.$$

Therefore, (3.13) may be written

$$Y_1 + 2 \int_0^T E(t)dt = \frac{1}{2} \int_0^T \int_{\Gamma_1} (m \cdot \nu)w'^2 d\Gamma dt - \frac{1}{2} \int_0^T a_{\Gamma_1}(w)dt$$

$$- \int_0^T \int_{\Gamma_1} (m \cdot \nu)g(w')m \cdot \nabla w d\Gamma dt + \frac{1}{2} \int_0^T \int_{\Gamma_0} (m \cdot \nu)(\Delta w)^2 d\Gamma dt. \tag{3.14}$$

3.1. A Priori Estimates. Introduce

$$\rho(t) = (w'(t), m \cdot \nabla w(t)). \tag{3.15}$$

Then

$$|\rho(t)| \leq \frac{1}{2} \int_\Omega w'^2 dx dy + \frac{1}{2} \int_\Omega (m \cdot \nabla w)^2 dx dy$$

$$\leq \frac{1}{2} \int_\Omega w'^2 dx dy + Ca(w) \leq CE(t) \tag{3.16}$$

for an appropriate constant C. From (3.12) we also have $Y_1 = \rho(T) - \rho(0)$, hence $\rho'(T) = dY_1/dT$, which may be calculated from (3.14). Writing t in place of T, we obtain from (3.14) the estimate (utilizing assumptions (1.15), (1.16))

$$\rho'(t) \leq -2E(t) + \frac{1}{2} \int_{\Gamma_1} (m \cdot \nu)w'^2 d\Gamma - \int_{\Gamma_1} (m \cdot \nu)g(w')m \cdot \nabla w d\Gamma. \tag{3.17}$$

For $\epsilon > 0$ define

$$F_\epsilon(t) = E(t) + \epsilon\rho(t)[E(t)]^{(p-1)/2}. \tag{3.18}$$

We are going to derive the estimate

$$F'_\epsilon(t) \le - \epsilon[E(t)]^{(p+1)/2}.$$

Henceforth we shall assume $p > 1$. (For $p = 1$ the calculations are similar, but simpler.) We then have

$$F'_\epsilon(t) = E'(t) + \epsilon\rho'(t)[E(t)]^{(p-1)/2} + \epsilon\frac{p-1}{2}\rho(t)[E(t)]^{(p-3)/2}E'(t). \tag{3.19}$$

Use of (3.16) and (3.17) in (3.19) gives

$$F'_\epsilon(t) \le [1 - \frac{\epsilon(p-1)}{2} C(E(0))^{(p-3)/2}]E'(t)$$

$$+ \epsilon(E(t))^{(p-1)/2}[-2E(t) + \frac{1}{2}\int_{\Gamma_1}(m\cdot\nu)w'^2 d\Gamma - \int_{\Gamma_1}(m\cdot\nu)g(w')m\cdot\nabla w d\Gamma]$$

$$= - [1 - \frac{\epsilon(p-1)}{2} C(E(0))^{(p-3)/2}]\int_{\Gamma_1}(m\cdot\nu)w'g(w')d\Gamma$$

$$+ \epsilon(E(t))^{(p-1)/2}[-2E(t) + \frac{1}{2}\int_{\Gamma_1}(m\cdot\nu)w'^2 d\Gamma - \int_{\Gamma_1}(m\cdot\nu)g(w')m\cdot\nabla w d\Gamma]. \tag{3.20}$$

Let us estimate the last term in (3.20). We have

$$|\int_{\Gamma_1}(m\cdot\nu)g(w')m\cdot\nabla w d\Gamma| \le \int_{\Gamma_1}(m\cdot\nu)\left|\frac{g(w')}{w'}\right||w'||m\cdot\nabla w|d\Gamma$$

$$\le \frac{1}{2}\int_{\Gamma_1}(m\cdot\nu)\left|\frac{g(w')}{w'}\right|(\frac{1}{\delta}|w'|^2 + \delta|m\cdot\nabla w|^2)d\Gamma$$

$$\le \frac{1}{2\delta}\int_{\Gamma_1}(m\cdot\nu)w'g(w')d\Gamma + \delta C_1\int_{\Gamma_1}(m\cdot\nu)\left|\frac{g(w')}{w'}\right||\nabla w|^2 d\Gamma. \tag{3.21}$$

Substituting (3.21) for the corresponding term in (3.20) yields

$$F'_\epsilon(t) \le - [1 - \frac{\epsilon(p-1)}{2} C(E(0))^{(p-3)/2} - \frac{\epsilon}{2\delta}(E(0))^{(p-1)/2}]\int_{\Gamma_1}(m\cdot\nu)w'g(w')d\Gamma$$

$$+ \epsilon((E(t))^{(p-1)/2}[-2E(t) + \frac{1}{2}\int_{\Gamma_1}(m\cdot\nu)w'^2 d\Gamma + \delta C_1\int_{\Gamma_1}(m\cdot\nu)\left|\frac{g(w')}{w'}\right||\nabla w|^2 d\Gamma]. \tag{3.22}$$

Introduce the sets

$$\Gamma_1^0 = \{X\in\Gamma_1| |w'(X)| \le 1\},$$

$$\Gamma_1^1 = \{X\in\Gamma_1| |w'(X)| > 1\}.$$

We estimate the last term in (3.22) as follows:

$$\int_{\Gamma_1}(m\cdot\nu)\left|\frac{g(w')}{w'}\right||\nabla w|^2 d\Gamma \le (\int_{\Gamma_1^0} + \int_{\Gamma_1^1})(m\cdot\nu)\left|\frac{g(w')}{w'}\right||\nabla w|^2 d\Gamma. \tag{3.23}$$

Since g' is bounded on bounded sets and $g(0) = 0$, there is a constant M such that

$$|g(\xi)| \le M|\xi|, \quad |\xi| \le 1.$$

Therefore

$$\int_{\Gamma_1^0}(m\cdot\nu)\left|\frac{g(w')}{w'}\right||\nabla w|^2 d\Gamma \le M\int_{\Gamma_1^0}(m\cdot\nu)|\nabla w|^2 d\Gamma. \tag{3.24}$$

As for the second term on the right side of (3.23), we apply the inequality

$$ab \le \frac{1}{\alpha}a^\alpha + \frac{1}{\beta}b^\beta, \quad a > 0, \ b > 0, \ \alpha > 1, \ \beta > 1, \ \frac{1}{\alpha}+\frac{1}{\beta}=1,$$

with

$$\alpha = \frac{q+1}{2}, \ \beta = \frac{q+1}{q-1}, \ a = (m\cdot\nu)^{1/\alpha}|\nabla w|^2, \ b = (m\cdot\nu)^{1/\beta}\left|\frac{g(w')}{w'}\right|.$$

We obtain

$$\int_{\Gamma_1^1}(m\cdot\nu)\left|\frac{g(w')}{w'}\right||\nabla w|^2 d\Gamma$$

$$\le \frac{2}{q+1}\int_{\Gamma_1^1}(m\cdot\nu)|\nabla w|^{q+1}d\Gamma + \frac{q-1}{q+1}\int_{\Gamma_1^1}(m\cdot\nu)\left|\frac{g(w')}{w'}\right|^{\frac{q+1}{q-1}}d\Gamma. \tag{3.25}$$

Assumption (3.3) implies

$$\left|\frac{g(\xi)}{\xi}\right|^{\frac{q+1}{q-1}} \le C_0^{2/(q-1)}\xi g(\xi), \quad |\xi| \ge 1. \tag{3.26}$$

Use of (3.26) in (3.25), combined with (3.24), results in the estimate

$$\int_{\Gamma_1}(m\cdot\nu)\left|\frac{g(w')}{w'}\right||\nabla w|^2 d\Gamma \le M\int_{\Gamma_1^0}(m\cdot\nu)|\nabla w|^2 d\Gamma$$

$$+ \frac{2}{q+1}\int_{\Gamma_1^1}(m\cdot\nu)|\nabla w|^{q+1}d\Gamma + C_2\int_{\Gamma_1^1}(m\cdot\nu)w'g(w')d\Gamma. \tag{3.27}$$

By the Sobolev Imbedding Theorem we have $H^{\frac{1}{2}}(\Gamma) \subset L^r(\Gamma)$ with continuous injection for every $r > 1$. Therefore

$$\int_{\Gamma_1}m\cdot\nu|\nabla w|^r d\Gamma \le (\text{const.})\|\nabla w\|^r_{L^r(\Gamma)} \le (\text{const.})\|\nabla w\|^r_{H^{\frac{1}{2}}(\Gamma)}$$

$$\le (\text{const.})[a(w)]^{r/2} \le (\text{const.})[E(t)]^{r/2}. \tag{3.28}$$

We apply (3.28) to the first two terms on the right side of (3.27) and obtain

$$M\int_{\Gamma_1^0}(m\cdot\nu)|\nabla w|^2 d\Gamma + \frac{2}{q+1}\int_{\Gamma_1^1}(m\cdot\nu)|\nabla w|^{q+1}d\Gamma$$

$$\le C_3[E(t) + (E(t))^{(q+1)/2}] \le C_3[1 + (E(0))^{(q-1)/2}]E(t). \tag{3.29}$$

It follows from (3.27) and (3.29) that

$$\int_{\Gamma_1}(m\cdot\nu)\left|\frac{g(w')}{w'}\right||\nabla w|^2 d\Gamma \le C_3[1 + (E(0))^{(q-1)/2}]E(t) + C_2\int_{\Gamma_1}(m\cdot\nu)w'g(w')d\Gamma. \tag{3.30}$$

Replacing the last term in (3.22) with (3.30) gives the estimate

$$F'_\epsilon(t) \le -[1 - \tfrac{\epsilon(p-1)}{2}C(E(0))^{(p-3)/2} - \tfrac{\epsilon}{2\delta}(E(0))^{(p-1)/2}$$

$$- \epsilon\delta C_1(E(0))^{(p-1)/2}]\int_{\Gamma_1}(m\cdot\nu)w'g(w')d\Gamma$$

$$+ \epsilon((E(t))^{(p-1)/2}\{[-2 + \delta C_2(1 + (E(0))^{(q-1)/2})]E(t) + \tfrac{1}{2}\int_{\Gamma_1}(m\cdot\nu)w'^2 d\Gamma\} \quad (3.31)$$

for some constants C, C_1 and C_2.

The last term in (3.31) is bounded above as follows. We have from (3.2)

$$(E(t))^{(p-1)/2}\int_{\Gamma_1'}(m\cdot\nu)w'^2 d\Gamma \le \tfrac{1}{c_0}(E(0))^{(p-1)/2}\int_{\Gamma_1}(m\cdot\nu)w'g(w')d\Gamma. \quad (3.32)$$

Also, by Hölder's inequality

$$\int_{\Gamma_1^0}(m\cdot\nu)w'^2 d\Gamma \le C_3[\int_{\Gamma_1^0}(m\cdot\nu)|w'|^{p+1}d\Gamma]^{2/p+1}.$$

Use of Hölder's inequality once again gives

$$(E(t))^{(p-1)/2}\int_{\Gamma_1^0}(m\cdot\nu)w'^2 d\Gamma \le C_3(E(t))^{(p-1)/2}[\int_{\Gamma_1^0}(m\cdot\nu)|w'|^{p+1}d\Gamma]^{2/p+1}$$

$$\le C_4\int_{\Gamma_1^0}(m\cdot\nu)|w'|^{p+1}d\Gamma + \tfrac{p-1}{p+1}(E(t))^{(p+1)/2}$$

$$\le C_4\int_{\Gamma_1^0}(m\cdot\nu)w'g(w')d\Gamma + \tfrac{p-1}{p+1}(E(t))^{(p+1)/2}. \quad (3.33)$$

for some constant C_4. Therefore, from (3.32), (3.33) we have

$$(E(t))^{(p-1)/2}\int_{\Gamma_1}(m\cdot\nu)w'^2 d\Gamma$$

$$\le C_4[1 + (E(0))^{(p-1)/2}]\int_{\Gamma_1}(m\cdot\nu)w'g(w')d\Gamma + \tfrac{p-1}{p+1}(E(t))^{(p+1)/2}.$$

Inserting the last inequality back into (3.31) gives an estimate of the form

$$F'_\epsilon(t) \le -[1 - \epsilon C_1 - \epsilon C_2(E(0))^{(p-3)/2}$$

$$- \epsilon(C_3 + \tfrac{1}{2\delta})(E(0))^{(p-1)/2}]\int_{\Gamma_1}(m\cdot\nu)w'g(w')d\Gamma$$

$$+ \epsilon(E(t))^{(p+1)/2}\{-2 + \delta C_4[1 + (E(0))^{(q-1)/2}] + \tfrac{p-1}{2(p+1)}\} \quad (3.34)$$

for some constants C_1 through C_4 and for $\epsilon > 0$, $\delta > 0$ and, say, $\delta \le 1$.

Choose $\delta > 0$ so small that

$$\delta C_4[1 + (E(0))^{(q-1)/2}] \le \tfrac{1}{2}.$$

Then choose $\epsilon > 0$ so small that

$$1 - \epsilon C_1 - \epsilon C_2 (E(0))^{(p-3)/2} - \epsilon (C_3 + \tfrac{1}{2\delta})(E(0))^{(p-1)/2} \geq 0.$$

We then obtain from (3.34)

$$F_\epsilon'(t) \leq - \epsilon (E(t))^{(p+1)/2} \tag{3.35}$$

as claimed.

Let $\beta > 0$, multiply (3.35) by $e^{-\beta t}$ and integrate from t to ∞. After an integration by parts we obtain

$$\epsilon \int_0^\infty e^{-\beta s} [E(s)]^{(p+1)/2} ds + \beta \int_t^\infty e^{-\beta s} F_\epsilon(s) ds \leq e^{-\beta t} F_\epsilon(t). \tag{3.36}$$

In view of (3.16) and (3.18) we have

$$\begin{cases} F_\epsilon(t) \leq [1 + \epsilon C (E(0))^{(p-1)/2}] E(t), \\ F_\epsilon(t) \geq [1 - \epsilon C (E(0))^{(p-1)/2}] E(t). \end{cases} \tag{3.37}$$

Therefore, $F_\epsilon(t) \geq 0$ if

$$1 - \epsilon C (E(0))^{(p-1)/2} \geq 0.$$

Letting $\beta \to 0$ in (3.36) and using (3.37) yields

$$\int_0^\infty [E(s)]^{(p+1)/2} ds \leq \frac{1}{\omega} E(t), \quad t \geq 0, \tag{3.38}$$

where

$$\omega = \epsilon [1 + \epsilon C (E(0))^{(p-1)/2}]^{-1}.$$

Proof of Corollary. We consider only the case $p > 1$. From (3.35), (3.37) we have

$$F_\epsilon'(t) \leq - \sigma [F_\epsilon(t)]^{(p+1)/2} \tag{3.39}$$

where

$$\sigma = \epsilon [1 + \epsilon C (E(0))^{(p-1)/2}]^{(p+1)/2}.$$

Upon integrating (3.36) we obtain (with $a = (p-1)/2$)

$$[F_\epsilon(t)]^a \leq (F_\epsilon(0))^a / [1 + a\sigma (F_\epsilon(0))^a] \tag{3.40}$$

provided $F_\epsilon(0) > 0$. Assume that

$$1 - \epsilon C (E(0))^{(p-1)/2} \geq \tfrac{1}{2}.$$

Then

$$\frac{1}{2} E(t) \leq F_\epsilon(t) \leq \frac{3}{2} E(t).$$ (3.41)

It follows from (3.40) and (3.41) that

$$E(t) \leq 3E(0)(1 + \lambda t)^{-2/(p-1)}$$ (3.42)

where

$$\lambda = \frac{a\sigma}{2}(E(0))^{(p-1)/2}.$$

Remark 3.3. The calculations leading to the estimate (3.4) do *not* utilize the assumption that $g(\cdot)$ is nondecreasing. Indeed, everything formally goes through assuming only that $g(\cdot)$ satisfies (3.2), (3.3), is locally Lipschitz continuous and $g(0) = 0$. On the other hand, the growth restriction (3.3), which was essential to the proof, should be unnecessary. An interesting question is whether utilizing monotonicity of g in a suitable manner would allow (3.3) to be dropped.

Remark 3.4. Stability results of a different nature, using nonlinear feedback at the boundary, were obtained in [7] in the case of the one dimensional wave equation. These results were greatly generalized in [6] to include a variety of second order systems of Petrowsky type, including wave equations in $\mathbb{R}^n$ and the plate equation (1.1), and a number of different possible boundary conditions. In both papers, monotone and possibly *multivalued* feedbacks $f(w')$ (important for applications) were considered. The point of both papers is to characterize the ω-limit set of the initial data and to establish strong convergence of solutions to the ω-limit set.

REFERENCES

[1] Brezis, H., *Operateurs Maximaux Monotones et Semi–Groups de Contractions dans les Espaces de Hilbert*, North Holland, Amsterdam, 1973.

[2] Browder, F., *Problemes Non–Lineaires*, Séminaire de Mathématiques Université de Montréal, 1965.

[3] Duvaut, G. and J. L. Lions, *Inequalities in Mechanics and Physics*, Springer–Verlag, Berlin, 1976.

[4] Lagnese, J. E., *Uniform boundary stabilization of homogeneous isotropic plates*, in Lecture Notes in Control and Information Sciences, Vol. 102, Springer–Verlag, Berlin, 1987, pp. 204–215.

[5] Lagnese, J. E. and J. L. Lions, *Modelling, Analysis and Control of Thin Plates*, R.M.A. Collection, Masson, Paris, 1988.

[6] Lasiecka, I., *Stabilization of wave and plate–like equations with nonlinear dissipation on the boundary*, to appear.

[7] Wang, H–K and G. Chen, *Asymptotic behavior of solutions of the one–dimensional wave equation with a nonlinear boundary stabilizer*, SIAM J. Control and Optimization, to appear.

[8] Zuazua, E., *Stabilization of the wave equation by nonlinear boundary feedback*, to appear.

Acknowledgements. This research was supported by the Air Force Office of Scientific Research through grant AFOSR–86–0162. The author wishes to thank E. Zuazua for useful suggestions concerning the proof of Theorem 2.

A GENERALIZATION OF A THEOREM OF DATKO AND PAZY

Walter Littman

University of Minnesota

1. Introduction.

A theorem of Datko, both in its original version [D] and in its generalization by Pazy [P] (stated below as Theorem 1) have proved very useful in proving exponential decay in a number of evolution problems.

Theorem 1. [P]. Let $S(t)$ be a C_0 semigroup on a Banach space B . If for some value of p , $1 \leq p < \infty$,

$$\int_0^\infty \|S(t)x\|^P dt < \infty \quad \text{for every} \quad x \quad \text{in} \quad B , \qquad [1]$$

then there are constants $M \geq 1$ and $\mu > 0$ such that

$$\|S(t)\| \leq Mc^{-\mu t} \quad \text{for} \quad t \geq 0 .$$

Datko's original theorem applied only to Hilbert spaces in the case $p = 2$.

Our aim is to prove the following generalization of Theorem 1.

Theorem 2. Let $S(t)$ be a C_0 semigroup on a Banach space B . Let $\alpha = \alpha(t)$ be a continuous strictly increasing function defined for $t \geq 0$, such that $\alpha(0) = 0$. If for every $x \in B$

$$\int_0^\infty \alpha(\|S(t)x\|) dt < \infty , \qquad [2]$$

then there exist positive constants M_1, μ , such that

$$\|S(t)\| \leq M_1 e^{-\mu t} \quad \text{for} \quad t \geq 0 . \qquad [3]$$

While Datko's proof used the idea of a Liapunov functional in Hilbert space, Pazy's proof (which is much simpler) is based on the fact that the integral in [1] defines a norm in a Banach space, making it possible to use such tools as the closed graph theorem. The integral in [2], on the other hand (without additional assumptions on $\alpha(t)$) cannot be used to define a Banach space or even a Frechet space.

Nevertheless it is still conceivable that an abstract approach is possible. However we shall follow a different route. In the proof of Theorem 2 we are going to use an explicit construction, which has the advantange that it could possibly be applied in other situations (such as nonlinear problems) where the abstract methods fail.

The explicit construction just referred to is an outgrowth of joint work with L. Markus [LM] where it was used to show non exponential decay in a particular situation involving an elastic beam.

2. Preparatory Lemmas

For the remainder of this paper we are going to assume that $S(t)$ is a C_0 semigroup. In that case there exist constants $M_2 \geq 1$ and a positive ω such that

$$\|S(t)\| \leq M_2 e^{\omega t} \quad \text{for} \quad t \geq 0 .$$

Also, $\alpha = \alpha(t)$ will be a nonnegative, strictly increasing function for $t \geq 0$ with $\alpha(0) = 0$.

Lemma 1. If for a particular $x \in B$

$$\int_0^\infty \alpha(\|S(t)x\|) dt < \infty$$

then $\|S(t)x\| \to 0$ as $t \to 0$.

Proof: We adapt the proof in [P]. We argue by contradiction. If the conclusion were false, we could find a $\delta > 0$ and a sequence of positive t_j $(j = 1, 2, \ldots)$ approaching infinity such that $\|S(t_j)x\| \geq \delta$. Without loss of generality we can assume that $t_{j+1} - t_j > \omega^{-1}$. If we set $\Delta_j = [t_j - \omega^{-1}, t_j]$, then $m(\Delta_j) = \omega^{-1} > 0$ (m = Lebesgue measure) and the intervals Δ_j do not overlap. For $t \in \Delta_j$, we have

$$\|S(t)x\| \geq \delta(M_2 e)^{-1}$$

and therefore

$$\int_0^\infty \alpha(\|S(t)x\|) dt \geq \sum_{i=1}^\infty \int_{\Delta_j} \alpha(\|S(t)x\|) dt$$

$$\geq \sum_{j=1}^\infty \int_{\Delta_j} \alpha(\delta(M_2 e)^{-1}) dt = \alpha(\delta(M_2 e)^{-1}) \sum_{i=1}^\infty m(\Delta_j) = \infty .$$

Since this is impossible, we must have $\|S(t)x\| \to 0$ as claimed.

Lemma 2. Let $S(t)$ be a C_0 semigroup on a Banach space B such that for all x in B and $t \geq 0$ we have:

    i) $\|S(t)\| \leq M$

    ii) $\|S(t)\| \geq 1$.

Then for any compact interval of the form $[0,b]$ there exists an element x in B such that $\|x\| = 1/M$ and

$$\frac{9}{10M^2} \leq \|S(t)x\| \leq 1 \quad \text{for} \quad t \in [0,b] .$$

Proof. There exists an element $x_1 \in B$ such that $\|x_1\| = 1$ and $\|S(b)x_1\| > \frac{9}{10}$. Now for $0 \leq t \leq b$,

$$\|S(b)x_1\| = \|S(b-t)S(t)x_1\| \leq \|S(b-t)\|\|S(t)x_1\| ,$$

$$\frac{9}{10} \leq M\|S(t)x_1\| ,$$

$$\frac{9}{10M} \leq \|S(t)x_1\| \leq M .$$

Setting $x = x_1/M$, we obtain the desired result.

3. Beginning of the Proof of Theorem 2

From Lemma 1 it follows that $\|S(t)x\| \to 0$ as $t \to 0$ for each x in B . Hence $\|S(t)x\|$ is bounded for all $t \geq 0$, for each x in B . By the uniform boundedness principle there exists a bound $M \geq 1$, independent of x , such that

$$\|S(t)\| \leq M \quad \text{for all} \quad t \geq 0 .$$

(Note that this part of the argument is nonconstructive. For a completely constructive proof, this would have to be an additional assumption).

To reach the conclusion of Theorem 2, it suffices to show the existence of a particular value $t_1 > 0$ such that $\|S(t_1)\| = \theta < 1$. For then we have

$$\|S(nt_1)x\| \leq \theta^n\|x\| .$$

Now given $t > 0$ pick n such that

$$nt_1 \leq t < (n+1)t_1 .$$

Then

$$\|S(t)x\| = \|S(t-nt_1)S(nt_1)x\|$$

$$\leq M\theta^n\|x\| = \frac{M}{\theta}\theta^{n+1}\|x\|$$

$$\le M\theta^{-1}\theta^{t/t_1}\|x\| = M_2 e^{-\omega t} ,$$

where $\omega = |\log \theta|/t_1$.

We are now going to prove Theorem 2 by assuming [3] does not hold and hence the hypotheses of Lemma 2 hold. We are going to use Lemma 2 to explicitly construct an $x \in B$ for which the integral in [2] is infinite.

4. Completion of the Proof of Theorem 2

Assume the conclusion of Theorem 2 is false. Then it follows from the preceding section that $\|S(t)\| \ge 1$ for $t \ge 0$. For simplicity we shall assume that $\alpha(t) \to \infty$ as $t \to \infty$. The minor modifications needed in the absence of this assumption will be obvious. Since α is strictly increasing, it has an inverse function $\alpha^{-1}(t)$ defined for $t \ge 0$. Now define

$$\psi(t) = \alpha^{-1}(1/1+t) \qquad t \ge 0 ,$$

so that $\alpha(\psi(t)) = 1/1+t$, and hence

$$\int_0^\infty \alpha(\psi(t)) \ dt = \infty .$$

Note that $\psi(t)$ is positive, strictly decreasing and $\psi(t) \to 0$ as $t \to \infty$.

Step 1. Pick $a_1 \ge 0$ so that

$$\psi(t) < \frac{1}{10M^2} \quad \text{for} \quad t \ge a_1 .$$

Then pick $b_1 > a_1$ in such away that, letting $I_1 = [a_1, b_1]$

$$\int_{I_1} \alpha(\psi(t)) \ge 1 .$$

Finally, using Lemma 1, pick x_1^0 so that

$$\frac{4}{10M^2} \le \|x_1(t)\| \le \frac{5}{10} \quad \text{on} \quad [0, b_1]$$

where $x_1(t) \equiv S(t)x_1^0$, and $\|x_1(0)\| = \frac{5}{10M}$.

Step 2. Pick $a_2 \ge b_1$ so that

i) $\|x_1(t)\| < \frac{1}{2!} \frac{1}{10^2}$ for $t \ge a_2$

ii) $\psi(t) < \frac{1}{2!} \frac{1}{10^2 M^2}$ for $t \ge a_2$.

Pick $b_2 > a_2$ such that (with $I_2 = [a_2, b_2]$)

iii) $\int_{I_2} \alpha(\varphi(t))dt \geq 1$

iv) Pick $x_2(t) = S(t)x_2^o$ such that

$$\frac{4}{10^2 M^2} \leq \|x_2(t)\| \leq \frac{5}{10^2} \quad \text{in} \quad [0,b_2] \quad \text{and} \quad \|x_2(0)\| = \frac{5}{10^2 M} .$$

<u>Step n</u> Pick $I_n = [a_n,b_n]$ to the right of I_{n-1} , such that

i) $\sum_{k=1}^{n-1} \frac{1}{k!}\|x_k(t)\| \leq \frac{1}{n!M^2 10^n}$ for $t \geq a_n$ (hence on I_n)

ii) $\psi(t) \leq \frac{1}{n!10^n M^2}$ for $t \geq a_n$ (hence on I_n)

iii) $\int_{I_n} \alpha(\varphi(t))dt \geq 1$

iv) Pick $x_n(t) = S(t)x_n^o$ such that

$$\frac{4}{10^n M^2} \leq \|x_n(t)\| \leq \frac{5}{10^n} \quad \text{on} \quad [0,b_n] , \quad \text{(hence on } I_j \text{ for } j \leq n)$$

and $\|x_n(0)\| = \frac{5}{10^n M}$.

Now define $x(t) = \sum_{k=1}^{\infty} \frac{1}{k!} x_k(t)$ for $0 \leq t < \infty$.

Since $\|x_n(0)\| = \frac{5}{10^n M}$, $\|x_n(t)\| \leq \frac{5}{10^n} < 1$, the series converges

uniformly and represents a continuous B space valued function $x(t) = S(t)x^o$, where $x^o = x(0)$.

<u>Estimating</u> $\|x(t)\|$ <u>on</u> I_n <u>from below</u>:

On I_n we have:

$$\|x(t)\| \geq \frac{1}{n!}\|x_n(t)\| - \sum_{k \neq n} \frac{1}{k!}\|x_k(t)\|$$

$$\sum_{k<n} \frac{1}{k!} \|x_k(t)\| \leq \frac{1}{n!M^2 10^n}$$

$$\sum_{k>n} \frac{1}{k!}\|x_k(t)\| \leq \frac{1}{(n+1)!} \frac{5}{10^{n+1}} + \frac{1}{(n+2)!} \frac{5}{10^{n+2}} + \cdots$$

$$\leq \frac{1}{(n+1)!} \frac{5}{10^{n+1}}[1 + \frac{1}{10} + \frac{1}{100} + \cdots]$$

$$\leq \frac{1}{(n+1)!} \frac{6}{10^{n+1}} \leq \frac{1}{(n+1)!} \frac{1}{10^n} .$$

Hence on I_n we have

$$\|x(t)\| \geq \frac{4}{10^n M^2} \frac{1}{n!} - \frac{1}{10^n M^2 n!} - \frac{1}{10^n (n+1)!}$$

$$\|x(t)\| \geq \frac{1}{10^n n!}[\frac{3}{M^2} - \frac{1}{n}] \geq \frac{1}{M^2 10^n n!}$$

provided $n > M^2$. Therefore

$$\int_0^T \alpha(\|x(t)\|)dt \geq \int_0^T \alpha(\psi(t))dt \geq \Sigma \int_{I_n} \alpha(\psi(t))dt$$

where the summation extends from the greatest integer $\leq M^2+1$ to the last n sucht hat I_n is to the left of T . As $T \to \infty$ the last term approaches infinity. This proves Theorem 2.

Acknowledgement: This research was partially supported by NSF grant 87-22402.

Walter Littman
Department of Mathematics
University of Minnesota
Minneapolis, MN 55455
U.S.A.

References

[D] R. Datko, Extending a theorem of A.M. Liapunov to Hilbert space, J. Math. Anal. and Appl. 32 (1970), 610-616.

[LM] W. Littman and L. Markus, Some recent results on control and stabilization of flexible structures, University of Minnesota Mathematics Report #87-139, (to appear in Proceedings of COMCON workshop held in Montpellier, France, Dec. 1987).

[P] A. Pazy, <u>Semigroups of Linear Operators and Applications to Partial Differential Equations</u>, Springer Verlag, New York, 1983 (p. 116).

THE DESTABILIZING EFFECT OF DELAYS ON CERTAIN VIBRATING SYSTEMS

R. DATKO

(GEORGETOWN UNIVERSITY)

Introduction

In this paper we shall summarize the results obtained in a series of recent papers. ([3], [4], [5], [7]). These concern the well-known phenomenon of destabilization due to time delays. In finite-dimensional vibrating systems (see e.g. [4]) it has long been known that a damped system with no time delays becomes unstable if time delays larger than a specified magnitude, which depend on the systems, occur in the damping terms. However these systems always preserve their stability properties for "small" delays. In fact for linear systems one can easily estimate the range where delays are permitted that preserve stability. For infinite dimensional vibrating systems the same behavior often does not occur, especially in the case of some well-known systems modelled by linear partial differential equations with nonhomogeneous boundary conditions whose undelayed versions are exponentially stable, but for whose delayed versions something very unpleasant happens. Namely that arbitrarily small time delays in the boundary values result in an unstable system. Since the boundary conditions in these systems are often viewed as feedback stabilizing controls, this implies that these systems are ill-posed in terms of feedback stabilization. Even some infinite dimensional vibrating systems which are damped when no feedback is present can be destabilized by small time delays in the feedback. In this introduction we shall give two simple motivating examples of one dimensional vibrating systems which are precursors of what occurs in infinite dimensions. In Section 1 we summarize the main results obtained by us to date and we present an example of a controlled vibrating system which exhibits the above destabilizing phenomenon.

Consider the two one dimensional vibrating systems:

$$\ddot{x}(t) + bx(t) = -\alpha \dot{x}(t), \tag{0.1}$$

$$\ddot{x}(t) + a\dot{x}(t) + bx(t) = -\alpha \dot{x}(t), \tag{0.2}$$

where a, b, α are fixed positive numbers. For purposes of exposition we view the right-hand sides of (0.1) and (0.2) as feedback controls on the velocity of the respective systems. Let $h \geq 0$ be fixed.

If we replace $\dot{x}(t)$ by $\dot{x}(t-h)$ on the right-hand sides of (0.1) and (0.2) and compute the characteristic equations of the resulting systems, i.e. the equations determining their spectra, we obtain

$$\mu^2 + b = -\alpha \mu e^{-\mu h} \tag{0.3}$$

and

$$\mu^2 + a\mu + b = -\alpha \mu e^{-\mu h}. \tag{0.4}$$

We first examine (0.3). We look for values of $h > 0$ where $\mu = i\omega$, $\omega > 0$, i.e. where

$$\ddot{x}(t) + bx(t) = -\alpha \dot{x}(t - h) \tag{0.5}$$

has periodic solutions. A simple calculation shows that this occurs if

$$\omega = \frac{\alpha + \sqrt{\alpha^2 + 4b}}{2}, \tag{0.6}$$

$$h = \frac{\pi}{\alpha + \sqrt{\alpha^2 + 4b}}. \tag{0.7}$$

If we fix α in (0.5) but permit the restoring force, b, to go to infinity we see that h tends to zero. If one observes that many infinite dimensional homogeneous vibrating systems are uncoupled systems of the form

$$\ddot{x}_n(t) + b_n x(t) = 0, \quad n = 1, 2, \cdots, \tag{0.8}$$

where $b_n \to \infty$ as $n \to \infty$, it is not difficult to imagine what might occur if we force (0.8) by a vector feedback which results in a system of the form

$$\ddot{x}_n(t) + b_n x_n(t) = -\alpha_n \sum_{j=1}^{\infty} \alpha_j \dot{x}_j(t - h), \tag{0.9}$$

$n = 1, 2, \cdots$. Naturally for each n (0.9) is not as simple as (0.5), but, pending a more difficult mathematical analysis, the limiting behavior of (0.5), vis á vis the existence of periodic solutions, as $b \to \infty$ is essentially the same as that of (0.9). The precise results are given in Section 1.

In the case of (0.2) if we change the right-hand side $-\alpha\dot{x}(t)$ to $-\alpha\dot{x}(t - h)$ we obtain the equation

$$\ddot{x}(t) + a\dot{x}(t) + bx(t) = -\alpha\dot{x}(t - h), \tag{0.10}$$

whose spectrum is given by (0.4). If we attempt to find an $h > 0$ such that (1.4) has a solution $\mu = i\omega$, $\omega > 0$, we find that this is impossible if $\alpha < a$ and that if $\alpha \geq a$ there is a solution

$$\omega = \frac{\sqrt{\alpha^2 - a^2} + \sqrt{\alpha^2 - a^2 + 4b}}{2} \tag{0.11}$$

$$h = \frac{\theta}{\omega}, \tag{0.12}$$

where $\frac{\pi}{2} < \theta \leq \pi$ satisfies

$$\cos\theta = -\frac{a}{\alpha}. \tag{0.13}$$

Thus if we fix α and a in (0.5), but let $b \to \infty$ we see that h in (0.12) tends to zero. On the other hand if $a > |\alpha|$ no matter how large the restoring force, b, in (0.5) becomes the system (0.5) remains uniformly exponentially stable for all $h \geq 0$. As we shall see in Section 1 this mimics the behavior of certain infinite dimemsional vibrating systems of the form

$$\ddot{x}(t) + B\dot{x}(t) + Ax(t) = -\alpha(\alpha, \dot{x}(t - h)), \tag{0.14}$$

where B is a positive symmetric operator in a Hilbert space H, A is a positive unbounded operator in H, α is vector in a Banach space $X \neq H$ and $(\alpha, \cdot)$ represents a suitable bilinear form.

I. THE SYSTEMS INVOLVED AND THEIR STABILITY PROPERTIES

<u>1.1 Notation and Conventions.</u> (i) $\mathbb{C}$ will denote the complex plane, $\mathbb{R}^+$ the non-negative and real line, $\mathbb{R}$ the real line and $\mathbb{N}$ the positive integers. The symbol l^p, $1 \leq p \leq \infty$, will denote the usual sequence spaces of complex numbers whose pth powers are summable. If $\alpha \in l^q$, $1 \leq q$, and $\frac{1}{p} + \frac{1}{q} = 1$, then $(\alpha, \cdot)$ will denote the linear mapping from $l^p \to \mathbb{C}$ determined by α. The identity mapping on any Banach space is denoted by I and the norm of all Banach spaces by $|\cdot|$.

(ii) Let $0 < \lambda_1 < \lambda_2 < \cdots$, be a sequence in $\mathbb{R}^+$ such that for some $\tau_0 \in \mathbb{R}^+$ $\tau_o < \lambda_1$ and $\tau_0 \leq \lambda_{k+1} - \lambda_k$ for all $k \in \mathbb{N}$. Let $\{e_k\}$ denote the unit vectors in l^p defined by

$$e_k = \{\delta_{kl}\}, \tag{1.1}$$

where δ_{kl} is the Kronecker delta. Let $A: l^p \to l^p$, $1 \leq p \leq \infty$ be the unbounded linear operator with dense domain defined by

$$A e_k = \lambda_k^2 e_k, \quad k \in N \tag{1.2}$$

The positive square root of A is given by relations

$$A^{\frac{1}{2}} e_k = \lambda_k e_k, \quad k \in \mathbb{N}, \tag{1.3}$$

(iii) Let $\alpha \in l^\infty$ be fixed. We shall assume $\alpha = \{a_k\}$ has real components and these satisfy the inequalities

$$0 < \tau_1 \leq |a_k| \leq \tau_2 < \infty, \quad h \in \mathbb{N}, \tag{1.4}$$

where τ_1 and τ_2 are fixed.

1.2 The Conservative Vibrating System and its Feedback.

In l_1 we consider the differential equations:

$$\ddot{y}(t) + Ay(t) = 0; \tag{1.5}$$

$$\ddot{y}(t) + Ay(t) = -(\alpha, \dot{y}(t))\alpha; \tag{1.6}$$

$$\ddot{y}(t) + Ay(t) = -(\alpha, \dot{y}(t - h))\alpha, \tag{1.7}$$

where $h \geq 0$ is fixed.

Equation (1.5) with initial conditions $y(0) = y_0$ and $\dot{y}(0) = y_1$ is our conservative vibrating system. Equation (1.6) is the feedback stabilized system. However, the solutions of this equation can only be considered in some weak sense the definition of which is given below. But first we must introduce one additional operator. Thus we define the linear operator $\alpha\alpha^T$ from $l^1 \to l^\infty$ by the infinite matrix where rows and columns satisfy

$$\alpha\alpha^T = \{\alpha_{ij}\}, \tag{1.8}$$

where

$$\alpha_{ij} = a_i a_j, \quad i, j \in \mathbb{N}.$$

Notice that if $y \in l^\infty$ then

$$\alpha\alpha^T y = (\alpha, y)\alpha \tag{1.9}$$

and if $(\mu^2 I + A)^{-1}$, $\mu \in \mathcal{C}$, exists, then

$$\alpha\alpha^T(\mu^2 I + A)^{-1}y = (\alpha, \mu^2 I + A)^{-1}y)\alpha \qquad (1.10)$$

and if $y \in l^1$

$$(\mu^2 I + A)^{-1}\alpha\alpha^T y = (\mu^2 I + A)^{-1}\alpha(\alpha, y). \qquad (1.11)$$

The following definition describes the sense in which (1.5), (1.6) and (1.7) have solutions in a weak sense.

Definition 1.1. Let $\dot{y}(0) = y_0$, $\dot{y}(0) = y_1$ and $\dot{y}(t) = \varphi(t)$, $t \in [-h, 0]$, be vectors in l^1. If the formal Laplace transforms of (1.5), (1.6) or (1.7) with these initial conditions represent the transforms of Lebesgue-measurable mappings from $\mathbb{R}^+$ into l^∞ then we define the inverse transforms as weak solutions of the respective differential equations.

Remarks 1.2. In the case of equation (1.5) we could actually consider strong solutions in l^1 with a dense subset of initial values in l^1. However, because α in equations (1.6) and (1.7) is in l^∞ and the generalized eigenvalues (to be defined below) of these equations possess the property that they are mapped by the operator A into l^∞ we have chosen to use the above definition of weak solution.

Definition 1.3. The generalized eigenvalues and generalized eigenvectors of (1.6) and (1.7) are respectively any $\lambda \in C$ and $y \in l^1$ which satisfy the following equations:

$$(\lambda^2 I + A)y = -\lambda(\alpha, y)\alpha; \qquad (1.12)$$

$$(\lambda^2 I + A)y = -\lambda e^{-\lambda h}(\alpha, y)\alpha. \qquad (1.13)$$

(Notice that $Ay \in l^\infty$ in both (1.12) and (1.13).)

Theorem 1.4. (i) The generalized eigenvalues of (1.6) satisfy the equation

$$\mu(\alpha, (\mu^2 I + A)^{-1}\alpha) = -1 \qquad (1.14)$$

and if (1.14) has any solutions they must lie in a left half plane

$$P = \{\mu: \text{Re } \mu \leq -\delta_0, \ \delta_0 > 0\} \qquad (1.15)$$

(ii) The generalized eigenvalues of (1.13) are given by the equation

$$\mu e^{-\mu h}((\alpha, (\mu^2 I + A)^{-1}\alpha, \alpha) = -1 \qquad (1.16)$$

and there exist solutions μ of (1.16), with $\text{Re } \mu > 0$, for h in $\mathbb{R}^+$ arbitrarily small.

The proofs of the statements in Theorem 1.4 are given in [6] and are based on the observation that if (1.12) or (1.13) is satisfied then there exists a nontrivial eigenvector x such that

$$x = (\mu^2 I + A)^{-1}\alpha. \qquad (1.17)$$

Theorem 1.5. For any y_0 and y_1 in l^1, with $Ay_0 \in l^\infty$, there exists a unique weak solution of (1.6), with y_0 and y_1 as initial conditions, and a $\delta_1 > 0$ such that the solution of (1.6) satisfies the inequality

$$|y(t)| \leq M \left[\|Ay_0\| + |y_1|\right] e^{-\delta_1 t}$$

for all $t \in \mathbb{R}^+$. The numbers δ_1 and M are independent of y_0 and y_1.

The proof of Theorem (1.5) is based on two observations. The first is that the complex function

$$\psi(\mu) = \frac{1}{1 + \mu(\alpha, \mu^2 I + A)^{-1}\alpha}$$ (1.18)

is uniformly bounded in some right complex plane $\operatorname{Re}\mu \geq -\delta_2, \delta_2 > 0$. (Notice that $\psi(\pm i\lambda_j) = 0$ for the $\{\lambda_j\}$ which define A in (1.2).) The proof of this statement may be found in [6]. The second observation is that if $y_0 \in l_1$, $y_1 \in l_1$ and $Ay_0 \in l^\infty$ the formal Laplace transform of (1.6) with these initial conditions may be given by the expression

$$y(\mu) = (\mu^2 I + A + \mu\alpha\alpha^T)^{-1}(\mu y_0 + y_1 + \alpha\alpha^T y_0)$$ (1.19)

and that

$$(\mu^2 I + A + \mu\alpha\alpha^T)^{-1}(\mu y_0 + \alpha\alpha^T y_0) = \frac{1}{\mu}\left[I - (\mu^2 I + A + \mu\alpha\alpha^T)^{-1}A\right] y_0.$$ (1.20)

These two observations coupled with a version of the Paley-Wiener Theorem are then used to establish the validity of Theorem 1.5.

Then using Theorems (1.4) and (1.5) we easily establish the fact that arbitrarily small delays $-(\alpha, \dot{y}(t-h)\alpha$ in the feedback $-(\alpha, \dot{y}(t))\alpha$ change the uniformly exponentially stable system (1.6) into the unstable system (1.7).

A question naturally arises as to the universality of the models (1.6) and (1.7). The one-dimensional wave equation with homogeneous Dirichlet boundary conditions on one end and a Neumann-type boundary condition $\frac{\partial u}{\partial x} = -\frac{\partial u}{\partial t}$ on the other end (see e.g. [3]) is one model. The one-dimensional Euler-Bernoulli beam equation with a variety of boundary conditions is another model (see e.g. [1] is another case in point). But in general, with the exception of special solutions (for example certain radial solutions of the wave equation on spheres in $\mathbf{R}^n$, $n \geq 2$) the general hyperbolic partial differential equations of Mathematical Physics do not satisfy (ii) in Notation and Conventions (see e.g. [2], Chapter V). However because of Theorem 1.5 we can consider subsystems of most of these equations which do satisfy our conditions and whose stabilizing feedbacks do not radically affect the remaining parts of the system. This is the content of the next result.

Theorem 1.6. Consider the coupled system

$$\ddot{y} + Ay = -(\dot{y}, \alpha)\alpha,$$ (1.21)

$$\ddot{x} + Bx = -(\dot{y}, \alpha)\beta,$$ (1.22)

where (1.21) satisfies the conditions in Notation and Conventions, the operator B generates a cosine group, $\beta \in l^\infty$ and the initial values y_0, y_1, x_0 and x_1 satisfy are such that Ay_0 and Bx_0 are in l^∞. Then the solutions of (1.21) decay exponentially in conformity with Theorem 1.5 and the solutions of (1.22) are uniformly bounded by constants of the form

$$M_1\left[\|Ay_0\| + |Bx_0| + |y_1| + |x_1|\right],$$ (1.23)

where M_1 is independent of x_0, y_0, x_1 and y_1.

The proof is based on the fact that the solutions of (1.22) can be written formally as

$$x(t) = \left[\cos B^{\frac{1}{2}}t\right] x_0 + B^{-\frac{1}{2}}\left[\sin B^{\frac{1}{2}}t\right] x_1 - \int_0^t \left[\sin B^{\frac{1}{2}}(t-\sigma)\right]\beta(\alpha, \dot{y}(\sigma))d\sigma$$ (1.24)

and that the Laplace transform of the integral term on the right side of (1.24) is

$$-(\mu^2 I + A)^{-1}[\mu(\alpha, y(\mu))\beta - (\alpha, y_0)\beta], \tag{1.25}$$

where $|\mu(\alpha, y(\mu))|$ is uniformly bounded on $\text{Re}\,\mu \geq -\delta_2$. (This follows from the boundedness of (1.18) and equation (1.20).)

Remark 1.7. Since many of the standard vibrating systems of Mathematical Physics can be decomposed into systems of the form (1.21) and (1.22) (see e.g. [2], Chapter V) we obtain some generality for Theorems 1.4 and 1.5. That is there exist a very general class of vibrating systems decomposable into coupled systems of the form (1.21) and (1.22), where (1.21) decays at a uniform rate and (1.22) is uniformly bounded and such that given $h \in R^+$ arbitrarily small the system

$$\ddot{y}(t) + Ay(t) = -(\alpha, \dot{y}(t - h))\alpha, \tag{1.26}$$

$$\ddot{x}(t) + Bx(t) = -(\alpha, \dot{y}(t - h))\beta \tag{1.27}$$

is unstable.

1.3. Some Damped Vibrating Systems.
In this section we shall briefly outline some results for damped vibrating systems which parallel the results in Section 1.2. The basic systems we consider are

$$\ddot{x}(t) + 2a\dot{x}(t) + (A + a^2 I)x(t) = 0, \tag{1.28}$$

$$\ddot{x}(t) + 2a\dot{x}(t) + (A + a^2 I)x(t) = -(\alpha, \dot{x}(t))\alpha, \tag{1.29}$$

$$\ddot{x}(t) + 2a\dot{x}(t) + (A + a^2 I)x(t) = -(\alpha, \dot{x}(t - h))\alpha, \tag{1.30}$$

where $a > 0$ is fixed.

Theorem 1.8. If the meromorphic function

$$\mu(\alpha, ((\mu + a)^2 I + A)^{-1}\alpha) = f(\mu) \tag{1.31}$$

has an infinite number of points $\{i\omega_n\}$, ω_n-real, $\lim_{n \to \infty} \omega_n = \infty$, such that $|f(i\omega_n)| = 1$ then, given any $\epsilon > 0$ there exist a nontrivial periodic solution of (1.30) for some $0 < h < \epsilon$.

Theorem 1.9. If the hypothesis of Theorem 1.8 is satisfied at each point $\mu_n = i\omega_n$ and

$$\text{Re}\,\frac{d}{d\mu}\left[\frac{1}{\mu}\ln(\alpha, (\mu(\mu + a)^2 I + A)^{-1}\alpha)\right] \neq 0, \tag{1.32}$$

then given any $\epsilon > 0$ there exists $0 < h < \epsilon$ such that (1.30) has an exponentially divergent solution.

Example 1.10. The system where the $\{\lambda_j\}$ in (1.2) are given by $\lambda_j = j\pi$ and a_j in $\alpha = \{a_j\}$ satisfy $|a_j| = \sqrt{2}$ satisfies the hypotheses of Theorems 1.8 and 1.9. For example (1.31) in this case is

$$f(\mu) = \frac{\mu}{\mu + a}\left[\coth(\mu + a) - \frac{1}{\mu + a}\right]. \tag{1.33}$$

Another example may be found in [3].

REFERENCES

[1] G. Chen, S. G. Krantz, D. W. Ma, C. E. Wayne, II. II. West, The Euler-Bernoulli beam equation with boundary dissipation, Operator Methods for Optimal Control Problems (New Orleans, La., 1986), 67–96, Lecture Notes in Pure and Appl. Math., 108, Marcel Dekker, New York, 1987.

[2] R. Courant and D. Hilbert, "Methods of Mathematical Physics", Vol. I, Interscience Publishers Inc., New York, 1966.

[3] R. Datko, J. Lagnese and M. P. Polis, "An example of the effect of time delays in boundary feedback of wave equations", SIAM J. Cont. and Optimization 24 (1986), 152–156.

[4] R. Datko, "Not all feedback stabilized hyperbolic systems are robust with respect to small time delays in their feedbacks", SIAM J. Cont. and Optimization 26 (1988), pp. 697–713.

[5] R. Datko, "A rank-one perturbation result on the spectra of certain operators", submitted for publication.

[6] R. Datko, "Exponential stability and rank-one perturbations of certain conservative second-order differential equations", submitted for publication.

[7] R. Datko, "Even some internally damped vibrating systems lose stability when small delays occur in their feedbacks", in preparation.

Address: R. Datko, Department of Mathematics, Georgetown University, Washington, D.C., 20057.

FEEDBACK STABILIZATION OF A VISCOELASTIC ROD

Wolfgang Desch
Universität Graz, Graz, Austria

Kenneth B. Hannsgen and Robert L. Wheeler
Virginia Polytechnic Institute and State University, Blacksburg, VA, USA

1. Introduction. We consider an input-output system arising from a class of models for vibrating mechanical structures containing viscoelastic materials. A complex-valued transfer function $T_0(s)$ for the open-loop system is determined formally, and we consider a family of closed-loop transfer functions $T_k(s)$, parametrized by feedback kernels k. We present conditions under which $T_k(s)$ belongs to H^∞ in a halfplane $\{\Re s > \sigma_0\}$, $\sigma_0 < 0$, so that the closed-loop system maps $L^2(\mathbb{R}^+, e^{-\sigma_0 t})$ to itself. The results are illustrated for the case of a hybrid system containing a viscoelastic wave equation that models the stabilization of torsional vibrations in a viscoelastic rod with control and observation acting through an attached inertial mass.

2. Statement of results. Let H be an infinite dimensional real Hilbert space, let Y denote the Hilbert space $\mathbb{R} \times H$, and let X be a third infinite dimensional Hilbert space. Let

$$D: X \supset \mathrm{dom}\, D \to Y, \quad D^*: Y \supset \mathrm{dom}\, D^* \to X$$

be closed linear operators that are adjoint to each other with $\ker D = \{0\}$, $\ker D^* = \{0\}$, and where the embeddings of $\mathrm{dom}\, D$ into X and $\mathrm{dom}\, D^*$ into Y are compact. Define $B: \mathbb{R} \to Y$ and $C: Y \to \mathbb{R}$ by

$$Bu = \begin{bmatrix} u \\ 0 \end{bmatrix}, \quad C \begin{bmatrix} u \\ v \end{bmatrix} = u.$$

Let $E > 0$, and let $a(t)$ be completely monotone on $(0, \infty)$ with Bernstein representation [7]

$$a(t) = \int_\delta^\infty e^{-\rho t} dg(\rho) = e^{-\delta t} \int_0^\infty e^{-\rho t} dg(\rho + \delta),$$

where $\delta > 0$ and g is nondecreasing on $[0, \infty)$ with $\delta \in \mathrm{supp}\{dg\}$. We assume that

$\int_0^1 a(t)\,dt < \infty$ so that $\int_1^\infty \rho^{-1}\,dg(\rho) < \infty$, and consider the system

$$u'(t) = v(t),$$

$$v'(t) = D\sigma(t) + Bf(t), \qquad (1)$$

$$\sigma(t) = -ED^*u - \int_0^t a(t-r)D^*v(r)\,dr + h(t),$$

on $\mathbb{R}^+$ with initial data $u(0) = u_0$, $v(0) = v_0$, inhomogeneity h (h often accounts for history effects caused by the state of the system before $t = 0$), and with $f: \mathbb{R}^+ \to \mathbb{R}$ given.

Recently, Desch and Miller [2] developed a semigroup framework to study the stabilization properties of (1) when f is a compact feedback. In particular, it follows from [2] that the essential growth rate of system (1) with $\dot{f}(t) \equiv 0$ is the same as the essential growth rate of (1) with the feedback $f(t) = -kCv(t)$, where k is a positive constant. The essential growth rate of a semigroup is the infimum of the exponential growth rates that can be obtained by restricting the semigroup to the complement of a finite dimensional subspace determined by eigenvalues of the generator. On the other hand, for the specific example of the viscoelastic wave equation discussed in Section 5, numerical calculations in [4] show that the best growth rate that can actually be obtained by feedbacks of the form $f(t) = -kCv(t)$ may be considerably larger than the essential growth rate of system (1). This paper is motivated by the question of designing feedbacks to closely approximate the essential growth rate of abstract systems having the form (1).

Consider the system (1) with $u(0) = v(0) = 0$, $h(t) \equiv 0$, and the scalar input-output mapping

$$T_0: f \to w \equiv Cv. \qquad (2)$$

We impose the feedback condition

$$f(t) = f_0(t) - k * w(t)$$

$$= f_0(t) - \int_0^t k(t-s)Cv(s)\,ds, \qquad (3)$$

where

$$k \in L^1_{\sigma_0} = L^1(\mathbb{R}^+, e^{-\sigma_0 t}) = \{\, k \in L^1(\mathbb{R}^+) : \int_0^\infty e^{-\sigma_0 t}|k(t)|\,dt < \infty \,\} \qquad (4)$$

and $-\delta < \sigma_0 < 0$. Then (2) becomes

$$\tau_k: f_0 \to w. \tag{5}$$

At this point the domains of τ_0 and τ_k are not specified. We shall derive formal expressions for these mappings in the transform domain, so that τ_0 and τ_k can be expressed as bounded mappings on suitable spaces L^2_σ by Laplace inversion.

3.<u>The transfer functions</u>. Formal application of the Laplace transform to (1) yields

$$\hat{w}(s) = sC[s^2 + s\hat{A}(s)DD^*]^{-1}B\hat{f}(s)$$
$$:= T_0(s)\hat{f}(s) \tag{2*}$$

for (2) and

$$\hat{w}(s) = \frac{T_0(s)}{1 + \hat{k}(s)T_0(s)}\hat{f}_0(s)$$
$$:= T_k(s)\hat{f}_0(s) \tag{5*}$$

for (5). Here $s\hat{A}(s) = E + s\hat{a}(s) = E + s\int_\delta^\infty (s+r)^{-1}dg(r)$ is analytic in $\{\Re s > -\delta\}$, and it is easy to check that $s\hat{A}(s) = 0$ has at most one solution in $\{\Re s > -\delta\}$ which, if it exists, is real and negative. Let σ_* denote this zero of $s\hat{A}(s)$ if such a zero exists; otherwise let $\sigma_* = -\delta$. Since DD^* is self-adjoint and positive, T_0 is meromorphic in $\{\Re s > \sigma_*\}$. Indeed, setting $\alpha(s) = [s\hat{A}(s)]^{1/2}$ (principal branch : $\alpha > 0$ if $s > 0$), $\beta(s) = s/\alpha(s)$, we can calculate

$$T_0(s) = \frac{1}{\hat{A}(s)}C[\frac{s}{\hat{A}(s)} + DD^*]^{-1}B$$
$$= \frac{1}{\alpha(s)}\beta(s)C[\beta^2(s) + DD^*]^{-1}B \tag{6}$$
$$:= \frac{1}{\alpha(s)}T_e(\beta(s)),$$

where $T_e(s)$ is the transfer function for the normalized purely elastic version ($a \equiv 0, E = 1$) of (2). T_e is meromorphic in $\mathbb{C}$ with poles at $z = \pm i\lambda_j$, where $\Lambda = \{\lambda_j^2, j = 1, 2, \dots\} \subset \mathbb{R}^+$ is the spectrum of DD^*.

4.<u>Stabilization</u>. Let $\sigma_1 = \max\{\sigma_*, \sigma_l\}$, where

$$\sigma_l = \begin{cases} -\infty, & \text{if } A'(0+) = -\infty \\ A'(0+)/2A(0+), & \text{if } A'(0+) > -\infty. \end{cases} \tag{7}$$

σ_1 is the essential growth rate of the semigroup determined by the system (1) with $f(t) \equiv 0$ as defined by Desch and Miller in [2]. Given $\sigma_1 < \sigma_0 < 0$, our object is to find $k \in L_{\sigma_0}^1$ such that $T_k \in H^\infty(\Sigma_0)$, $\Sigma_0 = \{\Re s > \sigma_0\}$. We first establish the analytic properties of T_0. Without loss of generality, we assume that T_0 has no poles on $\{\Re s = \sigma_0\}$.

PROPOSITION 1. T_0 has at most finitely many poles in Σ_0 and none in $\{\Re s \geq 0\}$. There exists $R < \infty$ such that T_0 is bounded in $\Sigma_0 \cap \{|s| \geq R\}$.

Proposition 1 is proved from the properties of the completely monotonic function $a(t)$, and, in particular, the estimates and asymptotic formulas developed in [2]. (See also, [4, Section 2].)

For $k \in L_{\sigma_0}^1$, we have $\lim_{|\tau|\to\infty} \hat k(\sigma + i\tau) = 0$, uniformly in $\{\sigma_0 \leq \sigma < \infty\}$. From (5*), the following consequence of Proposition 1 is then obvious.

PROPOSITION 2. If $k \in L_{\sigma_0}^1$ and

(i) $\hat k(s)\hat T_0(s) \neq -1$ in Σ_0,

(ii) $\hat k(s) \neq 0$ at the poles of T_0 in Σ_0,

then $T_k \in H^\infty(\Sigma_0)$.

The interpolation problem set in the hypothesis of Proposition 2 is familiar in systems theory. We can represent τ_k schematically as part of a system

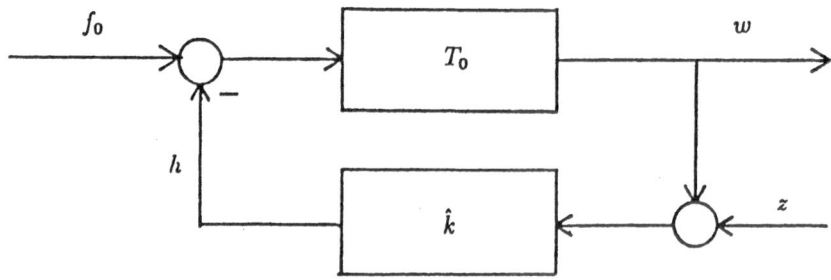

with transfer matrix

$$\Gamma = \begin{bmatrix} T_0(1 - T_0 Q) & T_0 Q \\ T_0 Q & Q \end{bmatrix}$$

$$= \begin{bmatrix} \frac{T_0}{1+\hat{k}T_0} & \frac{\hat{k}T_0}{1+\hat{k}T_0} \\ \frac{\hat{k}T_0}{1+\hat{k}T_0} & \frac{\hat{k}}{1+\hat{k}T_0} \end{bmatrix} = \begin{bmatrix} T_k & \hat{k}T_k \\ \hat{k}T_k & \hat{k}(1 - \hat{k}T_k) \end{bmatrix}$$

$(Q = \hat{k}(1 + \hat{k}T_0)^{-1})$, for which stabilizing schemes are known; see, for example [8], or [3, Chapters 4 and 6] and the references therein.

Note that we have required $k \in L^1_{\sigma_0}$. Stabilization of the scalar transfer function T_0 would only require $\hat{k}$ to be meromorphic in Σ_0 with $1 + \hat{k}T_0$ bounded away from zero near ∞ in Σ_0, as well as (i) and (ii) of Proposition 2. When the additional hypothesis $k \in L^1_{\sigma_0}$ also holds, one can show that system (1) with the feedback (3) defines a semigroup which is a perturbation of the semigroup constructed by Desch and Miller in [2] for the open loop system (1), and that this semigroup has growth rate no greater than σ_0. Thus, in addition to the stabilized scalar input-output transfer function T_k being exponentially stable with rate at most σ_0, the feedback system (1) with (3) is also internally stable with exponential rate at most σ_0. Our example will illustrate these ideas for the specific case of the viscoelastic wave equation. For additional work on the relationship between input-output stability and exponential stability of infinite dimensional systems, see the recent paper by Curtain [1].

5.Torsional vibrations in a viscoelastic rod. The system

$$U_{tt}(x,t) = EU_{xx}(x,t) + \int_0^t a(t-r)U_{txx}(x,r)\,dr + F_0(x,t) \quad (0 < x < 1),$$

$$U(0,t) = 0,$$

$$IU_{tt}(1,t) + EU_x(1,t) + \int_0^t a(t-r)U_{tx}(1,r)\,dr \tag{8}$$

$$= f_0(t) - \int_0^t k(t-r)U_t(1,r)\,dr,$$

$(I > 0, U(x,0) = U_t(x,0) = 0)$ models the stabilization of torsional vibrations in a viscoelastic rod (with linear mass density 1) by a feedback mechanism that acts through a rigid mass with moment of inertia I attached to the end $x = 1$. The corresponding system in which the integrated feedback is replaced by the instantaneous feedback $-kU_t(1,t)$ $(k > 0)$ was studied in [4] (see also [2]). System (8) is a special case of (1) where

$$DD^{\cdot} = \begin{bmatrix} 0 & \delta_1 \partial_x / I \\ 0 & -\partial_x^2 \end{bmatrix}$$

(δ_1 = point evaluation at 1) on the domain

$$\left\{ \begin{bmatrix} V_r \\ V_f \end{bmatrix} \in \mathbb{R} \times W^{2,2}(0,1) \subset \mathbb{R} \times L^2(0,1) := Y \mid V_f(0) = V_f'(1) = 0, V_r = V_f(1) \right\}.$$

(Here we have used the norm $\left\| \begin{bmatrix} V_r \\ V_f \end{bmatrix} \right\|^2 = I V_r^2 + \int_0^1 V_f^2(x)\, dx$ on Y to make $DD^{\cdot}$ self-adjoint.)

Using formal Laplace transforms we get the solution formula (cf [4, (2.11)])

$$(u_t)^\wedge(x,s) = -\frac{1}{\Delta(s)} \int_0^1 G(x,y,s)\hat{F}_0(y,s)\, dy + \frac{\sinh \beta(s) x}{\Delta(s)} f_0(s), \tag{9}$$

for (8) where $\Delta(s) = [Is + \hat{k}(s)] \sinh \beta(s) + \alpha(s) \cosh \beta(s)$,

$$G(x,y,s) = -\sinh \beta(s) x \left[\alpha(s)^{-1} (\hat{k}(s)+Is) \sinh \beta(s)(1-y) + \cosh \beta(s)(1-y) \right] \quad (0 \le x < y < 1),$$

with $G(x,y,s) = G(y,x,s)$ for $0 \le y < x \le 1$. Setting $F_0 \equiv 0$ in (9) we get

$$T_k(s) = \frac{\sinh \beta(s)}{\Delta(s)}.$$

On the other hand, if we allow a nonzero body force term F_0 in (8), and if $\hat{k}$ has poles in Σ_0, then (9) shows that $\|u_t(\cdot, t)\|_2$ need not decay like $e^{\sigma_0 t}$, even if $T_k \in H^\infty(\Sigma_0)$.

Finally, consider the pure boundary stabilization problem without the rigid mass ($I = 0$) consisting of the first two equations in (8) together with the feedback

$$EU_x(1,t) + \int_0^t a(t-r)U_{tx}(1,r)\, dr$$

$$= f_0(t) - \int_0^t U_t(1, t-r)\, dK(r). \tag{10}$$

This problem with $dK(t) = k_0 \delta(t)$, δ = point mass at 0, is also considered in [4]. Here we observe that even though systems of this type are not covered by the semigroup setting described above, one may still obtain the transfer functions T_0 and T_K for the scalar input-output mappings τ_0 and τ_K from input to output $U_t(1,t)$, and then use Laplace transfrom methods to investigate the stability of the system. For this system

$$T_0(s) = \frac{1}{\alpha(s)} \frac{\sinh \beta(s)}{\cosh \beta(s)} = \frac{1}{\alpha(s)} \frac{1 - e^{-2\beta(s)}}{1 + e^{-2\beta(s)}},$$

and T_K is defined as in (5^*). When $a(0^+) < \infty$, the theory of local analyticity of Laplace transforms [5, Proposition 2.3], may be used to show that $\alpha(s)$ is the Laplace-Stieltjes transform $\hat{K}(s)$ of a measure $dK(t) = \sqrt{E}\delta(t) + k(t)dt$ where $k \in L^1_{\sigma_0}$ for any $\sigma_0 > \sigma_*$. For this K, $1 + \hat{K}(s)T_0(s) \neq 0$ in Σ_{σ_0}, so $T_K \in H^\infty(\Sigma_{\sigma_0})$ for any $\sigma_0 > \sigma_*$, and all of the poles of T_0 have been removed at once. We comment that this method is nothing more than impedence matching, and should be viewed as stabilization by absorbing boundary conditions (cf [6]). In fact, the same technique applies whenever there exists a constant c such that $1 + cT_e(s) \neq 0$ in $\mathbb{C}$ where T_e is the transfer function for the corresponding elastic open-loop system since we can take $\hat{K}(s) = c\alpha(s)$ and use (6).

REFERENCES

1. R. F. Curtain, Equivalence of input-output stability and exponential stability for infinite dimensional systems, TW-276 University of Groningen, preprint.
2. W. Desch and R. K. Miller, Exponential stabilization of Volterra integral equations with singular kernels, J. Integral Equations Appl., to appear.
3. B. A. Francis, A Course in H_∞ Control Theory, Lecture Notes in Control and Information Sciences, Vol. 88, Springer-Verlag, Berlin, 1987.
4. K. B. Hannsgen, Y. Renardy and R. L. Wheeler, Effectiveness and robustness with respect to time delays of boundary feedback stabilization in one-dimensional viscoelasticity, SIAM J. on Control and Optimization, 26 (1988), 1200-1234.
5. G. S. Jordan, O. J. Staffans and R. L. Wheeler, Local analyticity in weighted L^1 - spaces and applications to stability problems for Volterra equations, Trans Amer. Math. Soc., 274 (1982), 749-782.
6. R. C. MacCamy, Absorbing boundaries for viscoelasticity, in Viscoelasticity and Rheology, A. S. Lodge, J. A. Nohel and M. Renardy, eds., Academic Press, New York, 1985, 323-344.
7. D. V. Widder, The Laplace Transform, Princeton University Press, Princeton, New Jersey, 1946.
8. G. Zames and B. A. Francis, Feedback, minimax sensitivity, and optimal robustness, IEEE Trans. Automat. Control, vol. AC-28 (1983), 585-601.

ACKNOWLEDGMENTS: The work of K.B.H. and R.L.W. was partially supported by the Air Force Office of Scientific Research under grant AFOSR-86-0085. Part of this work was carried out during a visit by R.L.W. to the Universität Graz supported by Fonds zur Förderung der Wissenschaftlichen Forschung, Austria, under project S3206.

CONTINUOUS DEPENDENCE OF THE ASYMPTOTICS
IN DISCRETE AND CONTINUOUS DYNAMICAL SYSTEMS

Jerome A. Goldstein   Gisèle Ruiz Rieder
Tulane University   Louisiana State University

Let T be a power bounded linear operator (i.e. $\sup_n \|T^n\| < \infty$) or let $T = \{T(t) : t \geq 0\}$ be a uniformly bounded strongly continuous semigroup of linear operators on a Banach space X. The mean ergodic theorem of von Neumann states that

$$\frac{1}{N+1}\sum_{n=0}^{N} T^n \varphi \to P\varphi$$

or

$$\frac{1}{\tau}\int_0^\tau T(t)\varphi dt \to P\varphi$$

as $N \to \infty$ or $\tau \to \infty$ for suitable vectors φ. Here the ergodic limit P is a certain bounded projection.

Our goal is to show that the projection P depends continuously on T (or, in the semigroup case, on the infinitesimal generator A of T) in a suitable sense. Thus, for the discrete dynamical system $\varphi_{n+1} = T(\varphi_n)$ or the differential equation $d\varphi(t)/dt = A(\varphi(t))$, the asymptotics (i.e., the behavior of φ_n or $\varphi(t)$ as n or t tends to infinity) depends continuously on the equation in a suitable sense.

Our results were motivated by some recent developments in the theory of Markov chains (cf. e.g. Seneta [3]). In this special case, X is finite dimensional and our results simplify greatly.

For concreteness we state one result explicitly.

THEOREM. For $j = 0,1$ let A_j be the infinitesimal generator of a (C_0) semigroup $T_j = \{T_j(t) : t \geq 0\}$ on a reflexive Banach space X (cf. [1]). Suppose $A_0 = A_1 B$ for some bounded operator B on X, Ran (A_0) is closed, and there is a dense set D in Dom $(A_o) \cap$ Dom (A_1) such that for $\varphi \in D$ and $t > 0$, we have $T_0(t)\varphi \in$ Dom (A_1) and $A_1 T_0(\cdot)\varphi \in L^1_{\text{loc}}([0,\infty); X)$. Then for all $\varphi \in X$,

$$\frac{1}{\tau}\int_0^\tau T_j(t)\varphi dt \to P_j\varphi$$

as $\tau \to \infty$ for $j = 0, 1$, and

$$(P_1 - P_0)\varphi = (P_1 - I)(B - I)P_0\varphi.$$

In particular, for all $\varphi \in X$,

$$\|(P_1 - P_0)\varphi\| \leq M\|(B - I)\varphi\|$$

where $M_j = \sup\{\|T_j(t)\| : t \geq 0\}$ for $j = 0, 1$ and $M = (M_1 + 1)M_0$. Moreover, we may take $M = M_0$ if X is a Hilbert space and $M_1 = 1$.

For A_0 fixed and $A_1 \to A_0$ in the sense that $B \to I$ in either the strong or uniform operation topology, then $P_1 \to P_0$ in the same topology.

More general results together with examples and proofs can be found in [2].

This work was partially supported by NSF grant No. DMS-8620148 (JAG) and LASER grant No. 86-LBR-016-04 (GRR).

REFERENCES

[1.] Goldstein, J. A., Semigroups of Linear Operators and Applications, Oxford University Press, 1985.

[2.] Goldstein, J. A., and Rieder, G. R., Continuous dependence of ergodic limits, in preparation.

[3.] Seneta, E., Perturbation of the stationary distribution measured by ergodicity coefficients, Adv. Appl. Prob. 20 (1988), 228-230.

Jerome A. Goldstein
Department of Mathematics
    and Quantum Theory Group
Tulane University
New Orleans, Louisiana 70118

Gisèle Ruiz Rieder
Department of Mathematics
Louisiana State University
Baton Rouge, Louisiana 70803

Persistency of Excitation and (lack of) Robustness in Adaptive Systems

William A. Sethares
University of Wisconsin

C. R. Johnson, Jr.
Cornell University

1. What a wonderful thing PE is!

It is impossible to read the literature on adaptive systems without being impressed with the significance of the PE (Persistence of Excitation) condition. Persistence of excitation leads to exponential stability, parameter convergence, identifiability, robustness to slow variation of the desired parameters, rejection of small disturbances, immunity from the effects of slight mismodeling, and protection from small nonlinearities. What a wonderful thing! Here are a few typical statements...

"... a persistently excited signal applied to a system in open loop operation will give identifiability" [1].

"Of the various conclusions drawn, the most general is that the various adaptive algorithms described will behave robustly, given satisfaction of a persistency of excitation condition..." [2].

"In the study of the behavior of adaptive filtering algorithms, persistence of excitation of the input process arises as a sufficient condition for convergence... of the parameter estimates." [3].

What exactly does PE mean? Persistence of excitation is usually defined as an intrinsic property of a vector sequence [3], [4] *without* reference to any particular algorithm. It is then used in conjunction with certain structural constraints in particular problem settings to demonstrate desirable behavior of the adaptive system. This approach has proven useful with a wide variety of algorithms: LMS, equation error ARMA identification, model reference adaptive control, output error identification, etc. As new and different algorithm forms are considered, however, this strategy fails. This paper considers one such system, the adaptive hybrid, for which persistently exciting inputs have no clear implications in terms of behavior and/or robustness. The investigation raises basic issues concerning the "true nature" of PE and its implications for robustness of adaptive systems.

2. The Adaptive Hybrid

The adaptive hybrid arises in telephony [5]. The purpose of the adaptation is to cancel the echoes generated in the translation of telephone signals through a hybrid device that converts 2 wire transmissions to 4 wire transmissions (and back again). Figure 1 shows a block diagram of the telephone system, where h and g represent the echo paths, and ĥ and ĝ represent the adaptive elements which attempt to cancel the echo paths. Note that *if* ĝ were equal to g and *if* ĥ were equal to h, then no near (far) end speech would be returned to the near (far) end. Unfortunately (as will be shown), ĝ = g and ĥ = h are not equilibria (even on average) of the system, and perfect echo cancellation is generically impossible.

At first, this may seem odd. Consider just the system at the near end (refer to figure 2). The signal x is the input and r is the error, which is used for standard Least Mean Squares (LMS) adaptation. Standard analysis of open loop LMS shows that if x is persistently exciting, then the parameter estimates ĥ will converge, on average, to a ball about h. Although it is easy to verify that x is PE (when v and w are sufficiently rich spectrally), the system can exhibit "bursting" behavior - brief periods of instability followed by long periods of good echo attenuation. The problem is that r is correlated with x through the action at the far end. The standard analytical approaches are misleading due to this feedback of the error into the input.

Since the same discussion also applies at the far end (where r plays the role of input and x plays the role of error), the near and far end systems cannot be analyzed in isolation. In fact, situations arise in which x is PE (thus "persistently exciting " the LMS at the near end) and r is PE (thus "persistently exciting" the LMS at the far end), yet the system undergoes repetitive bursts of destabilization [6] where the loop gain (g- ĝ)(h- ĥ) becomes larger than unity. This causes the signals to oscillate wildly until the system (hopefully) restabilizes. This bursting is clearly undesirable. It has been observed in real

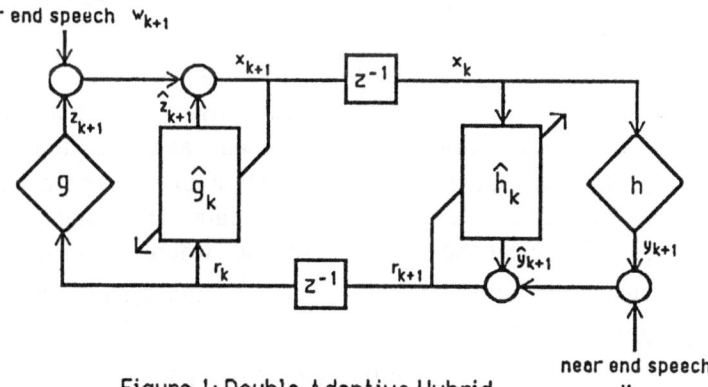

Figure 1: Double Adaptive Hybrid

time lab tests [6] and in simulation studies. Figure 3 shows a simple simulation that exhibits repetitive bursts.

Such behavior forces a reinterpretation of the implications of PE. For the adaptive hybrid of figure 1, which utilizes standard LMS adaptation inside a feedback loop, it is not enough that each input (w and v) be PE, it is not enough that some vector function of w and v be PE, it is not enough that the internal signals x and r be PE, it is not enough that some vector function of the x and r be PE (nor any combination of the above conditions). Persistency of excitation thus has no direct or obvious bearing on the behavior of the adaptive hybrid.

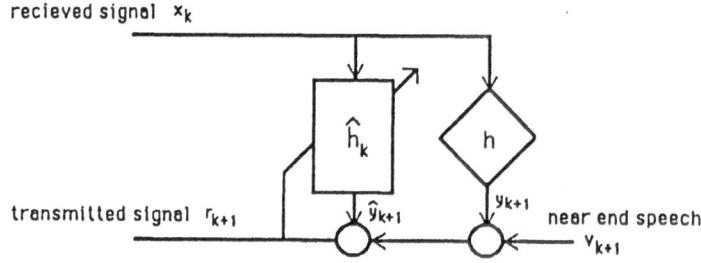

Figure 2: Simplified Adaptive Hybrid

In order to investigate the system concretely, consider the following three simple models. The first (see figure 2) corresponds to a single adaptive parameter at the near (or far) end, and ignores the feedback path. The system is identical to LMS adaptation and the law governing the evolution of the adaptive parameter is

$$\hat{h}_{k+1} = \hat{h}_k + \mu \, x_k \, r_{k+1} \qquad (2.1)$$

where μ is the stepsize, $r_{k+1} = \hat{y}_{k+1} - y_{k+1} + v_{k+1}$ is the error sequence, $y_{k+1} = h$ x_k is the near end echo source, and $\hat{y}_{k+1} = \hat{h}_k \, x_k$ is the output of the adaptive filter. Introducing the error variable $\tilde{h}_k = h - \hat{h}_k$, and recognizing that $r_{k+1} = \tilde{h}_k$ $x_k + v_{k+1}$, this can be rewritten

$$\tilde{h}_{k+1} = (1 - \mu \, x_k^2) \, \tilde{h}_k - \mu \, x_k \, v_{k+1}. \qquad (2.2)$$

The second model, the single adaptive hybrid of figure 4, assumes that there is one adjustable parameter at the near end, but that there is only a (nonadaptive) echo path at the far end, modeled by a small constant attenuation α. Using the LMS update leads to the same parameter estimate as in (2.1), but now $x_k = \alpha \, r_k + w_k$ is the echo from the far end, where w_k is the far end speech. Again, introducing the error variable $\tilde{h}_k = h - \hat{h}_k$, the system can be described by (2.2) in conjunction with

$$x_{k+1} = \alpha \, \tilde{h}_k \, x_k + \alpha \, v_{k+1} + w_{k+1} . \qquad (2.3)$$

The third situation, the double adaptive hybrid of figure 1, supposes that there are single parameter adaptive elements at each end. With all notation as above, the system equations are

$$\bar{h}_{k+1} = (1 - \mu\, x_k^2)\, \bar{h}_k - \mu\, x_k\, v_{k+1} \tag{2.4}$$

$$x_{k+1} = \tilde{g}_k\, r_k + w_{k+1} \tag{2.5}$$

$$\tilde{g}_{k+1} = (1 - \mu\, r_k^2)\, \tilde{g}_k - \mu\, r_k\, w_{k+1} \tag{2.6}$$

$$r_{k+1} = \bar{h}_k\, x_k + v_{k+1} \tag{2.7}$$

where $\tilde{g}_k = g - \hat{g}_k$.

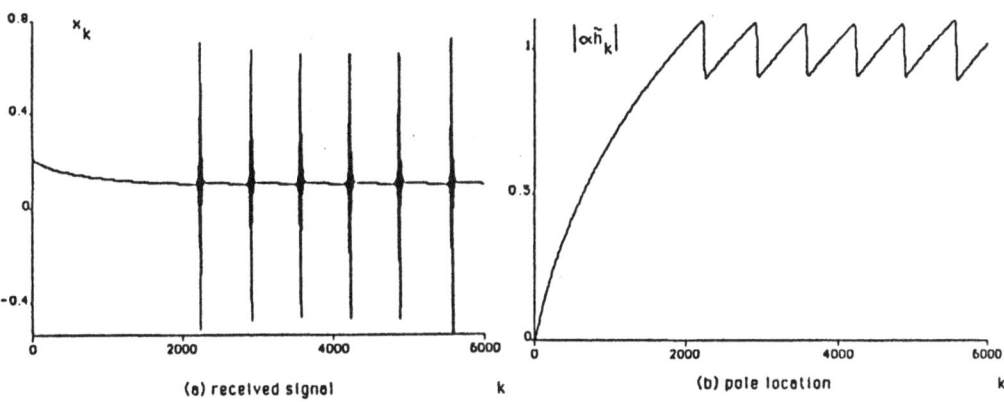

(a) received signal   k   (b) pole location   k

Figure 3: Bursting in the single adaptive hybrid

3. A Reevaluation of PE?

Convergence of the parameters of an adaptive system implies that the most significant dynamics of the unknown plant are matched, and implies that the prediction error (the difference between the actual output and the estimated output) converges to zero (in the ideal, or exact matching case). What properties of the input signals, the algorithm, and the problem setting imply such parameter convergence?

The idea of persistence of excitation originally grew out of an examination of the equation error adaptive algorithm as the key input dependent condition that implies asymptotic convergence of the parameter estimates to their desired values in the ideal setting [7], [8]. It was later realized [9] that this convergence could be characterized with an exponential rate, which implies a degree of robustness to suitably small disturbances in the nonideal case. Mathematically, PE (in this context) is a requirement that all eigenvalues of a certain matrix (which is a sliding summed outer product of the input/regressor sequence) be strictly positive. This condition has an intuitively appealing meaning in terms of a persistent spanning property of the regressor and a simple meaning in terms of the spectral complexity of the input sequence. This condition will be called PE for LMS, or the *standard* PE condition.

In 1986, most of the adaptive community would have agreed with this two part definition: PE is the full rank summed outer product condition on the input/regressor sequence outlined above, and PE implies exponential asymptotic convergence of the parameter estimates to their desired values in the ideal setting. This two part definition appeared to be generic in the sense that algorithms other than equation error identification can be interpreted within this framework. For instance, the "output error" forms [10] can be shown to be exponentially convergent in the ideal case if the input is PE for LMS and if a certain transfer function is strictly positive real. One way to interpret this is that the "PE for output error" condition can be translated to the PE for LMS condition plus a requirement on the function in the adaptive mechanism that converts the external signals into the regressor sequence. Unfortunately, such translation is not always possible.

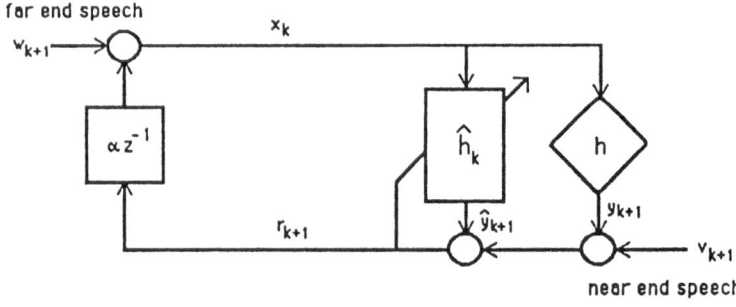

Figure 4: Single Adaptive Hybrid

One such situation is the Signed Regressor (SR) algorithm [11] in which the standard LMS update term $X_k e_k$ is replaced by $sgn(X_k) e_k$. The parameter estimates of the SR algorithm can be guaranteed to be exponentially convergent in the ideal setting if a condition on the eigenvalues of the summed outer product of the regressor and the *signed* regressor is fulfilled. Since this is a condition on the input/regressor sequence that implies exponential asymptotic stability of the error system, the condition is called PE for SR. The condition is similar to, but *not* equivalent to the PE for LMS condition. This forced a reevaluation of the twofold "definition" of PE. The consolidation was; PE is some condition on the input/regressor sequence that implies exponential convergence of the parameter estimates to their true values in the ideal setting.

Investigation of the signed error algorithm (in which the standard LMS update term $X_k e_k$ is replaced by $X_k sgn(e_k)$) revealed that even this consolidated concept of PE is inadequate [12]. The parameter estimates of the signed error algorithm were shown to be *linearly* convergent to a small ball about their desired values. They are not convergent to the true value even in the ideal case, nor is the convergence exponential. Yet the convergence is robust to small nonidealities, where robust means that small disturbances do not cause large changes in the behavior of the algorithm. Once again, the "definition" of PE must be modified; PE is some condition on the input/regressor sequence that implies that the parameter estimates converge to a small region about the true parameterization, and this convergence is suitable to offer an appreciable degree of robustness to small nonidealities.

The realization that the PE for LMS condition is not a generic robustifier for all adaptive algorithms seems, in retrospect, obvious. Yet it was not until 1987, in an adaptive control setting, in which the obvious was stated [13]. The thesis of [13] in this context is to relate PE to the Uniform Asymptotic Stability (UAS) of the adaptive system. Two examples are given. In one, the PE for LMS condition is not necessary for UAS parameter convergence. In the other, the PE for LMS condition is not sufficient for UAS convergence. PE is then defined with respect to a particular algorithm as the condition on the input/regressor sequence causes UAS of the algorithm about its equilibrium point. This is stated in terms of the uniform decrease of a Lyapunov function over a window of fixed finite length. Like the previous two "consolidations" of the concept of PE, this formulation will be found to be lacking when more adaptive schemes are considered.

Using the adaptive hybrid as a test case, we identify several issues that must be considered in the reevaluation of the term PE:

(1) It is imprecise to say "this signal is PE" without reference to a particular algorithm and a particular setting in which the algorithm will be applied.

(2) PE is a local, and not a global concept.

(3) PE should be defined with respect to a particular equilibrium set or trajectory, which may itself depend on the input sequence, the stepsize, or other features of the algorithm.

(4) The "equilibrium set" may only be an "average" stationary set.

4. What Can PE Tell Us?

The regressor sequence X_k is called persistently exciting for LMS (PE for LMS) if there are positive a,b, and t such that

$$bI > \frac{1}{t} \sum_{j=k}^{k+t-1} X_k X_k^T > aI \quad \text{for every k.} \tag{4.1}$$

This condition is used in [14] to show that the parameter estimate errors φ of the LMS algorithm

$$\varphi_{k+1} = \varphi_k - \mu X_k e_k \tag{4.2}$$

are globally exponentially stable to $\varphi=0$ when the X_k sequence is PE for LMS and the stepsize μ is small enough. Note that e_k, the measured error, is equal to $X_k^T \varphi_k$ in the ideal case. The exponential nature of this stability is then used to demonstrate that the algorithm is robust to suitably small disturbances. This result is directly applicable to the first model of the adaptive hybrid (figure 2). With x_k fulfilling the PE for LMS condition, $\tilde{h}_k$ will converge exponentially to a ball about 0, implying that $\hat{h}_k$ converges to a ball about h, where the radius of this ball is proportional to the magnitude of $\mu x_k v_k$. Consequently, there is no way that this system can exhibit the observed bursting. The model has been simplified too far to capture the bursting behavior. This emphasizes the danger of analyzing the adaptive element separately from the rest of the system, especially when the adaptive terms are embedded in a feedback loop.

The second model, incorporating a nonadaptive hybrid at the far end, is capable of reproducing the bursting. Figure 3 shows a simulation in which the far end transmission is silent ($w_k=0$) while the near end transmission is a constant ($v_k=v$ for every k). Irrespective of h, the algorithm drives $\alpha\hat{h}_k$ (the loop gain) to something greater than unity, which causes (2.4) to be momentarily unstable. This causes x_k to grow wildly until the contraction term in (2.3) becomes dominant and restabilizes the system. This bursting cycle is examined in detail in [6]. Notice that the excitation sequence x_k is PE for LMS, fulfilling definition (4.1). Yet the implications of the PE style theorems are false - there is no exponential stability to an average equilibrium.

Perhaps the most fundamental of the PE style results is boundedness of the parameter estimates in the presence of small disturbances. We now construct an example of a bounded (and even ε small) w_k and v_k which drive the parameter estimates $\hat{h}_k$ to infinity. This shows that, for the adaptive hybrid of figure 3, even the bounded state implications of PE are false.

Example 1: Suppose that the inputs are $w_k = \frac{1}{k} - \alpha\varepsilon$ and

$$v_k = \varepsilon - \frac{h}{k-1} + \frac{\mu\varepsilon}{k-1} \sum_{i=1}^{k-1} \frac{1}{i} \quad \text{where } \varepsilon \text{ is an arbitrary positive constant, and suppose}$$

that $\hat{h}$ is initialized at 0 while x is initialized at 1. Then the recursion pair

(2.1), (2.3) has the solution $x_k = \frac{1}{k}$ and $\hat{h}_k = \mu\epsilon \sum\limits_{i=1}^{k-1} \frac{1}{i}$. Thus $\hat{h}_k$ diverges to infinity as k approaches infinity.

Proof: Substituting the values of w_k, v_k, x_k and $\hat{h}_k$, in (2.1) and (2.3) yields $r_{k+1} = \hat{y}_{k+1} - y_{k+1} + v_{k+1} = h\ x_k - \hat{h}_k\ x_k + v_{k+1} = \epsilon$, which can then be substituted into the definitions of y_k and $\hat{y}_k$ to verify the result. Note that as k-> ∞, y_k ->0 and $\hat{y}_k$ ->0, while v_k -> ε and w_k -> - α ε.

It is not reasonable to suppose that divergence such as this is generic (quite the opposite, in fact, since this example was concocted to demonstrate a very controlled form of instability). Rather, the example shows that all hope of a bounded input bounded state result based on PE-like conditions involving only the input/regressor sequence is guaranteed to fail. Indeed, w_k and v_k, which converge to constants, each independently fulfill the one dimensional PE for LMS condition. Thus, for the adaptive hybrid system, the standard PE condition has no clear implications in terms of parameter convergence or stability. What factors *do* contribute to the stability of the adaptive hybrid system?

5. Input-Dependent Equilibria

For the LMS algorithm of (4.2), there is a clearly discernable equilibrium point at φ=0 which corresponds to $e_k = X_k^T\varphi_k = 0$. Any equilibrium h*, x* for the single adaptive hybrid of (2.2)-(2.3), however, must satisfy

$$h^* = (1 - \mu\ x^{*2})\ h^* - \mu\ x^*\ v_{k+1}. \tag{5.1}$$
$$x^* = \alpha\ h^*\ x^* + \alpha\ v_{k+1} + w_{k+1}. \tag{5.2}$$

for every k. This is only possible for very special input sequences v and w (i.e. constant), and, in general, there will be no equilibria. Thus parameter convergence is generically impossible. The discussion now branches in two directions: the remainder of this section examines the input dependence and stability properties of the equilibrium points when they do exist, the following section considers the more general situation in which v and w are averagable (for instance, periodic). In this latter case, although there may be no equilibrium point, there may be an averaged equilibrium which can shed light on the behavior of the algorithm. For constant "D.C." inputs, we have

Example 2: Suppose $v_k = v > 0$ and $w_k = w > 0$ for every k. Then there is an equilibrium at $h^* = -\frac{v}{w}$, $x^* = w$. If $|\alpha h^*| = |\frac{\alpha v}{w}| < 1$, then the equilibrium is stable. If $|\frac{\alpha v}{w}| > 1$, then the equilibrium is unstable.

Proof: The equilibrium can be verified directly by substituting into (5.1) and (5.2). Introducing the variables $z_k = h^* - \hat{h}_k$ and $y_k = x^* - x_k$, (5.1) and (5.2) can be linearized about $z=y=0$ with transition matrix

$$A(v,w) = \begin{pmatrix} 1-\mu w^2 & uv(1-1/w)+2\mu w \\ \alpha w & -\alpha v/w \end{pmatrix}. \tag{5.3}$$

With $|\alpha h^*| < 1$ and with μ small enough so that $\mu w << 1$, the largest eigenvalue is bounded above by $1 - \frac{1}{2}\mu w$, and the equilibrium is stable. When $|\alpha h^*| > 1$, the largest eigenvalue can be bounded below by $|\alpha h^*| - \mu w$, which is larger than unity for μ small enough, implying the instability of the equilibrium.

When the equilibrium is unstable, the system has a tendency to drive the parameter estimates towards the region of closed loop stability. This manifests itself in simulations as the bursting phenomenon. Example 2 shows explicitly that the equilibrium point of the algorithm depends on the inputs to the system. Note that the equilibrium point $h^*=0$ (corresponding to $\hat{h}_k = h$) occurs only when the input at the near end is identically zero. Thus convergence of the parameter estimate error to zero can only be expected in very nongeneric circumstances. Moreover, this equilibrium h^*, x^* may be stable or unstable depending on the values assumed by the inputs. Thus the existence, the value assumed, and the stability of the equilibria are dependent on the input sequences. This is in sharp contrast to standard LMS which has a single fixed global equilibrium, irrespective of the particular input sequences.

6. "Averaged" Equilibria

Even when there is no equilibrium point, there may be an averaged equilibrium. That is, even when (5.1) and (5.2) have no solution, an average solution may exist. This will occur, for instance, when the v and w are periodic signals.

To pursue this idea, suppose that v is p_v periodic and that w is p_w periodic. Then the v,w pair is $p=LCM(p_v,p_w)$ periodic (LCM= Least Common Multiple). There will be a stable p-periodic solution for h and x whenever the linearized p-periodic transition matrix

$$A = \prod_{k=1}^{p} A(v_k, w_k) \tag{6.1}$$

has all eigenvalues less than one in magnitude, where $A(v_k, w_k)$ is the linearization about the periodic trajectory h^*_k, x^*_k. This parallels the construction of $a(v,w)$ from (5.3) above. Thus A determines the stability of the solutions about the periodic trajectory h^*_k, x^*_k, and one may think of A as representing the average behavior of the system. When **A** is unstable (eigenvalues larger than unity), then there will be no stable periodic solution.

Example 3: With two periodic inputs $v_{even}=1.1$, $v_{odd}=0.9$, $w_{even}=1$, and $w_{odd}=1.5$, there is a two periodic equilibrium trajectory $h^*_{even}=-.74$, $h^*_{odd}=-.76$, $x^*_{even}=.94$, $x^*_{odd}=1.6$. It is easy to verify that A has all eigenvalues within the unit circle.

The situation with periodic inputs parallels the situation with constant inputs; both the stability of the equilibrium trajectories and the trajectories themselves are dependent on the particular input sequence. Of particular interest among periodic inputs are sinusoids. This case (where v and w are sinusoids of rationally related frequencies) has been investigated in detail in [15], where it is shown that the stability of the averaged equilibrium depends not only on the magnitudes of the sinusoids, but also on the frequencies. These same dependencies are also found in the double adaptive hybrid model.

7. When the Excitation is the Disturbance

The double adaptive hybrid of figure 1 and equations (2.4)-(2.7) is susceptible to all the problems of the simpler single adaptive hybrid of the previous sections, including input dependent (averaged) equilibria and the potential for bursting.

The "correct solution" $\hat{g} = g$ and $\hat{h} = h$, for instance, is not an equilibrium, nor even an averaged equilibrium. Substituting $\tilde{h} = \tilde{g} = 0$ into (2.4) and (2.6) yields $\mu x_k v_{k+1}=\mu r_k w_{k+1}=0$, while (2.5) and (2.7) yield $x_{k+1}=w_{k+1}$ and $r_{k+1}=v_{k+1}$. Clearly, this can only hold for very special inputs v and w. In general, the existence and stability of an average trajectory can be determined exactly as in the single hybrid case by finding the eigenvalues of the linearized map about the equilibrium trajectory.

The essential paradox is this. The input at the far end provides excitation for the near end adaptive algorithm. The input at the near end provides excitation for the far end adaptive algorithm. Meanwhile, the input at the near end acts as a disturbance to the algorithm at the far end, and the input at the far end acts as a disturbance to the algorithm at the near end. This odd structure, in which the signal provides excitation to one of the adaptive subsystems while it simultaneously acts as a disturbance on another piece of the adaptive system, shows that "PE for the adaptive hybrid" is not a straightforward generalization of the PE for LMS condition.

8. Conclusions

The adaptive hybrid system of figure 1 presents compelling evidence that our understanding of the behavior of adaptive systems is far from complete. The inclusion of even the simplest of adaptive filters inside a feedback loop renders useless many standard methods of analyzing the performance of adaptive systems. Worse, it also renders useless the intuitions and expectations that have arisen as a result of those analytic methods.

What does it mean to persistently excite an algorithm?

This investigation of the adaptive hybrid has revealed several issues which are crucial to an understanding of the behavior, stability, and robustness of adaptive systems. Consider the following four points:

(1) PE for an algorithm involves conditions on the input/regressor sequence, and other features of the algorithm.

(2) There is some stationary point, set, trajectory, or an "average" stationary point, set, or trajectory, which may depend on the input/regressor sequence, the unknown plant, the stepsize, and other features of the algorithm and the problem setting. Designate this point, set, or trajectory as $\varphi*$.

(3) There is a region of attraction about $\varphi*$.

(4) The persistently excited algorithm should be "robust" to all forms of suitably small perturbations including noise in the signals, slow motion of $\varphi*$ over time, and mismodeling errors.

The proposed "concept behind" PE can now be stated:

PE is a condition on the features of an algorithm and its inputs (1) which causes a particular set (2) to have behaviors (3) and (4).

Acknowledgements

The authors would like to thank L. Praly for several insightful comments, and M. A. Poubelle for her careful handling of earlier versions of this paper. William A. Sethares is with the Department of Electrical and Computer Engineering, University of Wisconsin, Madison, WI 53706. C. R. Johnson, Jr., may be found at the School of Electrical Engineering, Cornell University, Ithaca, NY 14853.

References

[1] Ljung L., and T. Soderstrom, *Theory and Practice of Recursive Identification*, MIT Press, Cambridge, MA., 1983.
[2] Anderson, B. D. O., and R. M. Johnstone, "Adaptive systems and time-varying plants," *Int. Journal Control*, vol. 37, pp. 367-377, 1983.

[3] Bitmead, R. R., "Persistence of excitation conditions and the convergence of adaptive systems," *IEEE Trans. on Information Theory*, vol. IT-30, no. 2, pp. 183-191, March 1984.

[4] Boyd S., and S. S. Sastry, "On parameter convergence in adaptive control," *Systems and Control Letters*, 3:311-319, 1983.

[5] Widrow, B., and S. D. Stearns, *Adaptive Signal Processing*, Prentice- Hall, 1985.

[6] W. A. Sethares, C. R. Johnson, Jr., C. Rohrs, "Bursting in Adaptive Hybrids," to appear in *IEEE Trans. on Communications*.

[7] Lion, P. M., "Rapid identification of linear and nonlinear systems," *Proc. of JACC*, 1966.

[8] Narendra, K. S. and L. E. McBride, "Multi-parameter self optimizating sustems using correlation techniques," *IEEE Trans. on Automatic Control*, vol.AC-9, pp. 31-39, January 1964.

[9] Anderson, B. D. O., "Exponential stability of linear equations arising in adaptive identification, *IEEE Trans. on Automatic Control*, vol. AC-22, February 1977.

[10] Johnson, C. R. Jr., "Adaptive IIR filtering: current results and open issues," *IEEE Trans. on Information Theory*, vol. IT-=30, no. 2, pp.237-250, March 1984.

[11] W.A. Sethares, I.M.Y. Mareels, B.D.O. Anderson, C.R. Johnson, Jr., "Excitation Conditions for Sign-Regressor LMS," *IEEE Trans. on Circuits and Systems*, June 1988.

[12] W. A. Sethares and C. R. Johnson, Jr., "A Comparison of Two Quantized State Adaptive Algorithms," to appear in *IEEE Trans. on Acoustics, Speech, and Signal Processing*.

[13] Narendra, K. S., and A. M. Annawamy, "Persistence of excitation in adaptive systems," *Int. Journal Control*, vol. 45, no. 1, pp. 127-160, 1987.

[14] Johnson, C. R. Jr., *Lectures in Adaptive Parameter Estimation*, Prentice Hall, 1987.

[15] Z. Ding, W. A. Sethares, and C. R. Johnson, Jr., "Frequency dependent bursting in adaptive echo cancellation and its prevention using double talk detectors," submitted to *Int. Journal of Adaptive Control and Signal Processing*..

ON COMPLEXITY IN SYNTHESIS OF FEEDBACK SYSTEMS

Michael K. Sain   Joseph L. Peczkowski
University of Notre Dame   Allied-Bendix Aerospace

Bostwick F. Wyman
The Ohio State University

Introduction. One basic number which characterizes complexity for a nonlinear
feedback system is the number of differential equations which it requires to describe
it. However, there is wide variation in systems having the same number of differential equations. A more detailed picture of the way in which the generalized system
velocity is related to the generalized system position is provided by the local pole
structure of the system. This structure carries not just the number of equations, as
the dimension of a state space, but also the module nature of the poles. In fact,
however, a more thorough assessment would be provided by the combined study of local
pole structure and local zero structure, inasmuch as poles and zeros may be regarded
as concrete evidence of the way in which integrations are related to each other, and
to inputs and outputs.

If we regard the complexity of the plant as fixed, then overall system complexity becomes a function of controller complexity. The fundamental, local, pole and
zero complexity attached to a given plant and a given set of closed loop system
specifications can be studied by means of the equivalent open loop controller. For
poles, such constraints have been investigated by Conte, Perdon, and Wyman (1), in a
context of the equation

$$T(z) = P(z)M(z), \qquad\qquad\qquad [1-1]$$

which is associated with the commutative diagram in Figure 1. This equation has
solutions if and only if im $T(z) \subset$ im $P(z)$; and the
symbols R(z), U(z), and Y(z), defined in Section 3,
are finite-dimensional vector spaces over k(z),
the field of transfer functions with coefficients
in a base field k. The results of (1) generalize
certain of the earlier studies by Wyman and Sain
(2, 3), where T(z) was an identity. For zeros, the
problem of Figure 1 has not yet been studied on the
module level. Previous works by Wyman and Sain

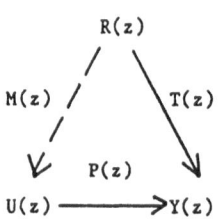

Figure 1

(4) and by Conte and Perdon (5) studied the composition of two given maps; but
neither of these addressed Figure 1, where T(z) and P(z) are given, and M(z) is the
unknown. Sain and Wyman (6) have examined fixed zeros in Equation [1-1] when M(z) is
given and P(z) is unknown. In a certain, somewhat intricate, sense, the study of this
paper can be phrased as a dual to that of (6), if we build upon the concepts of (7).
The dual approach would be approximately equal in length to that which we take here.
We choose the approach of Sections 3-6, in the sequel, because of the special
interest attached to the novel concept of free zeros, and because of the insights

associated with the corresponding vantage point of thought.

In Section 2, we provide further motivation and illustrations for nonlinear feedback system design. Section 3 discusses the finitely generated, torsion module theory for poles and zeros, and Section 4 introduces the new concept of free zeros. Section 5 develops and characterizes a module Z of fixed zeros associated with solutions M(z) to Figure 1, and Section 6 explains how this module relates to finitely generated, torsion zeros of such solutions. Extensions and conclusions are provided in Section 7.

Local Nonlinear Synthesis. A nonlinear model (8) of a simple turbojet engine is shown in Figure 2. It is representative, on a small scale, of the kind of nonlinear plant with which designers of turbine engines and turbine engine controls currently deal in practice. In essence, it is a computer simulation, typically constructed by engine manufacturers and provided to control designers and manufacturers. The model of Figure 2 incorporates nonlinear steady state and dynamical relationships between three inputs and six outputs. From one point of view, this nonlinear turbojet model consists of three integrators, nine nonlinear functions or schedules, including five bi-variant schedules, six multipliers, three dividers, and nine designated summations. Inputs to the plant model are fuel flow (W_f), exhaust nozzle area (A_j), and turbine vane position (β). Outputs from the plant model are engine speed (N), turbine temperature (T_4), engine thrust (FN), tailpipe pressure (P_5), compressor pressure (P_3), and tailpipe temperature (T_5).

At the most elementary level, we may count the number of integrators in Figure 2, in this case three. This provides the dimension of the local state space, if we believe that the plant is locally controllable and observable. It is quite clear, however, that the single integer three does little to connote the preponderance of detail which remains in the figure. For a fixed triple of inputs ($\underline{W}_f$, $\underline{A}_j$, $\underline{\beta}$), which lead to a corresponding state vector $\underline{x}$ and output vector $\underline{y}$, one may adjoin to the dimension (three) of the local state space X a local dynamics matrix $\underline{A}$: X → X which is a function of the level of operation. Equipped with the operation $\underline{A}$, X becomes a module over the ring R[$\underline{A}$], if R denotes the field of real numbers. The R[$\underline{A}$]-module X contains a great deal more information about Figure 2, and its complexity, than does the integer three. Yet we know, from the classical theory of systems having just one input and one output, that this pole information is at most just half of the story. To complete the dynamical picture, and to give a more thorough description of the way in which the integrators relate to one another, and to the inputs and outputs, we can append the corresponding module information for zeros. As will be seen in subsequent sections, zero modules are more varied in structure and character than are pole modules.

When the plant is fixed, and when the specified feedback system performance has been laid in, the remaining pole and zero complexity is that of the controller. A typical instance of such a controller (9) is depicted in Figure 3, where we can detect the same classifications of model characteristics which we examined for the

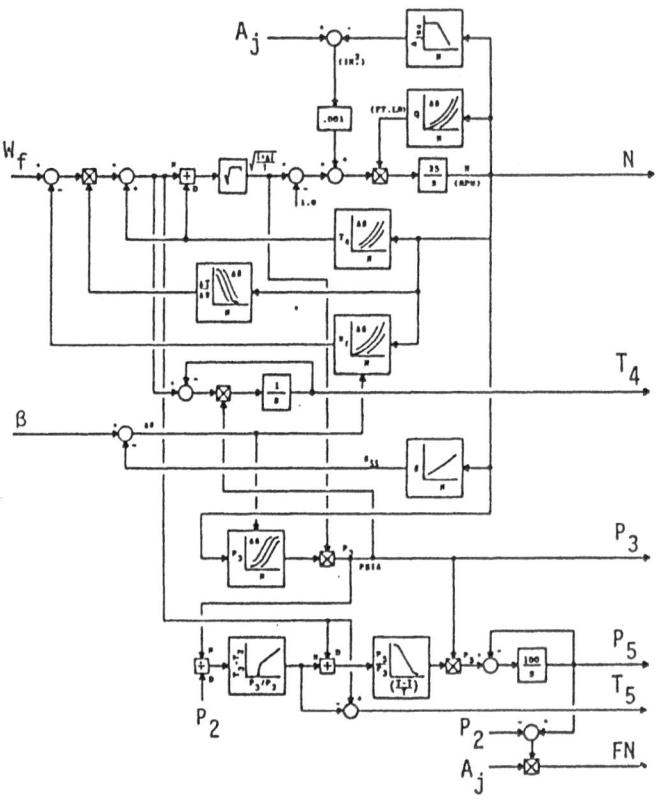

Figure 2. Nonlinear Turbojet Model

plant. Pole and zero modules for controllers may be approached through the idea of
an equivalent open loop. This is the idea of Figure 1.

Finitely Generated, Torsion Poles and Zeros. For a field k, let k[z] be the
ring of polynomials in z with coefficients in k, and let k(z) be its quotient field.
Suppose that R, U, and Y are finite dimensional vector spaces over k. Inasmuch as
k[z] is a k-vector space also, the k-bilinear tensor product can be used to construct
k[z]-modules

$$R[z] = k[z] \otimes_k R, \qquad\qquad [3\text{-}1a]$$

$$U[z] = k[z] \otimes_k U, \qquad\qquad [3\text{-}1b]$$

$$Y[z] = k[z] \otimes_k Y. \qquad\qquad [3\text{-}1c]$$

Also, the field of fractions k(z) can be viewed as a vector space over k. Then the
same type of tensor product yields k(z)-vector spaces

$$R(z) = k(z) \otimes_k R, \qquad\qquad [3\text{-}2a]$$

$$U(z) = k(z) \otimes_k U, \qquad\qquad [3\text{-}2b]$$

$$Y(z) = k(z) \otimes_k Y. \qquad\qquad [3\text{-}2c]$$

The purpose of this section is to discuss poles and zeros of k(z)-linear maps, in a

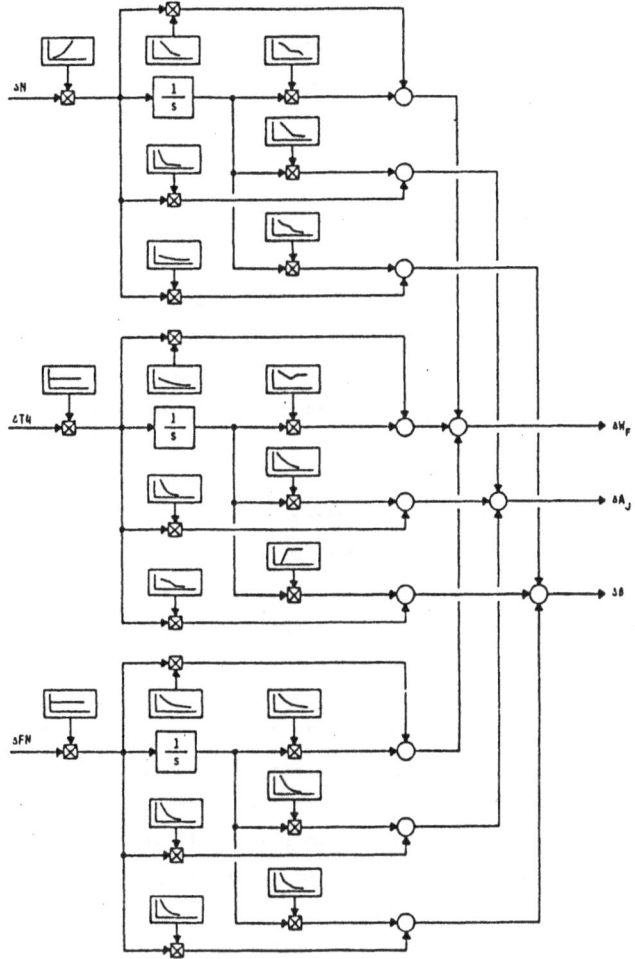

NONLINEAR CONTROLLER g

Figure 3. Scheduled Nonlinear Controller

module theoretic sense. An example of such a map is $P(z) : U(z) \rightarrow Y(z)$.

   We associate with each such map a pole module $P\big(P(z)\big)$, and define it to be the $k[z]$-factor module given by

$$P\big(P(z)\big) = \frac{PU[z]}{Y[z] \cap PU[z]} \, . \qquad\qquad [3\text{-}3]$$

As a $k[z]$-module, $P\big(P(z)\big)$ is finitely generated, because $U[z]$ is finitely generated. It is a torsion module because every element in $PU[z]$, even if not in $Y[z]$, can be made so by scalar multiplication. Finitely generated, torsion $k[z]$-modules correspond to state spaces of finite dimension over k. Considered as such a space, $P(P(z))$ = X_P, the usual state space of a minimal realization of $P(z)$; and the scalar module action z : $P(P(z)) \to P(P(z))$ defines a pole-state dynamics matrix A_P, which is k-linear on X_P into X_P. We refer to a module such as the pole module $P(P(z))$ as being of classical, finite-dimensional, state-space type. More briefly, we can call it a module of state-space type.

No matter what the nature of the kernel ker $P(z)$, or the cokernel coker $P(z)$, the pole module continues to be of state-space type. This is not true, in general, for the k[z]-zeros of $P(z)$. However, there is a useful module containing zeros of state-space type.

The finitely generated, torsion zero module $Z(P(z))$ associated with $P(z)$ was defined by Wyman and Sain (2) in 1981 to be the k[z]-factor module of form

$$Z(P(z)) = \frac{P^{-1}Y[z] + U[z]}{\ker P(z) + U[z]} . \qquad [3-4]$$

Notice that U[z] and Y[z] are finitely generated, and that elements in ker $P(z)$ are equivalent to zero. Accordingly, the zero module [3-4] is finitely generated. Moreover $Z(P(z))$ is torsion, for any element in U(z) can be brought into U[z] by means of scalar multiplication. The key to this definition giving a finitely generated module lies in the presence of ker $P(z)$ as a summand in the denominator of [3-4]. As a k-vector space, $Z(P(z))$ = X_Z has finite dimension. The module action z : $Z(P(z)) \to Z(P(z))$ defines a zero-state dynamics matrix A_Z : $X_Z \to X_Z$, which is k-linear. The space X_Z is not the usual state space of a minimal realization. To distinguish the cases, it is quite helpful to refer to X_Z as the zero-state space and to X_P as the pole-state space. Through the use of isomorphisms, it is sometimes possible to visualize zero spaces as subspaces or factor spaces of pole spaces. However, this thinking is k-linear, and does not extend to submodules and factor modules. Generally speaking, it seems best to regard zero spaces and pole spaces as separate spaces.

The k[z]-factor module

$$\frac{Y[z] \cap PU(z)}{Y[z] \cap PU[z]} = Z(P(z)) \qquad [3-5]$$

is a useful isomorphic form of the zero module definition [3-4]. In this work, unless otherwise indicated, we shall regard [3-5] as an identity. The k[z]-linear isomorphism of [3-5] can be deduced from the commutative diagram

$$P^{-1}Y[z] \cap (\ker P(z)+U[z]) \longrightarrow P^{-1}Y[z]$$

$$P(z) \Big\downarrow \qquad\qquad\qquad \Big\downarrow P(z)$$

$$Y[z] \cap PU[z] \longrightarrow Y[z] \cap PU(z)$$

Figure 4

of Figure 4, where the rows are natural inclusions and the columns are restrictions of $P(z)$. Notice that the first row is related to [3-4] by a standard isomorphism.

The pole module $P(P(z))$ and the zero module $Z(P(z))$ convey the meaning of, and

carry the structure for, k[z]-poles and k[z]-zeros of state-space type. For the
problem of interest in this paper, however, we shall make use of certain zeros which
are not torsion. These turn out to be free zeros.

Free Zeros. Poles and zeros of state-space type have been reviewed in Section
3, for a k(z)-linear map such as P(z) : U(z) → Y(z). The zero of state-space type
is, of course, quite familiar because of its resemblance to a pole. Such zeros have
received the lion's share of attention in the literature. Moreover, when U and Y
are one-dimensional, it certainly seems that this notion of zero is completely
adequate. When multiple inputs and multiple outputs arise, however, we must be ready
to deal with zeros which are not of state-space type. Some of these may be understood as finitely generated zeros which are not torsion. This section explores such
zeros, which turn out to be free.

The basic module in our discussion is the k[z]-factor module

$$Z_\Omega(P(z)) = \frac{Y[z]}{Y[z] \cap PU[z]} \, , \tag{4-1}$$

which is finitely generated, because Y[z] is finitely generated. If P(z) has zero
cokernel, as a k(z)-linear map, then

$$Y[z] \cap PU(z) = Y[z]; \tag{4-2}$$

and the state-space zero module Z(P(z)) of [3-5] is identical to $Z_\Omega(P(z))$. Clearly,
therefore, the module [4-1] is not torsion-free. On the other hand, if coker P(z) is
nonzero, it turns out that $Z_\Omega(P(z))$ is not a torsion module either. The precise
character of [4-1] is the subject of this section. For the purposes of calculating
$Z_\Omega(P(z))$, one could as well replace P(z) : U(z) → Y(z) with the restriction

$$\tilde{P}(z) : P^{-1}Y[z] \cap U[z] \to Y[z], \tag{4-3}$$

which is k[z]-linear. In classical realization theory (10), U[z] and Y[z] are
frequently denoted ΩU and ΩY, respectively; and the subscript Ω is also used to
denote k[z]-linear maps between such modules. For this reason, we have chosen the
Ω-subscript in [4-1]; and it is convenient to refer to $Z_\Omega(P(z))$ as the omega zero
module of P(z).

When coker P(z) is not equal to zero, so that [3-5] and [4-1] are distinct
modules, the structure of $Z_\Omega(P(z))$ can be studied by means of the commutative
diagram

$$
\begin{array}{ccc}
Y[z] \cap PU[z] & \longrightarrow & Y[z] \cap PU(z) \\
1 \downarrow & & \downarrow \\
Y[z] \cap PU[z] & \longrightarrow & Y[z]
\end{array}
$$

Figure 5

of Figure 5, where the rows and second column are natural inclusions, and column one
is a restriction of the identity. This diagram infers the existence of a monic,
k[z]-linear map from Z(P(z)) into $Z_\Omega(P(z))$. In fact, this is a natural inclusion,
so that Z(P(z)) is a submodule of $Z_\Omega(P(z))$. The cokernel of this inclusion is thus

$$\frac{Y[z]}{Y[z] \cap PU[z]} \bigg/ \frac{Y[z] \cap PU(z)}{Y[z] \cap PU[z]} \approx \frac{Y[z]}{Y[z] \cap PU(z)} \, . \tag{4-4}$$

We summarize this idea, with certain of its consequences, in the following theorem.

Theorem 1

Let $Z_\Omega(P(z))$ be the finitely generated, omega zero module of $P(z)$, as defined in [4-1]; and let $Z(P(z))$ be its module of state-space zeros, as in [3-5]. Define the finitely generated, $k[z]$-factor module

$$\Omega(P(z)) = \frac{Y[z]}{Y[z] \cap PU(z)} \, . \tag{4-5}$$

Then there is a short exact sequence

$$0 \to Z(P(z)) \xrightarrow{\alpha(z)} Z_\Omega(P(z)) \to \Omega(P(z)) \to 0 \tag{4-6}$$

of $k[z]$-modules and $k[z]$-linear maps, with $\alpha(z)$ the map induced from the diagram in Figure 5. In fact, $\Omega(P(z))$ is a free module; and the sequence [4-6] splits, so that

$$Z_\Omega(P(z)) \approx \Omega(P(z)) \oplus Z(P(z)). \tag{4-7}$$

Proof: The exact sequence [4-6] has been established in the prelude to the theorem. Suppose now that $\Omega(P(z))$ is not torsion-free. Then there exists a $y(z)$ in $Y[z]$, not in $PU(z)$, such that $p(z)y(z)$ is a member of $PU(z)$. But

$$\frac{1}{p(z)} \, PU(z) \subset PU(z), \tag{4-8}$$

which gives a contradiction. So [4-5] is torsion-free. Because $\Omega(P(z))$ is finitely generated, and because $k[z]$ is a principal ideal domain, it then follows that $\Omega(P(z))$ is a free, $k[z]$-module. The epimorphism onto this free module then ensures that $Z_\Omega(P(z))$ splits into the sum [4-7].

This first theorem establishes that omega zeros of $P(z)$ can be represented by a direct sum of free zeros with state-space zeros. Free zeros are connected with coker $P(z)$, and play a central role in the synthesis of multivariable, dynamical systems. The next section discusses them as fixed zeros in transfer function equations.

Fixed Zeros. Consider a scalar equation of transfer functions

$$t(z) = p(z)m(z), \tag{5-1}$$

and write the given fractions in relatively prime form

$$t(z) = \frac{n_t(z)}{d_t(z)} \, , \qquad p(z) = \frac{n_p(z)}{d_p(z)} \, . \tag{5-2}$$

It follows that

$$m(z) = \frac{n_t(z)d_p(z)}{d_t(z)n_p(z)} \, , \tag{5-3}$$

so that the state-space zeros of $m(z)$ are comprised of those zeros of $t(z)$ which are not zeros of $p(z)$, combined with those poles of $p(z)$ which are not poles of $t(z)$. In this section, we extend this basic idea to the multivariable case, with the aid of the notion of free zeros. The vehicle for discussion is a module of fixed zeros.

By the fixed zero module, we shall mean the $k[z]$-factor module

$$Z = \frac{PU[z] + TR[z]}{TR[z]} \, , \tag{5-4}$$

in which $T(z) : R(z) \to Y(z)$ and $P(z) : U(z) \to Y(z)$ are $k(z)$-linear maps. It is apparent that [5-4] is a finitely generated module, because $U[z]$ and $R[z]$ are

finitely generated modules. But Z is not in general a torsion module, and so the
fixed zero module is not of state-space type. Our purpose in this section is to
explain and characterize the module Z as a generalization of the notions of [5-1]
through [5-3].

Now consider the module [5-4] of fixed zeros in greater detail. In particular,
we form a commutative diagram

$$
\begin{array}{ccc}
Y[z] \cap TR[z] & \longrightarrow & TR[z] \\
\downarrow & & \downarrow \\
Y[z] \cap \{PU[z]+TR[z]\} & \longrightarrow & PU[z] + TR[z]
\end{array}
$$

Figure 6

of natural inclusions, as in Figure 6. Observe that column two has a cokernel which
is Z, as in [5-4]. Next consider a $k(z)$-linear map

$$[T(z) \quad P(z)] : R(z) \oplus U(z) \to Y(z). \tag{5-5}$$

The pole module of the map [5-5] is given by

$$P\big([T(z) \quad P(z)]\big) = \frac{[T \quad P](R[z] \oplus U[z])}{Y[z] \cap [T \quad P](R[z] \oplus U[z])}, \tag{5-6}$$

which we recognize as the quotient module induced from row two of Figure 6. Row one,
of the same diagram, on the other hand, induces a quotient module

$$\frac{TR[z]}{Y[z] \cap TR[z]} = P\big(T(z)\big), \tag{5-7}$$

the pole module of $T(z)$. The diagram, moreover, induces a monic, $k[z]$-linear map on
$P\big(T(z)\big)$ into $P\big([T(z) \; P(z)]\big)$, with a cokernel that we shall denote by X_p. On the
intuitive level, X_p is a module of poles of the map [5-5] which are not poles of
$T(z)$. It is a precise presentation of the more intuitive "poles of $P(z)$ which are
not poles of $T(z)$," and generalizes one-half of the analysis in [5-1] through [5-3].
Return now to the composite map [5-5], and form the omega zero module

$$Z_\Omega\big([T(z) \quad P(z)]\big) = \frac{Y[z]}{Y[z] \cap [T \quad P](R[z] \oplus U[z])}. \tag{5-8}$$

Next calculate the omega zeros of $T(z)$ in the manner

$$Z_\Omega\big(T(z)\big) = \frac{Y[z]}{Y[z] \cap TR[z]}. \tag{5-9}$$

By the inclusion $TR[z] \subset TR[z] + PU[z]$, there is an epic, $k[z]$-linear map on $Z_\Omega\big(T(z)\big)$
onto $Z_\Omega\big([T(z) \; P(z)]\big)$, with kernel isomorphic to

$$\frac{Y[z] \cap \big(TR[z]+PU[z]\big)}{Y[z] \cap TR[z]}. \tag{5-10}$$

In Figure 6, however, we see that [5-10] is just the cokernel of column one, which we
denote by Z_p. The module Z_p describes "omega zeros of $T(z)$ which are not omega zeros
of [5-5]." It is a precise statement of the more intuitive "zeros of $T(z)$ which are
not zeros of $P(z)$." We summarize in a second theorem

Theorem 2

Let im $T(z) \subset$ im $P(z)$, and let Z be the module [5-4] of fixed zeros associated
with solutions $M(z)$ to the transfer function equation $T(z) = P(z)M(z)$. Then there

exist modules X_p and Z_p, and appropriate $k[z]$-linear maps, such that the following three short sequences are exact:

$$0 \to P(T(z)) \to P([T(z) \quad P(z)]) \to X_p \to 0; \qquad [5\text{-}11a]$$

$$0 \to Z_p \to Z_\Omega(T(z)) \to Z_\Omega([T(z) \quad P(z)]) \to 0 ; \qquad [5\text{-}11b]$$

$$0 \to Z_p \to Z \to X_p \to 0. \qquad [5\text{-}11c]$$

Proof: Sequences [5-11a] and [5-11b] are in the prelude to the theorem. Sequence [5-11c] follows from the diagram of Figure 6, extended to include the induced third column, so that all three columns are monomorphisms.

Discussion: The module X_p, as a factor module of the finitely generated, torsion pole module of [5-5], is itself finitely generated and torsion. The omega zero behavior of Z thus appears in the sense of a submodule, by injection of Z_p. The clarity and brevity of the description of Z in Theorem 2 seems to be largely a consequence of the use of omega zeros in place of state-space zeros. In the section following, we investigate the relationship of the module Z of fixed zeros to the modules $Z(M(z))$ of solutions $M(z)$.

State-Space Zero Constraints. The sections preceding have reviewed the module theoretic ideas associated with classical poles and zeros, of state-space type, for $k(z)$-linear maps; and they have introduced the new notions of omega zeros of a transfer function, and free zeros. A module Z of fixed zeros, independent of $M(z)$, has been characterized in terms of certain poles and omega zeros of $T(z)$ and of the composite map $[T(z) \ P(z)]$. This module is the precise generalization of the classical ideas of [5-1] through [5-3]. This section introduces the relationship between Z and $Z(M(z))$, for $M(z)$ a solution to the equation $T(z) = P(z)M(z)$.

Theorem 3

Let im $T(z) \subset$ im $P(z)$; and let $M(z)$ satisfy the equation $T(z) = P(z)M(z)$. If Z is the fixed zero module [5-4], then there exists a $k[z]$-linear map $\beta(z) : Z(M(z)) \to$ Z, with kernel

$$\ker \beta(z) = \frac{U[z] \cap \{MR[z] + MR(z) \cap \ker P(z)\}}{U[z] \cap MR[z]} \qquad [6\text{-}1]$$

and cokernel

$$\operatorname{coker} \beta(z) = \frac{PU[z]}{P(U[z] \cap MR(z)) + PU[z] \cap TR[z]} . \qquad [6\text{-}2]$$

Proof: Consider the commutative diagram of Figure 7.

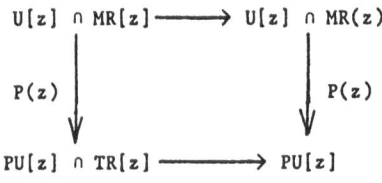

Figure 7

The rows are natural inclusions; and the columns are restrictions of $P(z)$. This diagram induces $\beta(z)$ with the action

$$\beta(z)[u(z)] = P(z)u(z) \bmod PU[z] \cap TR[z]. \qquad [6\text{-}3]$$

Representation of the kernel follows from

$$P^{-1}\bigl(PU[z] \cap TR[z]\bigr) \cap U[z] \cap MR(z)$$

$$= \bigl(U[z] + \ker P(z)\bigr) \cap \bigl(MR[z] + \ker P(z)\bigr) \cap U[z] \cap MR(z)$$

$$= \bigl(MR[z] + \ker P(z)\bigr) \cap U[z] \cap MR(z); \qquad\qquad [6-4]$$

then

$$\text{im } \beta(z) = \frac{P\bigl(U[z] \cap MR(z)\bigr) + PU[z] \cap TR[z]}{PU[z] \cap TR[z]}; \qquad\qquad [6-5]$$

and the cokernel follows by isomorphism.

It follows from Theorem 3 that the zero module $Z\bigl(M(z)\bigr)$ fits into a short exact sequence

$$0 \rightarrow \ker \beta(z) \rightarrow Z\bigl(M(z)\bigr) \rightarrow \text{im } \beta(z) \rightarrow 0. \qquad\qquad [6-6]$$

<u>Conclusions</u>. The work reported above has been mo ivated by the problem of synthesizing nonlinear feedback systems (8, 9). Because nonlinear plants offer dif-different opportunities for control as they move from region to region, the complexity of their control models, as well as the detail of any controller realizations greatly exceeds a mere count of the number of integrators. See, for instance, Section 2. If we regard the complexity of the plant as fixed, then overall system complexity is a direct function of that of the controller.

This paper proposes a local measure of complexity for these situations, in terms of the pole and zero modules of the equivalent open loop controller. The most basic inquiry about such modules involves fixed constraints which are enforced by the plant dynamics and closed loop specifications. For poles, such constraints have been studied by Conte, Perdon, and Wyman (1). For zeros, the needed prior work was not available. Thus, in this paper we propose and characterize a module of fixed zeros in solutions for equivalent open loop controllers. Unlike the case for poles, fixed zeros have their simplest and most intuitive characterization in terms of omega zero modules, which may be viewed as direct sums of zeros of classical, state-space, type (finitely generated, torsion) with zeros of a new type (finitely generated, free). Differences from the pole study also occur in the way in which the fixed zeros relate to solution zeros.

By a comparison of (1) with the results of this investigation, it would seem that the proposed measure of module theoretic complexity is feasible. Moreover, it appears that more design freedom exists for zeros than for poles, in the sense that we can address both factors of omega zeros: torsion and free.

The results of Sections 3-6 remain true in more general contexts. Let 0 be any subring of k(z) which contains k, has field of fractions k(z), and is a principal ideal domain. Instances of such rings are readily at hand, and include localizations of k[z] and discrete valuation rings. An example of the latter is O_∞, the subring of proper transfer functions. Further illustrations can be formed by intersecting discrete valuation rings, as in the case of O_{ps}, the subring of transfer functions which are proper and stable, when k = R, the real numbers. If we replace k[z] by such an 0, certain adjustments are made in Equations [3-1], which change to

$$\Omega_0 R = 0 \otimes_k R, \qquad\qquad\qquad [7\text{-}1a]$$

$$\Omega_0 U = 0 \otimes_k U, \qquad\qquad\qquad [7\text{-}1b]$$

$$\Omega_0 Y = 0 \otimes_k Y. \qquad\qquad\qquad [7\text{-}1c]$$

In the rest of the discussion, replace $R[z]$ by $\Omega_0 R$, and so forth. The physical interpretation varies with the ring.

The complexity of a feedback system has for a long time been locally assessed by the state spaces and dynamics matrices associated with various operating lines. Such an assessment is equivalent to that provided by the local pole modules. This paper proposes to augment such pictures with the corresponding local zero modules. For a fixed performance of the feedback system, together with a fixed plant, such a study can be implemented by means of the equivalent open loop controller. A finitely generated module of fixed zeros in such controllers is defined and studied. Unlike the previously studied module of fixed poles, this module is in general not torsion and thus does not always admit a classical, state-space interpretation. In fact, some of the zeros are free. If corresponding adjustments in the definition of zeros of a transfer function are made, however, then the fixed zeros have natural interpretation and serve as a springboard to determine solutions of least complexity.

Acknowledgements. The work of the first author was supported by The Ohio State University's Distinguished Visiting Professorship, by the University of Notre Dame's Frank M. Freimann Chair in Electrical Engineering, and by the National Science Foundation under Grant ECS-84-05714.

References.

1. G. Conte, A. M. Perdon, and B. F. Wyman, "Fixed Poles in Transfer Function Equations," SIAM Journal on Control and Optimization, Vol. 26, No. 2, pp. 356-368, March 1988.
2. B. F. Wyman and M. K. Sain, "The Zero Module and Essential Inverse Systems," IEEE Transactions on Circuits and Systems, Vol. CAS-27, No. 2, pp. 112-126, February 1981.
3. B. F. Wyman and M. K. Sain, "On the Design of Pole Modules for Inverse Systems," IEEE Transactions on Circuits and Systems, Vol. CAS-31, No. 10, pp. 977-988, October 1985.
4. B. F. Wyman and M. K. Sain, "Exact Sequences for Pole-Zero Cancellation," Proceedings International Symposium on the Mathematical Theory of Networks and Systems, pp. 278-280, August 1981.
5. G. Conte and A. M. Perdon, "Zeros of Cascade Compositions," in Frequency Domain and State Space Methods for Linear Systems, C. I. Byrnes and A. Lindquist, eds. North-Holland: Elsevier, 1986, pp. 23-34.
6. M. K. Sain and B. F. Wyman, "The Fixed Zero Constraint in Dynamical System Performance," International Symposium on the Mathematical Theory of Networks and Systems, June 1987, in press.
7. B. F. Wyman and M. K. Sain, "On Dual Zero Spaces and Inverse Systems," IEEE Transactions on Automatic Control, Vol. AC-31, No. 11, pp. 1053-1055, November 1986.
8. M. K. Sain and J. L. Peczkowski, "Nonlinear Control by Coordinated Feedback Synthesis, with Gas Turbine Applications," Proceedings American Control Conference, pp. 1121-1128, June 1985.
9. J. L. Peczkowski and M. K. Sain, "Synthesis of System Responses: A Nonlinear Multivariable Control Design Approach," Proceedings American Control Conference. pp. 1322-1329, June 1985.
10. M. K. Sain, Introduction to Algebraic System Theory. New York: Academic Press, May 1981, Chapter 7.

ROBUSTNESS IN THE PRESENCE OF PARAMETRIC UNCERTAINTY
AND UNMODELED DYNAMICS

Michael K.H. Fan

University of Maryland

John C. Doyle

California Institute of Technology

André L. Tits

University of Maryland

Introduction. The structured singular value (μ), introduced by Doyle [1] allows to analyze robust stability and performance of linear systems affected by parametric as well as dynamic uncertainty. While exact computation of μ can be prohibitively complex, an efficiently computable upper bound was obtained in [2], yielding a practical sufficient condition for robust stability and performance.

In this note, the results of [2] are used to study the case of state space models of the form

$$\dot{x} = (A_0 + \sum_{i=1}^{m} \delta_i A_i)x \tag{1}$$

where the A_i's are $n \times n$ real matrices and the δ_i's are uncertain real parameters. The case where the A_i's have low rank is given special attention. When the A_i's all have rank one, (1) is equivalent to the model used by Qiu and Davison [3], which itself generalizes that used by Yedavalli [4]. By means of two examples, we compare our bound to those proposed in [3] and [4].

Preliminaries. Throughout the note, given any square complex matrix M, we denote by $\bar{\sigma}(M)$ its largest singular value and by M^H its complex conjugate transpose. Given any Hermitian matrix A, we denote by $\bar{\lambda}(A)$ its largest eigenvalue. Given any integer k, I_k denotes the $k \times k$ identity matrix and O_k the $k \times k$ zero matrix. Finally, j will denote $\sqrt{-1}$.

Given a $p \times p$ complex matrix M and positive integers $k_1, \ldots, k_m$, with $\sum_{q=1}^{m} k_q = p$, consider the family of block diagonal $p \times p$ matrices (In this note we consider only parametric perturbations)

$$\mathcal{X} = \{\text{block diag }(\delta_1 I_{k_1}, \ldots, \delta_m I_{k_m}) : \delta_q \in \mathbb{R}\} .$$

Definition 1. [1] The *structured singular value* $\mu_{\mathcal{X}}(M)$ of a complex $p \times p$ matrix M with respect to $\mathcal{X}$ is 0 if there is no Δ in $\mathcal{X}$ such that $\det(I - \Delta M) = 0$, and

$$\mu_{\mathcal{X}}(M) = \left(\min_{\Delta \in \mathcal{X}} \{\bar{\sigma}(\Delta) : \det(I - \Delta M) = 0\} \right)^{-1}$$

otherwise. $\square$

Exact computation of the structured singular value is generally intractable. In [2], the following computable upper bound was obtained.

Fact 1. [2] For any matrix M and $\mathcal{X}$,

$$\mu_{\mathcal{X}}(M) \le \hat{\mu}_{\mathcal{X}}(M) := \inf_{D \in \mathcal{D}_{\mathcal{X}}} \nu_{\mathcal{X}}(DMD^{-1})$$

where

$$\mathcal{D}_{\mathcal{X}} = \left\{ \text{block diag}(D_1, \ldots, D_m) : 0 < D_q = D_q^H \in \mathbb{C}^{k_q \times k_q} \right\}$$

and where, for any matrix A and $\mathcal{X}$, $\nu_{\mathcal{X}}(A)$ is defined by

$$\nu_{\mathcal{X}}(A) = \sqrt{\max\left\{0, \inf_{G \in \mathcal{G}_{\mathcal{X}}} \overline{\lambda}[A^H A + j(GA - A^H G)]\right\}}$$

with

$$\mathcal{G}_{\mathcal{X}} = \left\{ \text{block diag}(G_1, \ldots, G_m) : G_q = G_q^H \in \mathbb{C}^{k_q \times k_q} \right\} .$$

□

An efficient algorithm for computing $\hat{\mu}_{\mathcal{X}}(M)$ is described in [5,6].

Main result. Following [7], one can easily show that system (1) is asymptotically stable for all $|\delta_i| \le \delta$ if, and only if,

$$\delta < \left(\sup_{\omega \ge 0} \mu_{\mathcal{X}}(H_1(j\omega))\right)^{-1}$$

where $H_1(s)$ is the transfer matrix defined by

$$H_1(s) = \begin{bmatrix} I \\ I \\ \vdots \\ I \end{bmatrix} (sI - A_0)^{-1} [A_1|A_2|\cdots|A_m]$$

and where

$$\mathcal{X} = \{\text{block diag}(\delta_1 I_n, \ldots, \delta_m I_n) : \delta_q \in \mathbb{R}\} .$$

However, whenever some A_i's are not of full rank, one can obtain a necessary and sufficient condition for robust stability involving a transfer matrix of lower size than H_1. To see this, decompose each A_i as

$$A_i = b_i c_i^T$$

where $b_i, c_i \in \mathbb{R}^{n \times r_i}$, with r_i the rank of A_i, and define

$$B = [b_1|\cdots|b_m]$$

and

$$C = \begin{bmatrix} c_1^T \\ \vdots \\ c_m^T \end{bmatrix} .$$

The following is then easily proven using [7].

Proposition 1. The system in (1) is asymptotically stable if, and only if,

$$\delta < \left(\sup_{\omega \geq 0} \mu_\mathcal{X}(H_2(j\omega)) \right)^{-1}$$

where $H_2(s)$ is the transfer matrix defined by

$$H_2(s) = C(sI - A)^{-1}B$$

and where

$$\mathcal{X} = \{ \text{block diag } (\delta_1 I_{r_1}, \ldots, \delta_m I_{r_m}) : \delta_q \in \mathbb{R} \} \ .$$

☐

Substituting for $\mu_\mathcal{X}$ its upper bound $\hat{\mu}_\mathcal{X}$ gives the following sufficient condition

Corollary 1. The system in (1) is asymptotically stable if

$$\delta < \left(\sup_{\omega \geq 0} \hat{\mu}_\mathcal{X}(H_2(j\omega)) \right)^{-1}$$

where $H_2(s)$ and $\mathcal{X}$ are defined as in Proposition 1. ☐

Models of the type (1) for which the A_i's have low rank are of practical importance. The case where all A_i's have rank one corresponds to the model used by Qiu and Davison,

$$\dot{x} = (A + B\Delta C)x \tag{2}$$

where Δ is uncertain. Yedavalli considered the model

$$\dot{x} = (A + \Delta)x \ ,$$

with Δ uncertain, which is clearly a special case of (2).

Numerical examples. We conclude this note by comparing on two examples the results obtained using Corollary 1 above to those obtained using the techniques of [3] and [4]. The examples are borrowed from [3,4,8].

Example 1. The following matrices were considered in [3,4].

$$A + B\Delta C = \begin{bmatrix} -3 & -2 \\ 1 & 0 \end{bmatrix} + \begin{bmatrix} 1 & 1 \\ 0 & 0 \end{bmatrix} \begin{bmatrix} \delta_1 & 0 \\ 0 & \delta_2 \end{bmatrix} \begin{bmatrix} 1 & 0 \\ 0 & 1 \end{bmatrix} = \begin{bmatrix} -3 + \delta_1 & -2 + \delta_2 \\ 1 & 0 \end{bmatrix} \ ,$$

$$A + B\Delta C = \begin{bmatrix} -3 & -2 \\ 1 & 0 \end{bmatrix} + \begin{bmatrix} 1 & 0 \\ 0 & 1 \end{bmatrix} \begin{bmatrix} \delta_1 & 0 \\ 0 & \delta_2 \end{bmatrix} \begin{bmatrix} 1 & 0 \\ 1 & 0 \end{bmatrix} = \begin{bmatrix} -3 + \delta_1 & -2 \\ 1 + \delta_2 & 0 \end{bmatrix} \ ,$$

$$A + B\Delta C = \begin{bmatrix} -3 & -2 \\ 1 & 0 \end{bmatrix} + \begin{bmatrix} 0 & 1 \\ 1 & 0 \end{bmatrix} \begin{bmatrix} \delta_1 & 0 \\ 0 & \delta_2 \end{bmatrix} \begin{bmatrix} 1 & 0 \\ 0 & 1 \end{bmatrix} = \begin{bmatrix} -3 & -2 + \delta_2 \\ 1 + \delta_1 & 0 \end{bmatrix},$$

$$A + B\Delta C = \begin{bmatrix} -3 & -2 \\ 1 & 0 \end{bmatrix} + \begin{bmatrix} 1 & 0 & 1 \\ 0 & 1 & 0 \end{bmatrix} \begin{bmatrix} \delta_1 & 0 & 0 \\ 0 & \delta_2 & 0 \\ 0 & 0 & \delta_3 \end{bmatrix} \begin{bmatrix} 1 & 0 \\ 1 & 0 \\ 0 & 1 \end{bmatrix} = \begin{bmatrix} -3 + \delta_1 & -2 + \delta_3 \\ 1 + \delta_2 & 0 \end{bmatrix}.$$

Bounds of δ given in [4] which guarantees robust stability were 1.0, 0.48, 0.5 and 0.317, respectively. Bounds given in [3] were 1.5201, 0.9150, 0.8108 and 0.6848, respectively. Our bounds are 2,1,1 and 1, respectively. In these cases, our bounds are also exact.

Example 2. Consider the following system [3,8].

$$\dot{x} = \begin{bmatrix} -1 & 0 \\ 0 & -2 \end{bmatrix} x + \begin{bmatrix} 7 & 8 \\ 12 & 14 \end{bmatrix} u, \quad y = \begin{bmatrix} 7 & -8 \\ -6 & 7 \end{bmatrix} x$$

with output feedback control

$$u = \begin{bmatrix} -k_1 & 0 \\ 0 & -k_2 \end{bmatrix} y .$$

The controller gains are subject to uncertainty such that $|\Delta k_1| \le \delta/2$ and $|\Delta k_2| \le \delta$. The nominal value of controller gains are $k_1 = k_2 = 1$. Corollary 1 implies that the closed loop system is stable if

$$\delta < \hat{\delta} := \left(\sup_{\omega \ge 0} \hat{\mu}_{\mathcal{X}}(H(j\omega)) \right)^{-1} \tag{3}$$

where $H(s) = C(sI - A)^{-1}B$, with

$$A = \begin{bmatrix} -1 & 0 \\ 0 & -2 \end{bmatrix} + \begin{bmatrix} 7 & 8 \\ 12 & 14 \end{bmatrix} \begin{bmatrix} -1 & 0 \\ 0 & -1 \end{bmatrix} \begin{bmatrix} 7 & -8 \\ -6 & 7 \end{bmatrix} = \begin{bmatrix} -2 & 0 \\ 0 & -4 \end{bmatrix}$$

$$B = \begin{bmatrix} 3.5 & 8 \\ 6 & 14 \end{bmatrix} \quad \text{and} \quad C = \begin{bmatrix} 7 & -8 \\ -6 & 7 \end{bmatrix} .$$

Solution of the optimization problem in (3) yields $\hat{\delta} = 0.0816$. This result agrees with that in [3]. It turns out to be an exact bound.

Acknowledgements. This research was supported in part by the National Science Foundation's Engineering Research Centers Program: NSF CDR 8803012 and by the National Science Foundation under Grant DMC-84-51515.

Authors' Addresses. Michael K.H. Fan is with the Systems Research Center, University of Maryland, College Park, MD 20742. John C. Doyle is with the Electrical Engineering Department, California Institute of Technology, Pasadena, CA 91125. André L. Tits is with the Systems Research Center and the Electrical Engineering Department, University of Maryland, College Park, MD 20742.

References.

[1] J.C. Doyle, "Analysis of Feedback Systems with Structured Uncertainties," *Proc. IEE-D* 129 (1982), 242–250.

[2] M.K.H. Fan, A.L. Tits & J.C. Doyle, "Robustness in the Presence of Joint Parametric Uncertainty and Unmodeled Dynamics," *Proc. of the 1988 American Control Conference*, Atlanta, Georgia (June 1988).

[3] L. Qiu & E.J. Davison, "New Pertubation Bounds for the Robust Stability of Linear State Space Models," *Proceedings of the 25th Conference on Decision and Control*, Athens, Greece (December 1986).

[4] R.K. Yedavalli, "Perturbation Bounds for Robust Stability in Linear State Space Models," *Internat. J. Control* 42 (1985), 1507–1517.

[5] M.K.H. Fan, "On Computing the Structured Singular Value of a Square Transfer Matrix," *(in preparation)* (1988).

[6] M.K.H. Fan, "User's Guide for MUSOL2: A Package for Computing the Structured Singular Value or Its Upper Bound," Systems Research Center, University of Maryland, Technical Research Report, SRC TR 87-204, 1988.

[7] J.C. Doyle, J.E. Wall & G. Stein, "Performance and Robustness Analysis for Structured Uncertainty," *Proc. 21st IEEE Conf. on Decision and Control*, Orlando, Florida (December 1982).

[8] W.H. Lee, "Robust Analysis for State Space Models," Alphatech Inc., Report TP-151, 1982.

Lecture Notes in Control and Information Sciences

Edited by M. Thoma and A. Wyner

Vol. 81: Stochastic Optimization
Proceedings of the International Conference,
Kiew, 1984
Edited by I. Arkin, A. Shiraev, R. Wets
X, 754 pages, 1986.

Vol. 82: Analysis and Algorithms
of Optimization Problems
Edited by K. Malanowski, K. Mizukami
VIII, 240 pages, 1986.

Vol. 83: Analysis and Optimization
of Systems
Proceedings of the Seventh International
Conference of Analysis and Optimization
of Systems
Antiba, June 26-27, 1986
Edited by A. Bensoussan, J. L. Lions
XVI, 901 pages, 1986.

Vol. 84: System Modelling
and Optimization
Proceedings of the 12th IFIP Conference
Budapest, Hungary, September 2–6, 1985
Edited by A. Prékopa, J. Szelezsán, B. Strazicky
XII, 1046 pages, 1986.

Vol. 85: Stochastic Processes
in Underwater Acoustics
Edited by Charles R. Baker
V, 205 pages, 1986.

Vol. 86: Time Series and
Linear Systems
Edited by Sergio Bittanti
XVII, 243 pages, 1986.

Vol. 87: Recent Advances in
System Modelling and
Optimization
Proceedings of the IFIP-WG 7/1
Working Conference
Santiago, Chile, August 27-31, 1984
Edited by L. Contesse, R. Correa, A. Weintraub
IV, 199 pages, 1987.

Vol. 88: Bruce A. Francis
A Course in H_∞ Control Theory
XI, 156 pages, 1987.

Vol. 88: Bruce A. Francis
A Course in H_∞ Control Theory
X, 150 pages, 1987.
Corrected - 1st printing 1987

Vol. 89: G. K. H. Pang/A. G. J. McFarlane
An Expert System Approach to
Computer-Aided Design
of Multivariable Systems
XII, 223 pages, 1987.

Vol. 90: Singular Perturbations
and Asymptotic Analysis
in Control Systems
Edited by P. Kokotovic,
A. Bensoussan, G. Blankenship
VI, 419 pages, 1987.

Vol. 91 Stochastic Modelling
and Filtering
Proceedings of the IFIP-WG 7/1
Working Conference
Rome, Italy, Decembre 10-14, 1984
Edited by A. Germani
IV, 209 pages, 1987.

Vol. 92: L. T. Grujić, A. A. Martynyuk,
M. Ribbens-Pavella
Large-Scale Systems Stability Under
Structural and Singular Perturbations
XV, 366 pages, 1987.

Vol. 93: K. Malanowski
Stability of Solutions to Convex
Problems of Optimization
IX, 137 pages, 1987.

Vol. 94: H. Krishna
Computational Complexity
of Bilinear Forms
Algebraic Coding Theory and
Applications to Digital
Communication Systems
XVIII, 166 pages, 1987.

Vol. 95: Optimal Control
Proceedings of the Conference on
Optimal Control and Variational Calculus
Oberwolfach, West-Germany, June 15-21, 1986
Edited by R. Bulirsch, A. Miele, J. Stoer
and K. H. Well
XII, 321 pages, 1987.

Vol. 96: H. J. Engelbert/W. Schmidt
Stochastic Differential Systems
Proceedings of the IFIP-WG 7/1
Working Conference
Eisenach, GDR, April 6-13, 1986
XII, 381 pages, 1987.

Lecture Notes in Control and Information Sciences

Edited by M. Thoma and A. Wyner

Vol. 97: I. Lasiecka/R. Triggiani (Eds.)
Control Problems for Systems
Described by Partial Differential Equations
and Applications
Proceedings of the IFIP-WG 7.2
Working Conference
Gainesville, Florida, February 3-6, 1986
VIII, 400 pages, 1987.

Vol. 98: A. Aloneftis
Stochastic Adaptive Control
Results and Simulation
XII, 120 pages, 1987.

Vol. 99: S. P. Bhattacharyya
Robust Stabilization Against
Structured Perturbations
IX, 172 pages, 1987.

Vol. 100: J. P. Zolésio (Editor)
Boundary Control and Boundary Variations
Proceedings of the IFIP WG 7.2 Conference
Nice, France, June 10-13, 1987
IV, 398 pages, 1988.

Vol. 101: P. E. Crouch,
A. J. van der Schaft
Variational and Hamiltonian
Control Systems
IV, 121 pages, 1987.

Vol. 102: F. Kappel, K. Kunisch,
W. Schappacher (Eds.)
Distributed Parameter Systems
Proceedings of the 3rd International Conference
Vorau, Styria, July 6–12, 1986
VII, 343 pages, 1987.

Vol. 103: P. Varaiya, A. B. Kurzhanski (Eds.)
Discrete Event Systems:
Models and Applications
IIASA Conference
Sopron, Hungary, August 3-7, 1987
IX, 282 pages, 1988.

Vol. 104: J. S. Freudenberg/D. P. Looze
Frequency Domain Properties of Scalar
and Multivariable Feedback Systems
VIII, 281 pages, 1988.

Vol. 105: Ch. I. Byrnes/A. Kurzhanski (Eds.)
Modelling and Adaptive Control
Proceedings of the IIASA Conference
Sopron, Hungary, July 1986
V, 379 pages, 1988.

Vol. 106: R. R. Mohler (Editor)
Nonlinear Time Series and
Signal Processing
V, 143 pages. 1988.

Vol. 107: Y. T. Tsay, L.-S. Shieh, St. Barnett
Structural Analysis and Design
of Multivariable Systems
An Algebraic Approach
VIII, 208 pages, 1988.

Vol. 108: K. J. Reinschke
Multivariable Control
A Graph-theoretic Approach
274 pages, 1988.

Vol. 109: M. Vukobratović/R. Stojić
Modern Aircraft Flight Control
VI, 288 pages, 1988.

Vol. 110: In preparation

Vol. 111: A. Bensoussan, J. L. Lions (Eds.)
Analysis and Optimization
of Systems
XIV, 1175 pages, 1988.

Vol. 112: Vojislav Kecman
State-Space Models of Lumped
and Distributed Systems
IX, 280 pages, 1988

Vol. 113: M. Iri, K. Yajima (Eds.)
System Modelling and Optimization
Proceedings of the 13th IFIP Conference
Tokyo, Japan, Aug. 31 – Sept. 4, 1987
IX, 787 pages, 1988.

Vol. 114: A. Bermúdez (Editor)
Control of Partial Differential Equations
Proceedings of the IFIP WG 7.2
Working Conference
Santiago de Compostela, Spain, July 6–9, 1987
IX, 318 pages, 1989

Vol. 115: H.J. Zwart
Geometric Theory for Infinite
Dimensional Systems
VIII, 156 pages, 1989.

Vol. 116: M.D. Mesarovic, Y. Takahara
Abstract Systems Theory
VIII, 439 pages, 1989

Lecture Notes in Control and Information Sciences

Edited by M. Thoma and A. Wyner

Vol. 117: K.J. Hunt
Stochastic Optimal Control Theory
with Application in Self-Tuning Control
X, 308 pages, 1989.

Vol. 118: L. Dai
Singular Control Systems
IX, 332 pages, 1989

Vol. 119: T. Basar, P. Bernhard
Differential Games and Applications
VII, 201 pages, 1989

Vol. 120: L. Trave, A. Titli, A. M. Tarras
Large Scale Systems:
Decentralization, Structure Constraints
and Fixed Modes
XIV, 384 pages, 1989

Vol. 121: A. Blaquière (Editor)
Modeling and Control of Systems
in Engineering, Quantum Mechanics,
Economics and Biosciences
Proceedings of the Bellman Continuum
Workshop 1988, June 13–14, Sophia Antipolis, France
XXVI, 519 pages, 1989

Vol. 122: J. Descusse, M. Fliess, A. Isidori,
D. Leborgne (Eds.)
New Trends in Nonlinear Control Theory
Proceedings of an International
Conference on Nonlinear Systems,
Nantes, France, June 13–17, 1988
VIII, 528 pages, 1989

Vol. 123: C. W. de Silva, A. G. J. MacFarlane
Knowledge-Based Control with
Application to Robots
X, 196 pages, 1989

Vol. 124: A. A. Bahnasawi, M. S. Mahmoud
Control of Partially-Known
Dynamical Systems
XI, 228 pages, 1989

Vol. 125: J. Simon (Ed.)
Control of Boundaries and Stabilization
Proceedings of the IFIP WG 7.2 Conference
Clermont Ferrand, France, June 20–23, 1988
IX, 266 pages, 1989

Vol. 126: N. Christopeit, K. Helmes
M. Kohlmann (Eds.)
Stochastic Differential Systems
Proceedings of the 4th Bad Honnef Conference
June 20–24, 1988
IX, 342 pages, 1989

Vol.127: C. Heij
Deterministic Identification
of Dynamical Systems
VI, 292 pages, 1989

Vol. 128: G. Einarsson, T. Ericson,
I. Ingemarsson, R. Johannesson,
K. Zigangirov, C.-E. Sundberg
Topics in Coding Theory
VII, 176 pages, 1989

Vol. 129: W. A.Porter, S. C. Kak (Eds.)
Advances in Communications and
Signal Processing,
VI, 376 pages, 1989.

Vol. 130: W. A. Porter, S. C. Kak,
J. L. Aravena (Eds.)
Advances in Computing and Control
VI, 367 pages, 1989